GRID INTEGRATION OF WIND ENERGY

GRID INTEGRATION OF WIND ENERGY

ONSHORE AND OFFSHORE CONVERSION SYSTEMS

Third Edition

Siegfried Heier

*Kassel University, Fraunhofer Institute for Wind Energy
and Energy System Technology (IWES) Kassel, Germany*

Translators: Gunther Roth
Adliswil, Switzerland

Rachel Waddington
UK

Originally published in the German language by Vieweg+Teubner, 65189 Wiesbaden, Germany, as "Siegfried Heier: Windkraftanlagen. 5. Auflage (5th Edition)" © Vieweg+Teubner | Springer Fachmedien Wiesbaden GmbH 2009

Springer Fachmedien is part of Springer Science+Business Media

This edition first published 2014
© 2014, John Wiley & Sons, Ltd

Registered office
John Wiley & Sons Ltd, The Atrium, Southern Gate, Chichester, West Sussex, PO19 8SQ, United Kingdom

For details of our global editorial offices, for customer services and for information about how to apply for permission to reuse the copyright material in this book please see our website at www.wiley.com.

Library of Congress Cataloging-in-Publication Data

Heier, Siegfried.
  [Windkraftanlagen im Netzbetrieb. English]
  Grid integration of wind energy / Siegfried Heier ; translated by Rachel Waddington. – Third editon.
     pages cm
  Translation of: Windkraftanlagen im Netzbetrieb.
   Includes bibliographical references and index.
  ISBN 978-1-119-96294-6 (hardback)
  1. Wind power plants. 2. Wind energy conversion systems. 3. Electric power systems. I. Title.
   TK1541.H3513 2014
   621.31'2136–dc23

                                                                              2013041087

A catalogue record for this book is available from the British Library.

ISBN: 978-1-119-96294-6

Typeset in 10/12pt Times by Laserwords Private Limited, Chennai, India

1   2014

Contents

Preface

In the long run, an ecologically sustainable energy supply can be guaranteed only by the integration of renewable resources. Besides water power, which is already well established, wind energy is by far the most technically advanced of all renewable power sources, and its economic breakthrough the closest. With a few exceptions, wind power will be used mostly for generating electricity.

Just three decades on from the 50 kW class machines of the mid-1980s, the development of wind turbines has led to production converters with outputs in the 3 MW range. Five to ten megawatt turbines are currently being launched in the market. In the development of these machines, successful techniques and innovations originating from small- and medium-sized turbines were carried over to larger ones, and this has led to a considerable improvement in the reliability of wind turbines. The technical availability currently achieves average values of approximately 98%. Furthermore, economic viability has increased enormously. As a result, wind energy has experienced an almost unbelievable upsurge and has already far exceeded the contributions of water power.

The rapid development of wind power has awakened strong public, political and scientific interest, and has triggered widespread discussion over the past few years, much of it concerning the degree to which nature, the environment and the electricity grid can withstand the impact of wind power.

If political requirements regarding the reduction of environmental pollution are to be met, long-term growth in the use of wind power must be the focus. Since obtaining electricity from the wind currently offers the most favourable technical and economic prospects of all the sources of renewable energy, it must be assigned the highest priority. Due to the fact that turbine sizes are still increasing, a high degree of grid penetration must be expected (regionally at any rate), meaning that the connection of wind turbines could come up against its technical limits. This is already the case today in some instances.

The objective of a forward-looking energy supply policy must therefore be to utilize the existing grid as well as possible ones for the supply of wind power. This is made possible by the use of turbines with good grid compatibility in connection with measures for grid reinforcement. In assessing grid effects, control operations and the electrical engineering design of wind turbines play a significant role. The themes developed in this work are therefore particularly concerned with this topic.

This edition of the book has been updated to cover important innovations in this rapidly changing technology in terms of energy converters, generators and controls, grid integration and development. Important additions were made especially in view of offshore use of wind energy. This has resulted in special importance being paid to network connections at sea and on land. The layout of the book has also been updated to achieve a consistent format, and a number of new illustrations have been included. A great deal of new material has also been added to cover changes in legislation.

This book is the result of a 37 years continuous work in research and development, especially as Head of Wind Energy Research and Professor at the University Kassel, in the Electrical Energy Supply Department of the Institut für Elektrische Energietechnik. Close cooperation with the Institut für Solare Energieversorgungstechnik (ISET) e.V. (now Fraunhofer Institut für Windenergiesysteme IWES, Kassel) brought with it a considerable broadening of the horizon of experience. My special thanks go to the founder of the ISET, Professor Dr Werner Kleinkauf. His suggestions and our technical discussions have had a considerable influence on this work.

The help and support of Ms Katherina Messoll, Dr.-Ing. Alejandro Gesino, Dipl.-Ing. Christof Dziendziol. Dipl.-Ing, Adit Ezzahraoui, Dr.-Ing. Gunter Arnold, Dr Boris Valov, Dipl.-Ing. Michael Durstewitz, Dr.-Ing. Martin Hoppe-Kilpper, Dipl.-Ing.Berthold Hahn, Dipl.-Ing. Martin Kraft, Dipl.-Ing. Volker Konig, Dipi.-Ing. Werner Döring, Dipl.-Ing. Bernd Gruss, Dr-Ing. Oliver Haas, Dr.-Ing Rajeh Saiju, Mr Thomas Donbecker, Mr Bernhard Siano, Mr Martin Nagelmilller, Ms Dipl.-Des. Renate Rothkegel, Frau Melanie Schmieder, Ms Anja Clark-Carina and Ms Judith Keuch have contributed greatly to the success of this book.

My grateful thanks also go to ENERCON GmbH for kindly granting permission to use the image of the wind turbine in the design of the front cover.

This book is intended not only for students in technical faculties. The procedural notes and experimental results will also be of great help to engineers both in theory and practice.

My special thanks must go to the publisher, John Wiley & Sons, Ltd and Laura Bell and Peter Mitchell for their readiness to publish this book and for the painstaking preparation involved.

I would like to thank my wife Hannelore for her assistance as adviser for the difficult formulation and for her understanding that was necessary for the creation of this work.

This book is dedicated to my grandchildren Serafin and Mila as well as my daughters Angela, Sandra and Tina.

The issue of the fifth revision marks the third decade of my future-oriented efforts in this sector and documents the headlong development of wind energy utilization. In this scientific and energy segment with its defining technology, successes have been achieved that open up optimistic perspectives for the future of energy supply.

Siegfried Heier, Kassel

Notation

a	Constant factor related to the pivot of the profile
a_a	Distance between point of application of lifting force and blade axis of rotation
a_p	Distance along blade axis between the points of application of torque and gravity
a_s	Blade deflection and slewing
A_1	Far-upstream cross-section of flow
A_2	Cross-section of flow at turbine
A_3	Broadening downstream cross-section of flow
A_{lt}	Long-term flicker factor
A_R	Rotor swept area
A_{st}	Short-term flicker factor
b	Acceleration of the rotor blade centre of gravity
b'	Acceleration of the rotor blade centre of gravity in the rotating coordinate system
b_c	Coriolis acceleration of rotor blade centre of gravity
b_{defl}	Blade bending in direction of deflection
b_o	Centripetal acceleration in the rotor head
b_R	Centripetal acceleration arising from ω_R
b_s	Bending acceleration of the rotor blade in the direction of deflection and slew
b_{slew}	Blade bending in direction of slew
$b\omega_A$	Centripetal acceleration arising from ω_A
c	Magnification factor for the initial short-circuit alternating current power or the maximum possible short-circuit current
c_a	Lift coefficient of blade profile
c_k	Capacitor bank capacitance
c_m	Torque coefficient of the turbine
c_p	Performance coefficient of the turbine
c_t	Torsional moment coefficient of blade profile ($t_B/4$-related)
c_w	Drag coefficient of blade profile
$\cos \varphi$	Power factor

$\cos \varphi_K$	Power factor in case of short-circuit
C	$= C(k)$, Theodorsen function
d	Half-profile depth
d_m	Average bearing diameter
dA_B	Blade element area
dF_A	Lift force on blade element
dF_{AW}	Resultant force on blade segment from lift and drag components
dF_{ax}	Axial force at blade element
dF_t	Tangential force at blade element
dF_W	Drag force on blade element
$d\dot{J}_{ax}$	Axial momentum losses by blade streaming
$d\dot{J}_t$	Change of tangential momentum of angular streaming
dM_L	Moment per unit of width during blade pitch adjustment due to acceleration of air mass and due to air damping
dM_{lift}	Torsional moment at blade element due to lifting forces
dM_T	Righting moment in direction of air flow on the blade element
dU	Voltage deviation, voltage drop
f	Frequency
f_1	Grid frequency
f_{2v}	Rotor current frequency (the vth harmonic) in asynchronous machines
f_G	Generator frequency
f_{L1}	Bearing coefficient dependent upon bearing type and loading
f_μ	Frequency of the μth subharmonic
f_v	Frequency of the vth harmonic
F_a	Axial force on bearing
F_N	Normal force
F_{Pr}	Force creating propeller moment
F_Q	Transverse force component
F_{St}	Actuating force on blade
F_Z	Centrifugal force
g_{L1}	Bearing load direction factor
i_{ABl}	Transmission ratio between adjustment mechanism and blade pitch adjustment
i_G	Total current (rotating pointer)
i_{G1}	Total current in phase 1
i_{G2}	Total current in phase 2
i_{G3}	Total current in phase 3
i_{Gd}	Total current in longitudinal direction of field coordinates
i_{Gq}	Total current in transverse direction of field coordinates
i_{MBl}	Transmission ratio between adjustment motor and blade rotation
$i_{MBl,rot}$	Transmission ratio between servomotor and rotor blade adjustment
$i_{MBl,lin-rot}$	Transmission ratio between servomotor and blade pitch adjustment in the case of direct motor drive
i_{ms}	Magnetizing current in the stator
i_R	Machine-side rotor current (rotating pointer)

i_{R1}	Machine-side rotor current in phase 1
i_{R2}	Machine-side rotor current in phase 2
i_{R3}	Machine-side rotor current in phase 3
i_{Rd}	Machine-side rotor current in longitudinal direction of field coordinates
$i_{Rd\,act}$	Actual value of i_{Rd}
$i_{Rd\,des}$	Desired value of i_{Rd}
i_{RN}	Grid-side rotor current (rotating pointer)
i_{Rq}	Machine-side rotor current in phase 1
$i_{Rq\,act}$	Actual value of i_{Rq}
$i_{Rq\,des}$	Desired value of i_{Rq}
I_0	No-load current in one machine phase
I_1	Stator current
I_1	Effective value of fundamental component current
I'_2	Rotor phase current acting on stator side
I_{an}	Starting current of asynchronous machines
I_E	Exciter current
I_{Fe}	Iron loss current in one machine phase
I_{St}	Electric current or hydraulic flow for blade pitch positioning
I_Z	Current of reactive power compensation system
I_μ	Magnetizing current in one machine phase
I_ν	Effective value of the νth harmonic current
J_B	Moment of inertia of blade during rotation around the hub
J_{Bl}	Moment of inertia of rotor blade when turned about its longitudinal axis
$J_{Bl(A)}$	Moment of inertia of rotor blade taken at the drive motor side
J_G	Moment of inertia of generator rotor
J_{LB}	Equivalent moment of inertia due to accelerated air mass
J_{Mot}	Moment of inertia of the drive motor
$J_{Mot(Bl)}$	Moment of inertia of the drive motor acting on the rotor blades
J_R	Moment of inertia of all rotating masses
J_{tot}	Total moment of inertia of blade pitch adjustment system
$J_{tot(A)}$	Total moment of inertia of the entire blade pitch adjustment system taken from the drive side
$J_{tot(Bl)}$	Total moment of inertia of the entire blade pitch adjustment system taken from the blade side
J_{trans}	Moment of inertia of transmission elements such as gears, couplings, etc., between drive motor and blade turning mechanism
k_A	Rate of change factor of the rotor displacement angle after falling out of step
k_d	Characteristic damping
k_{DB}	Coefficient of structural and aerodynamic damping
k_{DK}	Coefficient of damping for the drive train
k_{RL}	Coefficient of friction for bearing friction at rotor blade during blade pitch adjustment
k_t	Ratio of the acceleration moments of the drive-train component to the entire rotor system (M_{BT}/M_{BR})
k_{THD}	Total harmonic distortion

k_{THD0}	Grid-state-dependent and grid short-circuit power-dependent output value of total harmonic distortion
k_{THD1}	Gradient of relative harmonic content
k_{THD2}	Elongation factor of relative harmonic content
k_{TS}	Torsional stiffness of the drive train
k_u	Harmonic distortion of voltage
k_U	Factor for the maximum magnification of generator moment
m	Number of phases of three-phase current windings
m_B	Mass of a rotor blade
m_{dyn}	$= M_{KD}/M_{KS\,max}$, dynamic increase in moment
M	Moment
M	Torque
M_A	Driving torque
M_{act}	Actual value of moment
M_{AG}	Driving torque at generator
M_{AM}	Motor start-up torque
M_{An}	External torque of blade pitch adjustment drive taking into account spring and damping characteristics
M_{AT}	Driving torque of drive train including losses
M_{AV}	Internal torque of blade pitch adjustment drive
M_{AW}	Wind turbine driving moment
M_{bend}	Torsional moment at rotor blade due to bending
M_{BG}	Acceleration moment at the generator
M_{Bl}	Rotor blade torsional moment during turning about blade longitudinal axis
$M_{Bl\,max}$	Maximum blade torsional moment in extreme situations
M_{Bln}	Blade torsional moment in normal operation
M_{BR}	Acceleration moment in rotor system
M_{BT}	Acceleration moment in drive train
M_{BW}	Acceleration moment on wind turbine
M_{Cz}	Coriolis moment in relation to the z axis
M_D	Damping moment of the synchronous machine
M_{des}	Desired value of torque
M_{frict}	Moment of friction of all blade bearings during blade pitch adjustment
M_K	Breakdown torque of asynchronous machine
M_K	Pull-out torque of synchronous machine
M_{KD}	Dynamic breakdown or pull-out torque
M_{KG}	Generator breakdown or pull-out torque
M_{KM}	Motor breakdown or pull-out torque
$M_{K\,max}$	Maximum breakdown or pull-out torque
$M''_{K\,max}$	Maximum moment of synchronous machines due to subtransient short-circuit currents in the damping winding
$M'_{K\,max}$	Maximum value of pulsating short-circuit moment of synchronous machines due to transient currents
M_{KS}	Static breakdown or pull-out torque
$M_{KS\,max}$	Static breakdown or pull-out torque at maximum excitation
$M_{KS\,min}$	Static breakdown or pull-out torque with no-load excitation

M_{Ku}	Coupling torque at generator
M_{L}	Moment with blade pitch adjustment by acceleration of air masses and air damping
M_{LB}	Moment with blade pitch adjustment due to acceleration of air masses
M_{LD}	Moment with blade pitch adjustment due to air damping
M_{lift}	Torsional moment at rotor blade due to lift forces
M_{max}	Maximum moment
M_{N}	Nominal moment
M_{NG}	Generator nominal moment
M_{NM}	Motor nominal moment
M_{Pr}	Propeller moment
M_{res}	Reserve moment during acceleration of the blade pitch adjustment mechanism
M_{RL}	Load-dependent moment of friction of a bearing
M_{RLk}	Load-dependent moment of friction of bearing k
M_{s}	Steady-state torque
M_{S}	Pull-up torque
M_{SM}	Pull-up torque of a motor (asynchronous machine)
M_{St}	Moment exerted upon blade by actuator
M_{T}	Righting moment in direction of air flow on blade profile
M_{TD}	Damping component of drive train moment
M_{teeter}	Torsional moment at the blade due to teetering of the rotor
M_{TT}	Torsionally elastic component of drive-train moment
M_{W}	Load torque
M_{WG}	(Electrical) load torque of the generator
n	Rotational speed
n_1	Speed of rotating field or synchronous speed
n_{A}	Number of turbines
n_{act}	Actual value of rotational speed
n_{AV}	Rotational speed of blade pitch adjustment drive
n_{des}	Desired value for speed
n_{KG}	Generator breakdown-torque speed (asynchronous machine)
n_{KM}	Motor breakdown-torque speed (asynchronous machine)
n_{NG}	Generator nominal speed (asynchronous machine)
n_{NM}	Motor nominal speed (asynchronous machine)
n_v	Speed of the harmonic field of ordinal number v
p_1	Number of pairs of poles in the stator
p_2	Number of pairs of poles in the rotor
P	Average value of power
P_{E}	Power of producer in the grid
P_{el}	Electrical generator power
P_{G}	Total active power in rotor and stator
P_{L}	Power of the load in the grid
P_{L0}	Equivalent static bearing loading
P_{mech}	Mechanical input power of generator
P_{N}	Nominal power
P_{O}	Moving air mass power

P_{Stn}	Power for normal positioning procedures
P_{Sts}	Power for fast positioning procedures
$P_{T\,act}$	Actual value of total active power
$P_{T\,des}$	Desired value of total power
P_W	Wind turbine power
$P_{W\,max}$	Maximum wind turbine power
P_δ	Air gap power of an electrical machine
P_σ	Standard deviation of power
Q_C	Compensation reactive power
Q_G	Total reactive power in rotor and stator
$Q_{T\,act}$	Actual value of total reactive power
$Q_{T\,des}$	Desired value of total reactive power
r	Radius of a blade element
r'	Radius of the rotor blade centre of mass
r_o	Distance between yaw and rotor blade fulcrums
R_1	Stator resistance of one machine phase
R'_2	Rotor resistance of one phase of an asynchronous machine transformed on the stator side
R_a	Outer radius of rotor blade
R_{grid}	Resistance of the connection elements between higher grid and point of common coupling
R_i	Inner radius of rotor blade
R_{L+T}	Ohmic resistance of lines and transformers
R_{L+T}	Resistance between wind turbine and point of common coupling
s	Slip (of an asynchronous machine)
s_K	Breakdown slip (asynchronous machine)
s_N	Nominal slip (asynchronous machine)
s_ν	Slip of the νth harmonic (asynchronous machine)
S_{grid}	Grid apparent power
S_k	Grid short-circuit power
S''_k	Initial value of alternating current short-circuit power
S_{load}	Load apparent power
S_p	Centre of gravity
S_{rG}	Generator rated apparent power
S_{rT}	Transformer rated apparent power
S_{supply}	Supply apparent power
t	Time
t_0	Time for rotor blade adjustment into a safe operating state
t_{0b}	Time for rotor blade adjustment into a safe operating state with pure acceleration processes
t_{Af}	Acceleration time of blade positioning drive in the case of fast positioning procedures
t_{APD}	Acceleration time of blade positioning drive system
t_{APDd}	Acceleration time of direct-driven blades
t_{APDz}	Acceleration time of z rotor blades adjusted by positioning drive system

t_B	Blade thickness
t_f	Duration of secondary effect of flicker
t_v	Delay time
t_v	Rotor blade adjustment time at constant speed
T_D	Time constant of damping of torque oscillation
T_E	Time constant of the exciter circuit
T_G	Generator acceleration time constant
T_n	$= 1/\omega_0 = 1/A_o$, time constant of the rotation speed integrator
T_R	Rotor system acceleration time constant
T_V	Time constant for the decaying dynamic pull-out torque to its steady-state value
T_W	Wind turbine acceleration time constant
T_ε	$= p/\omega_0$, time constant of integrator for the determination of the angle of torsion (generator side)
u_G	Total voltage (rotating pointer) corresponds to stator voltage ($u_T = u_S$)
u_{Gq}	Total voltage in quadrature-axis direction of the field coordinates ($u_{Tq} = u_{Sq}$)
u_{kASM}	Magnification factor of the short-circuit power of asynchronous machines
u_{kSM}	Magnification factor of the short-circuit power of synchronous machines
u_N	Nominal voltage
u_{R1}	Machine-side rotor voltage in phase 1
u_{R2}	Machine-side rotor voltage in phase 2
u_{R3}	Machine-side rotor voltage in phase 3
u_{Rd}	Machine-side rotor voltage in direct-axis direction of the field coordinates
u_{Rq}	Machine-side rotor voltage in quadrature direction of the field coordinates
u_{S1}	Stator voltage in phase 1
u_{S2}	Stator voltage in phase 2
u_{S3}	Stator voltage in phase 3
u_{Sq}	Stator voltage in quadrature direction of the field coordinates
$u_{\mu VT}$	Compatibility level of the μth subharmonic specific to the fundamental component
u_ν	Harmonic voltage of the νth-order specific to the fundamental component
$u_{\nu VT}$	Compatibility level of the νth harmonic specific to the fundamental component
U_0	Direct voltage component
U_1	Effective value of fundamental component voltage, grid voltage
U_1	Grid voltage
U_C	Reference conductor voltage of capacitor banks
U_{di}	Ideal direct voltage
U_g	Direct current link voltage
U_{Gen}	Generator voltage
U_{Grid}	Grid voltage
U_i	Induced machine voltage
U_{Mot}	Motor voltage
U_p	Rotary-field (internal) voltage of a synchronous machine
$U_{R\,max}$	Maximum rotor voltage
U_ν	Effective value of the νth harmonic voltage
U_Z	Ignition impulse voltage

v	Wind speed
$\bar{v}$	Average wind speed
v_0	Rotational speed of the rotor head
v_1	Undisturbed far-upstream wind speed
v_2	Wind speed at the turbine
v_{2ax}	Axial component of the decelerated wind speed at the rotor blade
v_{2t}	Tangential component of the decelerated wind speed at the rotor blade
v_3	Decelerated wind speed far downstream of the turbine
v_r	Resultant wind speed
v_u	Peripheral speed
v_w	Local wind speed
v_σ	Standard deviation of wind speed
V_a	Airstream volume element
W_W	Energy drawn by wind turbine
X_d''	Subtransient reactance of a synchronous machine
$X_{1\sigma}$	Leakage reactance of one stator phase
$X_{2\sigma}'$	Leakage reactance of one rotor phase specific to the stator
X_d	Synchronous direct-axis reactance of a synchronous machine
X_d'	Transient reactance of a synchronous machine
X_G	Leakage reactance from stator and rotor of an asynchronous machine
X_{grid}	Reactance of the connection elements between grid and point of common coupling
X_h	Main reactance of one machine phase
X_{L+T}	Reactance between wind turbine and point of common coupling
X_{L+T}	Reactance of lines and transformers
X_q	Synchronous quadrature reactance of a synchronous machine
X_u	Reactance of the frequency converter valves
z	Number of rotor blades
z_a	Number of driven rotor blades
Z_C	Impedance of the capacitor bank
Z_k	Grid impedance (in short-circuit)
Z_{load}	Load impedance
α	Local profile flow angle at the rotor blade
α	Trigger or firing angle of thyristors
α_0	Ignition angle output value
α_{max}	Maximum thyristor trigger angle/thyristor inverter stability limit
$\alpha_{max\ 15}$	Maximum thyristor trigger angle/thyristor inverter stability limit at 15% voltage drop
β	Rotor blade pitch
$\dot{\beta}$	Adjustment speed of the rotor blade
$\dot{\beta}_n$	Normal adjustment speed of the rotor blade
$\dot{\beta}_s$	Fast adjustment speed of the rotor blade
$\ddot{\beta}$	Adjustment acceleration of the rotor blade
γ	Cone angle
ν	Ordinal number of one harmonic (integer)

δ	Angle between plane of rotation and resultant air flow velocity
Δ_n	Adjustment range of rotation speed
ΔP	Power fluctuation range
ΔU	Voltage drop
ΔU_r	Voltage rise
$\Delta U_{r\,perm}$	Permissible voltage rise
Δv	Fluctuation range of wind velocity
$\Delta \varepsilon$	Angle of torsion
ε_G	Generator rotor angle of rotation
$\dot{\varepsilon}_G$	Angular velocity of the generator rotor
ε_W	Wind turbine angle of rotation
$\dot{\varepsilon}_W$	Angular velocity of the turbine
η_{ABl}	Transfer efficiency between positioning system and rotor blade
θ	Rotor displacement angle (electrical) of a synchronous machine
θ	Angle (mechanical) between plane of rotation and chord
$\theta_{0.7}$	Angle between plane of rotation and chord at the rotor blade given at 0.7 times the radius
θ_B	Angle between $\theta_{0.7}$ and the chord of the rotor blade
λ	Load angle (electrical) in asynchronous machines between fixed grid slip voltage and load-dependent slip voltage
λ	Line angle
λ	Tip speed ratio (mechanical) of the rotor blade tip speed to wind velocity
λ_N	Tip speed ratio in nominal operating state
μ	Noninteger factor of subharmonics
v	Ordinal number of harmonics
ρ	Air density
φ_{Gen}	Phase angle of generator current (input angle)
φ_{Mot}	Phase angle of motor current
φ_v	Phase displacement angle of the vth harmonic
ψ	Position of rotor in relation to tower
ψ_{Bl}	Rotor blade position in relation to tower
ω	Resultant angular velocity from azimuth yaw and turbine rotation
ω_0	Steady-state grid (angular) frequency
ω_1	Angular velocity of the stator rotating field in two-pole winding design ($p = 1$)
ω_2	Angular velocity of the rotor rotating field in two-pole winding design ($p = 1$)
ω_A	Yaw control angular velocity
ω_{Bl}	Angular velocity of blade rotation about its longitudinal axis
ω_{BV}	Design value for angular velocity of blade pitch adjustment system
ω_G	Angular velocity of the generator rotor
ω_{mech}	Angular velocity of the mechanical rotation of the generator rotor
ω_{Mot}	Angular velocity of the servomotor
ω_N	Nominal angular velocity
ω_R	Angular velocity of the turbine rotor (vector quantity)
ω_{st}	Angular velocity of actuator
ω_W	Angular velocity of wind turbine
ω_v	Angular frequency of the vth harmonic

1

Wind Energy Power Plants

Rising pollution levels and worrying changes in climate, arising in great part from energy-producing processes, demand the reduction of ever-increasing environmentally damaging emissions. The generation of electricity – particularly by the use of renewable resources – offers considerable scope for the reduction of such emissions. In this context, the immense potentials of solar and wind energy, in addition to the worldwide use of hydropower, are of great importance. Their potential is, however, subject to transient processes of nature. Following intensive development work and introductory steps, the conversion systems needed to exploit these power sources are still in the primary phase of large-scale technical application. For example, in Germany around 8% of electricity is already being provided by wind turbines. However in the German provinces Mecklenburg-Western-Pomerania, Schleswig-Holstein, Brandenburg and Saxony-Anhalt there are about 50% wind power feed in. In Germany more power is supplied by wind energy than by hydroelectric plants.

These environmentally friendly technologies in particular require a suitable development period to establish themselves in a marketplace of high technical standards.

The worldwide potential of wind power means that its contribution to electricity production can be of significant proportions. In many countries, the technical potential and – once established – the economically usable potential of wind power far exceeds electricity consumption. Good prospects and economically attractive expectations for the use of wind power are, however, inextricably linked to the incorporation of this weather-dependent power source into existing power supply structures, or the modification of such structures to take account of changed supply conditions.

1.1 Wind Turbine Structures

In the case of hydro, gas or steam, and diesel power stations (among others) the delivery of energy can be regulated and adjusted to match demand by end users (Figure 1.1(a)). In contrast, the conversion system of a wind turbine is subject to external forces (Figure 1.1(b)). The delivery of energy can be affected by changes in wind speed, by machine-dependent factors such as disruption of the airstream around the tower or by load variations on the consumer side in weak grids.

Grid Integration of Wind Energy: Onshore and Offshore Conversion Systems, Third Edition. Siegfried Heier.
© 2014 John Wiley & Sons, Ltd. Published 2014 by John Wiley & Sons, Ltd.

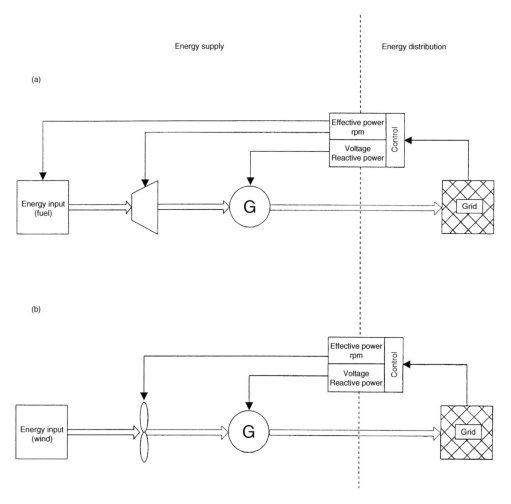

Figure 1.1 Energy delivery and control in electrical supply systems: (a) diesel generators, etc., and (b) wind turbines

The principal components of a modern wind turbine are the tower, the rotor, the nacelle (which accommodates the transmission mechanisms and the generator) and – for horizontal-axis devices – the yaw systems for steering in response to changes in wind direction. Switchgear and protection systems, lines, and maybe also transformers and grids, are required for supplying end users or power storage systems. In response to external influences, a unit for operational control and regulation must adapt the flow of energy in the system to the demands placed upon it. The next two figures show the arrangement of the components in the nacelle and the differences between mechanical–electrical converters in the modern form of wind turbines. Figure 1.2 shows the conventional drive train design in the form of a geared transmission with a high-speed generator. Figure 1.3, by contrast, shows the gearless variant with the generator being driven directly from the turbine. These pictures represent the basis for the functional relationships and considerations of the system.

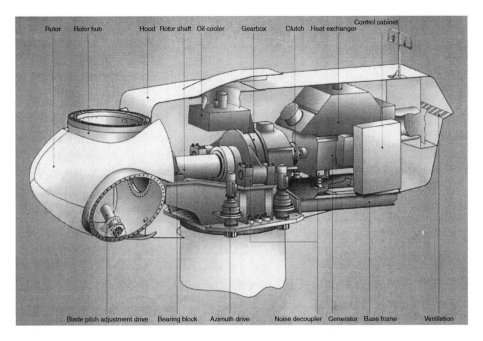

Figure 1.2 Nacelle of a wind turbine with a gearbox and high-speed 1.5 MW generator (TW 1.5 GE/Tacke). Reproduced by permission of Tacke Windenergie

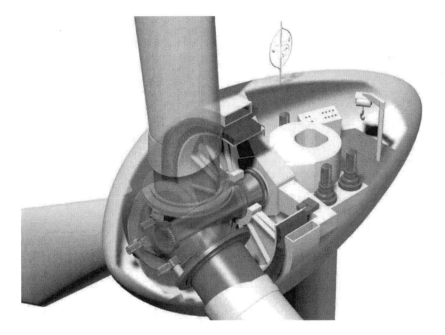

Figure 1.3 Schematic structure of a gearless wind turbine (Enercon E66, 70 m rotor diameter, 1.8/2 MW nominal output). Reproduced by permission of Enercon

Following a brief glance back into history, developmental stages and different wind turbine designs and systems will be briefly highlighted and the processes of mechanical–electrical power conversion explained. Moreover, particular importance is assigned to the interconnection of wind turbines to form wind farms and their combined effect in grid connection.

1.2   A Brief History

For thousands of years, mankind has been fascinated by the challenge of mastering the wind. The dream of defying Aeolos[1] and taming the might of the storm held generations of inventors under its spell. To attain limitless mobility by using the forces of Nature, thereby expanding the horizons of the then known world, was a challenge even in antiquity. Thus, sailing and shipbuilding were constantly pursued and developed despite doldrum, hurricane, tornado and shipwreck. Progress could only be achieved by employing innovative technologies. These, together with an unbridled lust for voyages of discovery, built up in the minds of sovereigns and scholars a mosaic of the world, the contours of which became ever more enclosed as time went by.

With wind-harnessing technology on land and on the sea, potentials could be realized and works undertaken that far outpaced any previously imagined bounds. For example, using only the power of animals and of the human arm, it would never have been possible for the Netherlands to achieve the drainage that it has through wind-powered pumping and land reclamation.

Archaeological discoveries relating to the use of wind energy predate the beginning of the modern era. Their origins lay in the Near and Middle East. Definite indications of windmills and their use, however, date only from the tenth century, in Persia [1.1]. The constructional techniques of the time made use of vertical axes to apply the drag principle of wind energy capture (Figure 1.4). Such mills were mostly found in the Arab countries. Presumably, news

Figure 1.4   Persian windmill (model)

[1] Aeolos: Greek god of the winds.

Figure 1.5   Sail windmill

of these machines reached Europe as a result of the Crusades. Here, however, horizontal-axis mills with tilted wings or sails (Figure 1.5) made their appearance in the early Middle Ages.

The use of wind energy in Western Europe on a large scale began predominantly in England and Holland in the Middle Ages. Technically mature post mills (Figure 1.6) and Dutch windmills (Figure 1.7) were used mostly for pumping water and for grinding. More than 200 000 (two hundred thousand) of these wooden machines were built throughout North-West Europe, representing by far the greatest proportion of energy capture by technical means in this region. At the beginning of the twentieth century, some 20 000 (twenty thousand) windmills were still in use in Germany.

From the nineteenth century onwards, mostly in the USA, the so-called 'western wheel' type of turbine became widespread (Figure 1.8). These multibladed fans were built of sheet steel, with around 20 blades, and were used mostly for irrigation. By the end of the 1930s, some 8 million units had been built and installed, representing an enormous economic potential.

1.3   Milestones of Development

The first attempt to use a wind turbine with aerodynamically formed rotor blades to generate electricity was made over half a century ago. Since then, besides the design and construction of large projects in the 1940s by the German engineers Kleinhenz [1.2] and Honnef [1.3], the pilot

Figure 1.6   Post mill

projects of the American Smith-Putnam (1250 kW nominal output, 53 m rotor diameter, 1941), the Gedser wind turbine in Denmark (200 kW nominal output, 24 m rotor diameter, 1957) and the technically trail-blazing Hütter W34 turbine (100 kW nominal output, 34 m rotor diameter, 1958) are worthy of mention (Figure 1.9).

The German constructor Allgaier started the first mass production of wind power plants in the early 1950s. They were designed to supply electricity to farmsteads lying far from the public grid. In coastal areas these turbines drove 10 kW generators; inland they were fitted with 6 kW units. Their aerodynamically formed blades of 10 m diameter could be pitched about the longitudinal axis so as to regulate the power taken from the wind. Even today, some of these

Figure 1.7   Dutch windmill

turbines (see Figure 1.10) are in operation with full functionality, after more than 50 years of service.

After the 1960s, cheaper fossil fuels made wind energy technology economically uninteresting, and it was only in the 1970s that it returned to the spotlight due to rising fuel prices. Some states then developed experimental plants in various output classes.

In particular in the USA, Sweden and the Federal Republic of Germany, turbines with outputs in the megawatt class have attracted most attention. Here, with the exception of the American MOD-2 (Figure 1.11) with five units and the Swedish–American WTS-4 (Figure 1.12) with five or two units, large converters such as the German GROWIAN (Figure 1.13), the Swedish WTS-75 AEOLUS model, the Danish Tvind turbine and the US MOD-5B variants in Hawaii were all one-offs. Despite many and varied teething troubles with the pilot installations, it was

Figure 1.8   American wind turbine

clear even then that technical solutions could be expected in the foreseeable future that would permit the reliable operation of large-scale wind turbines. Second-generation megawatt-class systems such as the WKA 60 (Figure 1.14) and the Aeolus II (Figure 1.15) have confirmed this expectation.

Mainly in the US state of California, but also in Denmark, Holland and the Federal Republic of Germany, considerable efforts were being made, independently of the development of large turbines, to use wind power to supply energy to the grid on a large scale. In the 1980s, wind turbines with total capacity of around 1500 MW were installed in California alone. In the initial phases, turbines of the 50 kW categories were used (Figure 1.16). Scaling-up the systems that were successful through the 100, 150 and 250 kW classes (Figures 1.17 and 1.19) and the 500/600 kW order of magnitude (Figures 1.18 and 1.20) has led to wind farms with turbines in the megawatt range (Figure 1.21).

Figure 1.9   Hütter W 34 turbine

This development has made the mass production of wind turbines possible. A considerable improvement of performance can thus be achieved. Progressively increasing turbine size (see Figures 1.22 to 1.25) using designs of widely differing types and costs has led to the development of machines in the 500 kW and megawatt classes that are remarkable for their high availability and good return-on-investment potential.

The individual manufacturers have chosen very different routes to market success in relation to this trend. NEG Micon has retained the classic Danish stall-regulated turbines with an asynchronous generator rigidly coupled to the grid in the power classes up to 1.5 MW (Figure 1.22). Bonus (Figure 1.23), Nordex (Figure 1.24) and Vestas (Figure 1.25) as well as GE/Tacke (Figure 1.26) have altered their turbine configuration in the different size classes, particularly with regard to the turbine regulation (stall or pitch) and generator systems (fixed-speed or variable-speed with a thyristor/ IGBT frequency converter). Currently 3 to 5 MW systems from all well-known manufacturers are being operated as prototypes or are available on the market.

One new development has been the trend towards gearless wind turbines. Several attempts have been made to introduce and establish in the market small, high-speed, horizontal-axis

Figure 1.10   Allgaier turbine

turbines with direct-drive generators. Up until now these attempts have met with limited success. Microturbines (Figure 1.27) with a permanent-magnet synchronous generator driven directly from the turbine are usually used as battery chargers. The success of such systems is rooted in their attractive design and low price as well as in the modern worldwide sales concept and the simple installation of the plants.

To some degree, companies that have entered into the production of wind generators at a later stage have been able to draw upon existing developments and techniques, thus allowing their first efforts to overtake the systems of established manufacturers. DeWind started its development (Figure 1.28) with a pitch-regulated 600 kW turbine and a variable-speed generator system (double-fed asynchronous machine), which could not have been produced at an economical cost a few years previously and which is currently favoured by most manufacturers. Then 1 and 2 MW systems of the same design followed.

Figure 1.11   MOD 2 in the Goodnoe Hills (USA): 2.5 MW nominal output, 91 m rotor diameter, 61 m hub height

Figure 1.12   WTS-4 turbine in Medicine Bow, USA.: 4 MW nominal output, 78 m rotor diameter, 80 m tower height

Figure 1.13   GROWIAN by Brunsbüttel/Dithmarschen, 3 MW capacity, 100 m rotor diameter, 100 m hub height

The development of wind power systems has largely been carried out by medium-sized companies. Smaller manufacturers, however, face financial limits in the development of MW systems. The 1.5 MW turbine MD 70/MD 77 (Figure 1.29), again with the double-fed asynchronous generator design, which was developed by pro + pro for the manufacturers BWU, Fuhrländer, Jacobs Energie (now REpower Systems) and Südwind / Nordex is opening up new developmental and market opportunities for smaller companies in the field of large-scale plants.

Vertical-axis rotors, so-called Darrieus turbines, are enchantingly simple in structure. In their basic form they have up until now mostly been built with gearing and generators at base level

Figure 1.14   WKA 60 in Kaiser-Wilhelm-Koog: 1.2 MW nominal output, 60 m rotor diameter, 50 m tower height

(Figure 1.30). Variants in the form of so-called H-Darrieus gearless turbines in the 300 kW class were first designed with rotating towers and large multiple generators at ground level (Figure 1.31(a)). Further development led to machines with fixed tripods and annular generators in the head (Figure 1.31(b)). These variants have not, however, been successful in establishing themselves widely in the wind power market.

The Enercon E 40 horizontal-axis turbine was the first system in the 500 kW class with a direct-drive generator to establish itself in the market with great success in a very short time. Figure 1.32 shows the schematic construction of the nacelle. The generator, specially developed for this model, connects directly to the turbine and needs no independent bearings. In this way, wear on mechanical components running at high speed is reduced to a minimum. Operational run times of 180 000 hours have been quoted for many years.

Figure 1.15   AEOLUS II near Wilhelmshaven: 3 MW nominal output, 80 m rotor diameter, 88 m tower height

   The gearless E 30, E 40, E 58, E 66 and E 112/E 126 models from Enercon were produced as a development of the stall-regulated geared models E15/E16 and E17/E18, by way of the E 32/E 33 variable-pitch turbines (Figure 1.33). In parallel, but with a slight delay, the conversion from thyristors to pulse inverters was accomplished. This configuration thus unites the advantages of variable speeds (and the associated reduction in drive-train loading) with those of a grid supply having substantially lower harmonic feedback.
   In comparison to the gearless designs with electrically excited synchronous generators, as shown in Figure 1.33(d) to (h), permanent-magnet machines permit the arrangement of higher numbers of poles around the rotor or stator. By using high-quality permanently magnetic

Figure 1.16   Wind farm in California with turbines in the 50/100 kW class

Figure 1.17   Wind farm in California with turbines in the 250 kW class

Figure 1.18   Wind farm in Wyoming with turbines in the 600 kW class

Figure 1.19   Wind farm in North Friesland with turbines of the 250 kW class

Figure 1.20 Wind farm on Fehmarn Island with turbines of the 500 kW class

(a) On land   (b) At sea. Reproduced by permission
                                         of GE Wind Energy

Figure 1.21 Wind farm with 1.5 MW turbines

(a) NTK 150/25   (b) NTK 300/31   (c) NTK 600–180/43   (d) NEG 1500/60

Figure 1.22 Size progression of stall-regulated turbines of the same design (fixed-speed, fixed-pitch machines) from NEG Micon / Nordtank. Reproduced by kind permission of NEG Micon

(a) 300 kW / 33-2   (b) 600 kW / 44-3   (c) 1 MW / 54   (d) 2 MW

Figure 1.23 Size progression of Bonus turbines: (a,b) fixed-speed, stall-controlled turbines; (c,d) active (combi-)stall turbines with a slight blade pitch adjustment

(a) N 27/29   (b) N 43 (600 kW)   (c) N 62 (1300 kW)   (d) N 80/90 (2500 kW).
(150/250 kW)   Reproduced by
                                                              permission of Nordex

Figure 1.24 Size progression of Nordex turbines: (a,b,c) fixed-speed, fixed-pitch machines; (d) a large-scale, variable-speed, variable-pitch unit

materials, relatively favourable construction sizes can thus be achieved (Figure 1.34) and very high efficiencies attained, particularly in the partial load range. Such a plant configuration of the 600 kW class (Figure 1.34(a)) has been able to achieve excellent returns over several years of fault-free operation. A 2 MW unit with such a generator design (Figure 1.34(b)) was designed with a medium-voltage generator of 4 kV system voltage.

A further possibility, which has been considered for large, slow-running turbines in particular, is the combination of a low-speed generator and a turbine-side gearbox, as shown in Figure 1.35. The single-stage gearbox turns the generator shaft at around eight times the turbine speed of approximately 100 revolutions per minute. Thus, even for units in the 5 MW range, generators in compact and technically favourable construction sizes of approximately 3 m diameter can be used.

Further large-scale turbines in the 5MW class with a rotor diameter of over 125m are REpower 5 M and 6 M and Siemens SWT 6-154 (Fig. 1.36). A double-fed asynchronous generator with medium-voltage isolation in the low-voltage range (950 V stator-side or 690 V rotor-side) is used in the Repower system. The Siemens turbine has a direct drive permanent excited synchronous generator.

In the following we consider various real operational situations, the essential differences between the systems involved and the resulting effects on supply to the grid, taking as a basis the functional structure of wind power machines and their influences.

(a) V 17 (55 kW)   (b) V 25 (225 kW)   (c) V 39 (500 kW)

(d) V 44 (600 kW)   (e) V 66 (1650 kW)   (f) V 112 (3 MW)

Figure 1.25   Size progression of Vestas turbines: (a) small, fixed-speed, fixed-pitch machine; (b,c,d) larger variable-pitch units; (d,e) machines with speed elasticity; or double-fed asynchronous generators; (f) machines with permanent excited synchronous generators.

1.4   Functional Structures of Wind Turbines

For the following consideration, which is mainly concerned with the mechanical interaction of electrical components and with interventions to modify output, we will draw upon the nacelle layout shown in Figure 1.2. With the correct design, the influences of the tower and of steering in response to changes in wind direction can be handled separately (Section 2.2.1) or treated as changes in wind velocity. The block diagram shown in Figure 1.37 (see page 28),

(a) TW 80 (80 kW,
21 m rotor diameter)

(b) TW 250 (250 kW,
26 m rotor diameter)

(c) TW 300 (300 kW)

(d) TW 600 (600 kW,
43 m rotor diameter)

(e) TW 1.5/1.5S (1500 kW,
65/70 m rotor diameter)

(f) GE 3.6 (3.6 MW, 100 m rotor
diameter). Reproduced by
permission of GE Wind Energy

Figure 1.26 Size progression of turbines from GE / Tacke: first (a,b) and second (c,d,e,f) generation machines, from fixed-speed, fixed-pitch turbines (a to d) to large-scale, pitch-controlled, variable-speed turbines (e,f)

Figure 1.27 Small system-compatible turbine from aerosmart. Reproduced by permission of Aerodyn Energiesystems GmbH

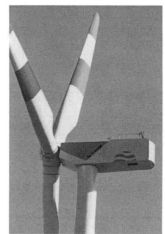

(a) DeWind 4 (600 kW,   (b) DeWind 6 (1000/ 1250kW,
    46/48 m rotor diameter)   60/62/64 m rotor diameter)

Figure 1.28 DeWind 4 (600 kW, 46/48 m rotor diameter). Reproduced by permission of DeWind

Figure 1.29   Joint development of the 1.5 MW MD 70/MD 77 turbine (70/77 m rotor diameter)

Figure 1.30   Fixed-speed 300 kW Darrieus unit with gearing and a conventional generator

(a) annular generator   (b) annular generator
at ground level   in head

Figure 1.31   Variable-speed 300 kW gearless H-Darrieus unit

which illustrates the links between the most important components and the associated energy conversion stages, may serve as the basis for later detailed observations. This diagram also gives an idea of how operation can be influenced by control and supervisory processes. Furthermore, the central position occupied by the generator is made particularly clear.

The following pages therefore explain the physical behavior of a wind energy extraction system and the conversion of this mechanical energy to electrical energy by means of generators. We examine how mechanical moments are handled in the drive unit when the generator is connected to the grid, the design of generators suitable for wind turbines and the combined effects of turbines and power supply grids, as well as the regulation of turbines in isolation and in grid operation, bearing in mind the conditions imposed by the grid and the consumer.

From Figure 1.37 (see page 28), the functional structures for entire wind energy conversion systems, or for particular types of wind energy converter as shown in Figure 1.38(a) and (b) (see page 29), can be further developed. Such simplified block diagrams can help us to understand how the principal components of pitch- or stall-regulated horizontal-axis wind energy converters work and interact.

Wind energy converters with variable-blade pitch (Figure 1.38(a)) allow direct control of the turbine. Figure 2.61 shows that by varying the blade pitch it is possible, firstly, to influence the power input or torque of the rotor, with a smaller blade pitch angle β (or greater ϑ) leading to a lower turbine output and a greater β leading to a higher turbine output (pitch regulation). Secondly, by a few degrees adjustment of the rotor blades, the profile can be brought more fully into stall when β is greater (active stall regulation) and the turbine power falls. A slight

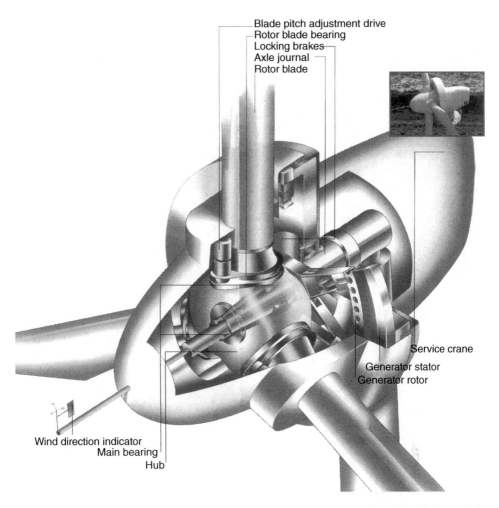

Blade pitch adjustment drive
Rotor blade bearing
Locking brakes
Axle journal
Rotor blade

Service crane
Generator stator
Generator rotor

Wind direction indicator
Main bearing
Hub

Figure 1.32 Schematic layout of the Enercon E 40 gearless turbine. Reproduced by kind permission of Enercon

reduction to the blade pitch angle, on the other hand, guides the rotor out of stall and power increases until laminar flow is achieved on the blade profiles. In this way, the speed of rotation, determined by integration of the difference between turbine torque and the generator's load torque, taking rotating masses (or mechanical time constants) into account, can be influenced at all performance levels – insofar as sufficient energy is available. The pitch control of a wind turbine therefore makes it possible to regulate energy extraction. In this way, adaptation to user needs (e.g. in standalone operation) can be achieved, as well as a measure of protection in storm conditions.

(a) E 15/16 (55 kW)	(b) E 17/18 (80 kW)	(c) E 32/33 (300 kW)	(d) E 30 (200 kW)
(e) E 40 (500 kW)	(f) E 58 (850 kW / 1000 kW)	(g) E 66 (1500 kW / 1800 kW)	(h) E 112 /E 126 (4,5 MW/6 MW/7,5 MW)

Figure 1.33 Enercon turbines from variable-speed geared models with thyristor inverters (a,b,c) to gearless configurations with pulse inverters (d,e,f,g,h); (a,b) with fixed and (c,d,e,f,g,h) with variable pitch. Reproduced by kind permission of Enercon

In (passive) stall-controlled converters (Figure 1.38(b)), the rotor speed is kept at an almost constant speed by the load torque of a rigidly coupled asynchronous (mains) generator, usually of large dimensions. When wind strength rises above nominal levels, the flow over the rotor blades achieves partial or even total stall – whence the so-called 'stall regulation'. The power take-up of the turbine is thereby passively (i.e. design-dependently) limited under full loading

(a) Genesys 600

(b) Zephyros Z72 (70.65 m rotor
diameter and 2 MW nominal
output) with medium-voltage
generator. Reproduced by
permission of Harakosan

Figure 1.34 Gearless wind turbines with permanent-magnet synchronous generator (46 m rotor diameter, 600 kW nominal output)

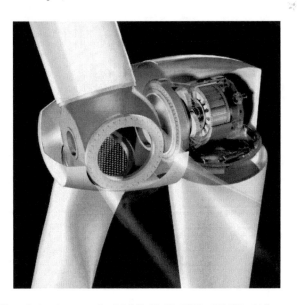

Figure 1.35 Nacelle of the large-scale Multibrid N 5000 (5 MW, 116 m rotor diameter) with single-stage gearing, integral hub and low-speed synchronous generator. Reproduced by permission of Multibrid Entwicklungsgesellschaft GmbH

(a)   (b)

Figure 1.36   Offshore turbines: (a) Repower offshore and onshore turbine 5M/6M, 5 MW/6 MW nominal output power, 126,5 m rotor diameter. *Source:* Repower; (b) Siemens offshore turbine SWT 6-154, 6 MW nominal output power, 154 m rotor diameter. *Source:* Siemens

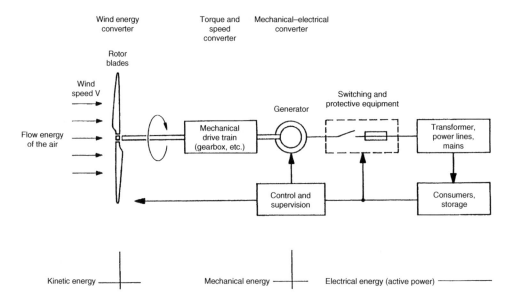

Figure 1.37   Functional chain and conversion stages of a wind energy converter

to values such that under operational wind speed conditions the nominal output of the generator is not significantly exceeded.

The use of variable-speed generators in both regulation systems allows the reduction of sudden load surges, and considerably extends the range of operation. The optimal power can be produced by adjusting the speed of the rotor to the desired speed. For example, it is also

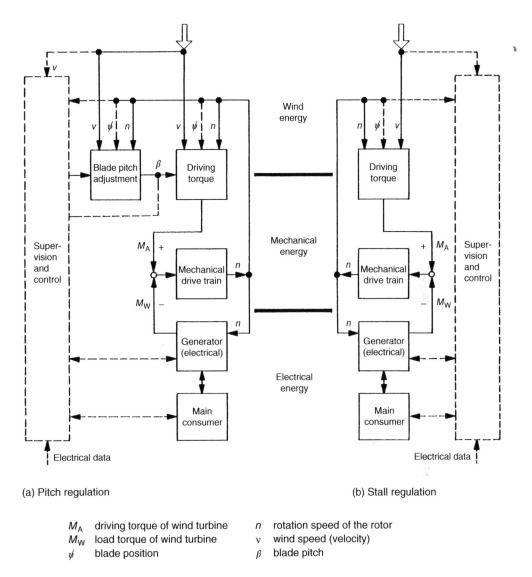

(a) Pitch regulation   (b) Stall regulation

M_A	driving torque of wind turbine	n	rotation speed of the rotor
M_W	load torque of wind turbine	v	wind speed (velocity)
ψ	blade position	β	blade pitch

Figure 1.38   Functional structure of a wind energy converter

possible, in cases where partially increased transitional loads must be handled, to influence the drive torque of stall-regulated turbines by varying the rotational speed of the generator.

A detailed treatment of the generator and associated discussions on the theme of turbine regulation require knowledge of the physical processes and a review of the mathematical laws governing the entire converter system. The following text should encompass this, insofar as is necessary. More detailed studies are also necessary regarding the combined effects of wind turbines working together with existing grid systems and the measures that must be applied to control these effects throughout the entire system.

The following chapters summarize the results of years of research and development. Through practical references achieved from completed projects, particular weight has been given to the usefulness of the results for plant conception and design.

References

[1.1] *Meyers Enzyklopädisches Lexikon*, Vol. 25.9, Completely Revised Edn, Bibliographisches Institut AG, Mannheim, 1979.
[1.2] Kleinhenz, F., Projekt eines Großwindkraftwerkes, *Der Bauingenieur*, 1942, 23/24.
[1.3] Honnef, H., *Windkraftwerke*, Vieweg, Braunschweig, 1932.
[1.4] Zephyros brochure, *Permanent Performance*.

2

Wind Energy Conversion Systems

As shown in Figure 1.37, the blades of a wind turbine rotor extract some of the flow energy from moving air, convert it into rotational energy and then deliver it via a mechanical drive unit (shafts, clutches and gears), as shown in Figures 1.2 and 1.3 respectively, to the rotor of a generator and thence to the stator of the same by mechanical–electrical conversion. The electrical energy from the generator is fed via a system of switching and protection devices, lines and if necessary transformers to the grid, consumers or an energy storage device. In the further discussion of the relevant system components, special attention must be given to the drive torque or performance properties, the structures based thereon and the actions necessary to limit turbine speed, together with reaction effects of the transmission on the turbine.

2.1 Drive Torque and Rotor Power

In contrast to the windmills of yesteryear, modern wind turbines used for generating electricity have relatively fast-running rotors. A few blades with high lift-to-drag ratio profiles that utilize lift attain much higher levels of efficiency than drag-type rotors. They also make it possible to alter the power of the turbine more quickly. Such turbines can, given sufficient wind, attain operating conditions akin to those of conventional power stations.

In examining the operational characteristics of a wind energy converter we must determine the variables influencing the turbine, beginning with the forces on the rotor blades or on small areas thereof, and thence derive the resulting drive torque and corresponding output power.

2.1.1 Inputs and outputs of a wind turbine

To model the power or torque properties of a wind turbine, we can consider the block labelled 'driving torque' in Figure 1.38(a) as a single structure having inputs and outputs as shown in Figure 2.1. These can be subdivided as follows:

- the independent input quantity 'wind speed', which determines the energy input but which can act at the same time as an interference quantity;

Grid Integration of Wind Energy: Onshore and Offshore Conversion Systems, Third Edition. Siegfried Heier.
© 2014 John Wiley & Sons, Ltd. Published 2014 by John Wiley & Sons, Ltd.

- machine-specific input quantities, arising particularly from rotor geometry and arrangement and
- the state variables 'turbine speed', 'rotor blade position' and 'rotor blade angle' arising from the transmission system of the complete wind-power unit, and with the aid of which
- turbine output quantities, such as 'power' or 'drive torque', may be controlled.

The power or torque of a wind turbine may be determined by several means. The best results might be obtained by considering the axial and radial momentum of the airstream, or of small-diameter infinite flow-tubes impinging on the rotor blades, or of elements thereof, thus allowing us to determine local flow conditions and the resulting forces or rotational action on the turbine blades. The following considers this briefly. Calculations based on circulation and vortex distribution over aerofoils [2.1], which allow the derivation of solutions by the Biot–Savart theorem, are not taken into consideration here.

2.1.2 Power extraction from the airstream

Within its effective region, the rotor of a wind turbine absorbs energy from the airstream, and can therefore influence its velocity. Figure 2.2 represents the flow that develops around a converter in an unrestricted airstream in response to prevailing transmission conditions, whereby the airstream is decelerated axially and deviated tangentially in the opposite direction to the rotation of the rotor.

From references [2.2, 2.3] and [2.4], the energy absorbed from an air volume V_a of cross-section A_1 and swirl-free speed of flow v_1 far upstream of the turbine, which results

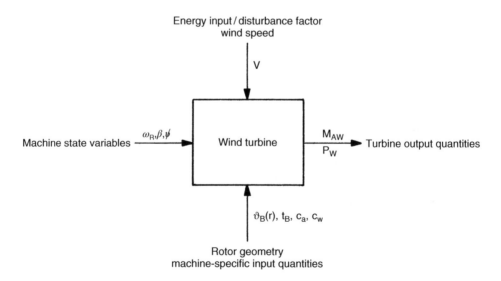

Figure 2.1   Wind turbine inputs and outputs

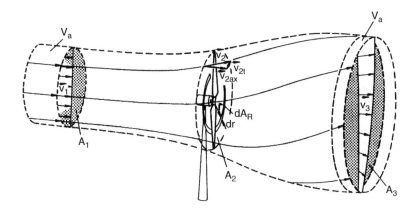

Figure 2.2   Airstream around the turbine

in a downstream reduction of flow speed to v_3 with a corresponding broadening of the
cross-sectional area (wake decay) to A_3, can be expressed as

$$W_w = V_a \frac{\rho}{2} \left(v_1^2 - v_3^2\right) \tag{2.1}$$

The wind turbine power may therefore be expressed as

$$P_w = \frac{dW_w}{dt} = d\frac{\left(V_a \frac{\rho}{2}\left(v_1^2 - v_3^2\right)\right)}{dt}. \tag{2.2}$$

An air volume flow in the rotor area $(A_2 = A_R)$ of

$$\frac{dV_a}{dt} = A_R v_2 \tag{2.3}$$

yields, in the quasi-steady state,

$$P_W = A_R \frac{\rho}{2}\left(v_1^2 - v_3^2\right) v_2. \tag{2.4}$$

The power absorption and operating conditions of a turbine are therefore determined by the
effective area A_R, by the wind speed and by the changes occurring to these quantities in the
field of flow of the rotor. The power of the turbine can thus be influenced by varying the flow
cross-sectional area and by changing flow conditions at the rotor system.

According to Betz [2.2], the maximum wind turbine power output

$$P_{W_{max}} = \frac{16}{27} A_R \frac{\rho}{2} v_1^3 \tag{2.5}$$

is obtained when

$$v_2 = \frac{2}{3}v_1 \quad \text{and} \quad v_3 = \frac{1}{3}v_1. \tag{2.6}$$

Under normal operating conditions up to the nominal output capacity, this reduction of wind speed is approached. When the rotor is idling, or running under light load, the value of v_2 approaches that of v_1.

The ratio of the power P_W absorbed by the turbine to that of the moving air mass

$$P_0 = A_R \frac{\rho}{2} v_1^3 \tag{2.7}$$

under smooth flow conditions at the turbine defines the dimensionless performance coefficient

$$c_p = \frac{P_W}{P_0}. \tag{2.8}$$

The above expression is based upon the assumption that tubular axial air mass transport only occurs from the leading side of the entry area A_1 to the exit area A_3. A more detailed examination of the turbine or rotor blades can be carried out using the modified blade element theory, by introducing a radial wind speed gradient and by taking into account any angular movement of the airstream.

2.1.3   Determining power or driving torque by the blade element method

If, instead of a circular section, we consider an annular section of radius r, width dr and area at the turbine of

$$dA_R = 2\pi r dr, \tag{2.9}$$

then the following is valid for the mass flow rate $d\dot{m}$ in front of, at and behind the rotor in a quasi-steady state:

$$d\dot{m}_1 = d\dot{m}_2 = d\dot{m}_3 \tag{2.10}$$

or

$$\rho\, dA_1 v_{1ax} = \rho\, dA_{2ax} = \rho\, dA_3 v_{3ax}. \tag{2.11}$$

The force that brakes the air axially from v_{1ax} to v_{3ax} may be derived from the loss of momentum from entry to exit by

$$d\dot{J}_{ax} = dF_{ax} = d\dot{m}\left(v_{1ax} - v_{3ax}\right). \tag{2.12}$$

In the rotor area, by application of Froude's theorem,

$$v_{2ax} = \frac{v_{1ax} + v_{3ax}}{2} \quad \text{or} \quad v_{1ax} - v_{3ax} = 2\left(v_{1ax} - v_{2ax}\right) \tag{2.13}$$

and the thrust of the air tubes

$$dF_{ax} = 4\pi\rho\, dr\ v_{2ax}\left(v_{1ax} - v_{2ax}\right) \tag{2.14}$$

can be obtained as a function of the axial wind speed on the rotor (to be determined). The tangential change of momentum may be determined in the same fashion:

$$d\dot{J}_t = v_{1t}\, d\dot{m}_1 - v_{3t}\, d\dot{m}_3. \tag{2.15}$$

When the air entering the flow tube is swirl-free, no other moment is applied to it. The tangential force, which brings the air into rotational flow, is derived as

$$dF_t = dJ_t = -v_{3t}\, d\dot{m}_3 = -v_{2t}\, d\dot{m}_2 \tag{2.16}$$

or, in applications,

$$dF_t = -2\pi \rho r\, dr\, v_{2t} v_{2ax}. \tag{2.17}$$

The force thus depends on the radius and the axial and tangential airstream at the turbine. The air, according to equations (2.14) and (2.17), exerts identical forces on the rotor blades.

For the sake of clarity, the physical processes will be shown for a single rotor blade. Multi-blade arrangements for fast-running turbines (e.g. with $z = 2$, 3 or 4 lift-type blades) can be handled by extension of this system, considering conditions at a single blade of z-fold depth. Depending on blade radius, Figure 2.2 shows that there is different flow behavior at the profile for different blade angles (Figure 2.3).

The combined effect of velocity components and the resultant forces are shown for a single blade element in Figure 2.4. Total values (forces, moments, power) are obtained by the integration of the corresponding values over the blade radius, or by summation of the components of individual blade sections.

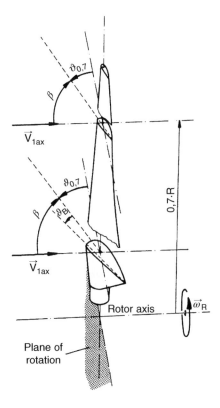

Figure 2.3   Definition of blade pitch angle

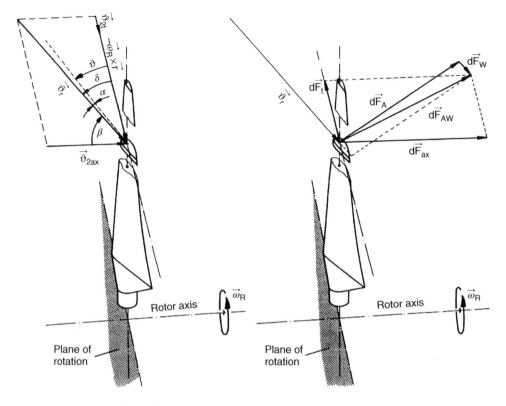

Figure 2.4   Airflow and forces on a rotor blade segment

A segment at radius r of a blade rotating with angular velocity ω_R experiences two airflows: that due to the wind deceleration across the swept area,

$$v_2 = v_{2ax} + v_{2t} \tag{2.18}$$

and that due to the speed of the rotating element at the given radius,

$$v = -\omega_R \times r. \tag{2.19}$$

If we disregard the cone angle then, in the direction of the resultant velocity component,

$$v_r = \sqrt{v_{2ax}^2 \left(\omega_R r + v_{2t}\right)^2}, \tag{2.20}$$

a drag of

$$dF_W = \frac{\rho}{2} t_B v_r^2 c_w \left(\alpha\right) dr \tag{2.21}$$

is exerted, acting against the movement of the blade, while an orthogonally directed lift of

$$dF_A = \frac{\rho}{2} t_B v_r^2 c_a \left(\alpha\right) dr \tag{2.22}$$

exerts a propulsive component. The force on the blade segment resulting from the lift and drag components is

$$dF_{AW} = dF_A + dF_W. \qquad (2.23)$$

Separating this into axial and tangential components leads, for z blades, to a drive torque-generating value of

$$dF_t = z\frac{\rho}{2}t_B v_r^2 \left(c_a \sin\delta - c_w \cos\delta\right) dr \qquad (2.24)$$

and an axial thrust on a rotor blade and on the hub of

$$dF_{ax} = z\frac{\rho}{2}t_B v_r^2 \left(c_a \cos\delta - c_w \sin\delta\right) dr. \qquad (2.25)$$

According to this, and for the respective wind and blade tip velocities, these forces are essentially dependent on:

- the pitch angle ϑ of the blade element with respect to the plane of rotation or β with respect to the rotor axis;
- the local angle of attack α between the resultant wind velocity and chord of the aerofoil;
- the lift and drag coefficients (c_a, c_w) for the blade profile (Figure 2.5), the Reynolds number and the roughness of the blade surface, which should be ignored here, as should skewed airflow impingement due to the cone angle, etc.

The drive torque of the wind turbine is therefore

$$M_{AW} = \int_{R_i}^{R_a} r \, dF_t \int_{R_i}^{R_a} rz\frac{\rho}{2}t_B v_r^2 \left(c_a \sin\delta - c_w \cos\delta\right) dr. \qquad (2.26)$$

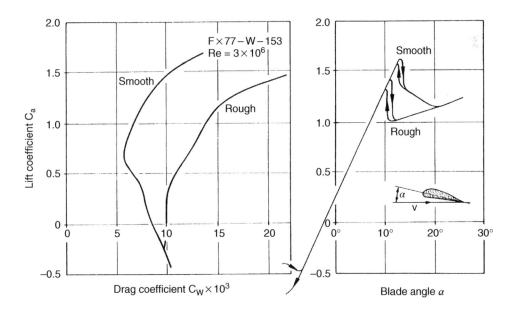

Figure 2.5   Lift and drag coefficients of a Wortmann aerofoil section [2.4]

To determine the forces, it is necessary to determine the local wind velocity v_2 at the rotor or the deceleration relative to v_1. By equating the axial and tangential forces to the corresponding losses in momentum, the components v_{2ax} and v_{2t} or the vector sum v_2, and thus the air flow acting on the rotor, may be derived.

Marginal vortex losses, due mainly to flow around the blade tips and roots, which cause a vortex field downstream of the rotor, can be allowed for by applying a correction factor of

$$F_S = \frac{2}{\pi} \arccos \left\{ \exp \left[-\frac{z}{2} \frac{1}{\sin \delta} \left(1 - \frac{r}{R_a} \right) \right] \right\} \qquad (2.27)$$

for the blade tips [2.5] and

$$F_N = \frac{2}{\pi} \arccos \left\{ \exp \left[-\frac{z}{2} \frac{1}{\sin \delta} \left(\frac{r}{R_i} - 1 \right) \right] \right\} \qquad (2.28)$$

for near the hub. The product of these two factors

$$F = F_S F_N \qquad (2.29)$$

allows the influence of the free peripheral eddies to be described with sufficient exactitude.

When the wind deceleration due to the turbine is very high, a turbulent vortex field develops in the downstream of the machine (turbulent wake state) [2.6]. However, we only need to allow for this state in a few machines, when the outer blade areas are running close to the cut-in speed. If necessary, various empirical approximations (see references [2.6] and [2.7]) may be used.

2.1.4 Simplifying the computation method

The method detailed above yields very good results that are in agreement with practice. The effort needed to determine operational conditions of the turbine or to examine the behavior of wind energy plants is, however, enormous. Very reliable estimates of operational performance can be arrived at by computational methods that give errors in the order of a few percent. The objective of further research was to considerably simplify the method, while keeping errors within predetermined limits. Starting with the method detailed above, various possibilities were examined with the objective of reducing computational effort while holding the errors within acceptable bounds (e.g. 2.5%) [2.8]. To this end, the complete method is compared with the chosen simplification for two wind turbines of different sizes (12.5 m and 60 m rotor diameter), divided into 50 blade elements.

This examination showed the following:

- To calculate the drop in wind velocity v_2 at the turbine, swirl effects can be ignored. Under operational conditions, the errors relative to the long method are less than 2.5%.
- To determine thrust for variable-pitch machines, it is possible to remain within the same margin of error while ignoring the drag component. For stall-regulated machines, values that are dependent upon the operating state must also be incorporated.
- Blade tip and hub losses can be approximated to the same level of exactitude using the given wind speed v_1 far upstream of the turbine instead of the iteratively determined wind speed

v_2 at the rotor in determining $\sin \delta$, or

$$\frac{1}{\sin \delta} = \sqrt{1 + \left(\frac{v_u}{v_2}\right)} \approx \sqrt{1 + \left(\frac{v_u}{v_1}\right)^2} \approx \sqrt{1 + \left(\frac{\omega_R r}{v_{1ax}}\right)^2}. \qquad (2.30)$$

The reduction factor is then determined just once for each blade element according to its own operating conditions at the beginning of the iteration.

- The torque, instead of being calculated by summing partial moments over 50 blade elements by the step-by-step method

$$M_{AW} = \int_{R_i}^{R_a} dM_{AW} \approx \sum_{k=0}^{n} \Delta M_k, \qquad (2.31)$$

may be determined using Simpson's rule

$$M_{AW} = \int_{R_i}^{R_a} dM_{AW} \approx \frac{R_i - R_a}{n} \left(\frac{dM_0}{dr} + 4\frac{dM_1}{dr} 2\frac{dM_2}{dr} + 2\frac{dM_{n-2}}{dr} + 4\frac{dM_{n-1}}{dr} + \frac{dM_n}{dr}\right) \qquad (2.32)$$

to roughly the same precision and may be made even faster by taking

$$n = 20 \qquad (2.33)$$

elements, i.e. 21 points. Then, because of the flow around the blade tips and roots, the boundary values

$$dM_0 = dM_n = 0. \qquad (2.34)$$

Using this simplification, a considerable reduction of computation time – to around 1% – may be achieved by:

- series expansion of the transcendent arctan and root functions;
- approximation of the aerofoil section with polar equations $c_A = f(\alpha)$ in the form of a polynomial to the third power;
- solving the resulting cubic equations by the cardanic method to determine the wind velocity v_{2ax} at the turbine. Time-consuming iterations thereby become superfluous.

The results can then be plotted, on the basis of the performance coefficient c_p in relation to the tip speed ratio λ, to give the usual $c_p - \lambda$ characteristics (see Section 2.2), as shown in Figure 2.6 (a) and (b). The tip speed ratio is obtained here from the quotient of the peripheral to the undisturbed wind velocity

$$\lambda = \frac{v_u}{v_1}. \qquad (2.35)$$

This two-dimensional representation of turbine power calculations shows the dependencies both upon blade pitch angle ϑ (or β) and also upon the speed n, which is proportional to the peripheral speed v_u and the wind speed v_1.

Further studies have shown that the result of the cardanic method under normal operating conditions is essentially limited to one of the three possible partial solutions. These conditions can thus be described in the form of a closed solution path for wind turbine power or torque calculation. The errors arising from this, using narrower element widths in comparison with blade element theory and including swirl, etc., are shown in Figure 2.7.

Because of the cubic approximation, the simplified method described here has its limits. Just below cut-in and just above cut-out wind speed, it compares unfavourably with the blade element method. However, blade element theory also yields results that differ considerably from measurements taken in these conditions, so such conditions can only be reconstituted with limited success.

The results of the calculation were shown as examples for two model turbines. Despite differences in size, number of blades, tip speed ratio, geometry, etc., hardly any difference can be seen in the calculated results within the operating ranges of the machines, indicated by continuous lines (see Figure 2.7(a) and (b)). Particularly for smaller machines, errors of more than 5% are few and occur mostly at very low and very high wind speeds. However, the conventional computation-intensive methods do not yield any exact error-margin data under these conditions either. It should therefore be possible to apply this method to high-speed wind turbines with variable pitch. For stall-regulated machines, however, some limitations should be expected, since working out the lift using a third-degree polynomial can yield significant errors with the high angles of attack used. Furthermore, the drag component should then be included in determining thrust.

The correct conditions for using the method described, and the resulting limitations to its validity, can thus only be defined after appropriate investigation. On the whole, however, with this considerably simplified method, a wide range of machines and operational areas can be handled. In examining complete systems, the physical characteristics of turbines in quasi-steady states can be taken into consideration. Thereafter the dynamic effects and the resulting deformations of system components can be included. These will be determined essentially by external effects and by whatever direct manipulations are performed on the turbine to limit and regulate its energy extraction performance. This forms the main theme of Section 2.2. In the text that follows we now examine simple methods of depicting turbine characteristics.

2.1.5 Modeling turbine characteristics

For studies centred on transient mechanical or electrical effects in wind energy machines, the frequency ranges of the components to be examined take on special significance. For simulations that target power fluctuations in wind energy units and transient effects in wind farms, the time constants between the pitch setting range (Section 2.3) and the rotor range (Section 2.5) are decisive. The resulting frequencies vary from 10 to 0.01 Hz. In modeling electrical effects, frequencies of around a thousand times this must be considered.

As the basis for the determination of rotor performance, Equations (2.4), (2.7) and (2.8) yield the relation

$$P_W = \frac{\rho}{2} c_p \left(\lambda, \vartheta \right) A_R v_w^3. \tag{2.36}$$

From this, the performance coefficient c_p can be determined according to Sections 2.1.3 or 2.1.4, etc. The most interesting torque levels of the rotor are thus obtained by the simple mechanical relation

$$M_W = \frac{P_W}{\omega_W}. \tag{2.37}$$

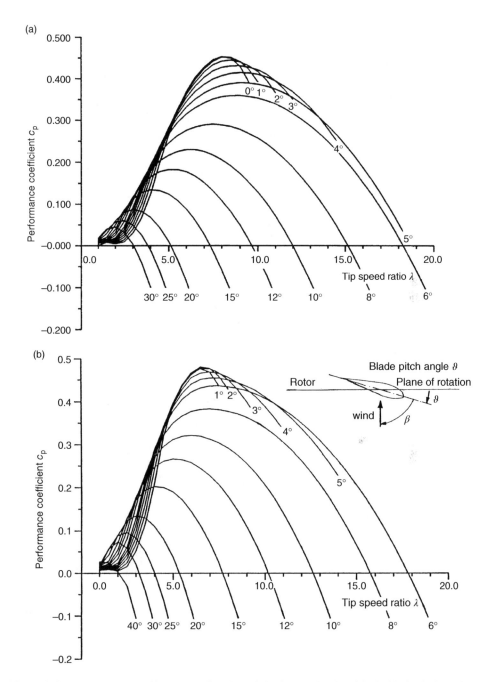

Figure 2.6 Performance coefficient as a function of the tip speed ratio with the blade pitch angle ϑ as a parameter, so-called $c_p - \lambda$ characteristics; v_{2ax} analytically with cubic equation, torque calculated over 21 points and integrated using Simpson's rule: (a) two-bladed turbine with 12.5 m rotor diameter, 20 kW nominal output (Aeroman 12.5); (b) three-bladed turbine with 60 m rotor diameter, 1.2 MW nominal output (WKA-60)

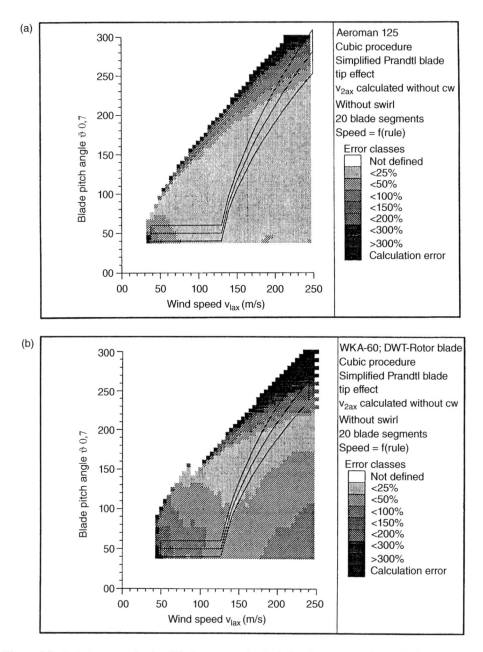

Figure 2.7 Relative errors in simplified rotor speed calculation for (a) a small two-bladed Aeromann 12.5 turbine and (b) a large three-bladed WKA-60 rotor

If the $c_p - \lambda$ performance characteristics (see Figure 2.6), derived from calculations or from direct measurement, are available, the turbine parameters can also be modeled from:

- data fields containing the group of curves derived from measurement or from calculation or
- analytical functions.

Both methods are explained briefly below to provide the user with simple simulation and design tools.

2.1.5.1   Determining the performance coefficient from data fields

If turbine characteristics as in Figure 2.6, or the data for plotting them, are available, data fields can be created by reading off the various values or entering them directly. These then form the basis for power computation in system simulations. The validity of the results thereof will then be dictated by that of the data. When enough data are available, linear interpolation can be used to arrive at intermediate values, although a quadratic derivation is usually better for ongoing processes.

To arrive at a complete data set for the operation of a turbine, however, it may be necessary to extend the characteristics plot. By extending the characteristic curves for small, or even negative, or large angles and by supplementing incomplete characteristics, undefined operating states can be avoided. Also, when the characteristics of a turbine in such areas cannot be forecast exactly, useful decisions and courses of action in extreme situations can be indicated or deduced. Furthermore, many interruptions to the program that would otherwise occur are largely avoided.

2.1.5.2   Approximating the performance coefficient by analytic functions

Groups of $c_p - \lambda$ curves obtained by measurement or by computation can also be approximated in closed form by nonlinear functions. Following reference [2.9], a model can be derived in the form

$$c_p = c_1 \left(c_2 - c_3 \vartheta - c_4 \vartheta^x - c_5 \right) e^{-c_6(\lambda, \vartheta)}. \tag{2.38}$$

According to reference [2.10],

$$c_1 = 0.5, \qquad c_2 = v_w/\omega_w, \qquad c_3 = 0, \qquad x = 2$$

$$c_4 = 0.022, \qquad c_5 = 5.6, \qquad c_6 = 0.17 v_w/\omega_w$$

can be used for the MOD 2 turbine, where v_w represents wind velocity and ω_w is the angular velocity of the turbine. On the other hand, procedures in reference [2.11] yield

$$c_1 = 0.5, \qquad c_2 = 116/\lambda_i, \qquad c_3 = 0.4$$

$$c_4 = 0, \qquad c_5 = 5 \qquad c_6 = 21/\lambda_i$$

and

$$\frac{1}{\lambda_i} = \frac{1}{\lambda + 0.08\vartheta} - \frac{0.035}{\vartheta^3 + 1}.$$

Figure 2.8 was obtained using slightly modified factors with an exponent $x = 1.5$.

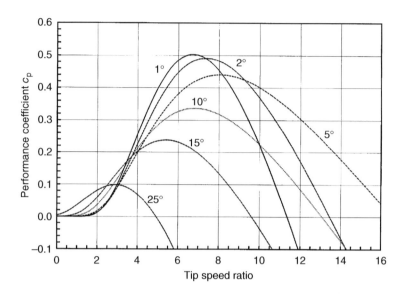

Figure 2.8 Analytical approximation of $c_\mathrm{p}-\lambda$ characteristics (blade pitch angle ϑ as the parameter)

According to the characteristics chosen, the above figures must be modified to obtain a close simulation of the machine in question. To manage this, however, demands a non-negligible investment of time and effort, even for those with a long experience of performing such approximations.

2.1.5.3   Wind-speed profiles, tower effects and transmission losses

The above sections assume a constant distribution of wind speed over the area of the rotor. However, wind-speed profiles result in different wind speeds at the blades nearest to the ground level compared to those at the top of blade travel, which in turn produce corresponding flow and power effects on the turbine (Figure 2.9). Furthermore, gusts and changes in wind speed do not affect the entire rotor at the same instant.

For wind speeds that lie in the operational range of the turbine and exceed about 4 m/s, the wind speed at a given height can be found from the relation

$$v_\mathrm{w}(h) = v_{10}\left(\frac{h}{h_{10}}\right)^{a},$$

where $0.14 \leq a \leq 0.17$ and where v_{10} is the determined or computed wind speed for $h_{10} = 10$ m, $h = h(\psi)$ corresponding to blade position and a is the so-called Hellman exponent.

In addition to the vertical speed gradient, influences due to tower wind-shadow or windbreak effects cause fluctuations in power and hence torque during blade rotation. Depending upon the number of blades (usually two or three), the height of the turbine, the positioning of the rotor upwind or downwind of the tower, the turbulence depending on tower diameter and the influence of the general surroundings, the results can be very different. The changes in power or torque can only be determined reliably if these influences can be well defined. Once the

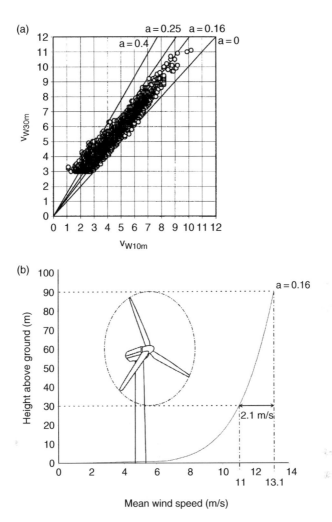

Figure 2.9   Wind-speed gradient: (a) Hellmann exponent (measured values); (b) vertical profile effects on turbines

variations in wind speed over the rotation of a blade have been determined or estimated, the variation in power or torque can be read off the performance characteristic. As a first approximation for research purposes, during a blade rotation the wind conditions at $r = 0.7R$, i.e. at 70% of the rotor radius, may be used.

   In single-blade turbines – as yet the exception and so not studied very closely – power/torque levels at the transmission oscillate widely. In symmetrically arranged multibladed rotors (given identical wind conditions at all blades) the torque produced at the transmission shaft is more or less constant. In these systems, however, flow disturbances also lead to power/torque fluctuations. These results are mostly generated from tower-created turbulence. Nonlinearities in the transmission system (c_p–λ characteristic) lead to further oscillations in the drive train. These

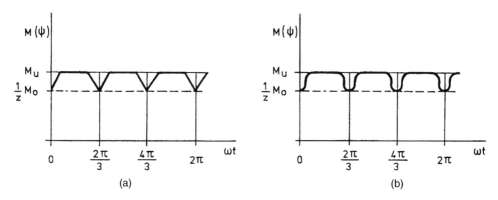

Figure 2.10   Pitch-dependent torque $(z = 3$ rotor blades): (a) ramp representation; (b) cosine representation

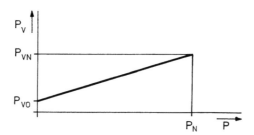

Figure 2.11   Power-dependent loss modeling

can be represented in simplified form using the pitch-dependent torque

$$M(\psi) = M_{\mathrm{u}} - \frac{1}{z}\left[M_{\mathrm{o}}\left(z\omega_{\mathrm{w}}t\right)\right]$$ (2.39)

where M_{u} represents torque under undisturbed wind distribution at the rotor and M_{o} is the oscillating component. This exerts an effect on the transmission shaft that is inversely proportional to the number of blades and has a recurrence frequency of $z\omega_{\mathrm{m}}$. This can be represented in ramp form or – closer to what actually happens – in cosine form (Figure 2.10(a) and (b)).

Mechanical and electrical losses depend mostly on performance and rotor speed. Effects due to thermal changes, etc., cannot easily be handled mathematically. For simple modeling, machine-dependent losses $P_{\mathrm{v}} = f(P)$ can be handled using constants (P_{vo}) and a power-dependent component (P_{vN}), as shown in Figure 2.11.

2.2   Turbines

The conversion of kinetic energy from the wind for technical applications is effected by means of a variety of turbine types [2.4] (Figure 2.12). Machine designs are divided into horizontal and vertical axis types. Depending on the way in which energy is extracted from the wind,

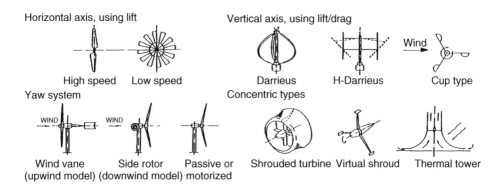

Figure 2.12   Main turbine types

wide variations are discernible between converters, depending on whether they use the drag developed at the surface of the moving parts or the lift exerted on the blades.

In turbines that use drag only, e.g. cup types, board constructions and other surfaces set against the wind, the energy derived is lower than that developed by lift types. Because of lower speeds of rotation, the use of such machines is limited essentially to driving mechanical devices. Their construction is usually simple and very heavy; they will not be considered here.

Turbines using concentrating or suction effects (Figure 2.13), which have, for example, been developed and tested in windy New Zealand, plus thermal turbines (Figure 2.14), a prototype of which has been constructed in sunny southern Spain, are the exceptions in wind turbine technology.

Figure 2.13   Vortex turbine

Figure 2.14   Thermal turbine

Wind power machines for generating electricity are produced in horizontal and vertical axis format. The turbines are so constructed that they can utilize the force of lift. Lift originates in the flow of air past the rotor blade, which causes an overpressure on the underside of the blade and an underpressure on the top. The tangential component of the lifting forces causes the rotor blade to rotate.

According to Betz [2.2], a lifting rotor can only extract 60% of the energy/power from an airstream. The remaining 40% of power must remain in the air flowing past (see Figure 2.2). Complete braking of the moving air mass to $v_3 = 0$ would result in air backing up at the turbine; the inflow of air would be halted and energy extraction would no longer be possible. In practice, after conversion losses, lower levels in the order of 45% are achieved. Therefore, in contrast to water-driven turbines, for example, no efficiency levels are quoted for wind turbines – rather the performance coefficient c_p is used. For operating machines, this figure gives the ratio of the power taken from the wind to that contained by it (Equation 2.8).

A deceleration of the air as in Figure 2.2 can occur just as well with many blades moving slowly as with a few blades rotating at high speed. Simple wood and sheet metal constructions only allow slow rotation with high blade counts (e.g. more than six). The torques to be transmitted are correspondingly high, and the constructions must, in consequence, be massive. A few, swiftly spinning blades (e.g. one to three) attain higher power extraction levels and thus better performance coefficients (Figure 2.15). Such figures are, however, only possible using correctly formed aerofoil sections having a low structural area and low vortex creation and which generate little resistance to rotational movement. In such machines, the torque transmitted at high speed is lower, allowing the transmission components to be correspondingly

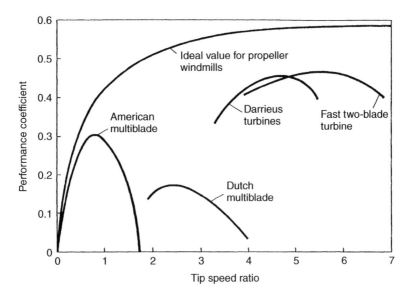

Figure 2.15   Performance coefficient as a function of tip speed ratio for various turbine types, compared with the ideal value [2.12]

lighter. The rpm, or so-called tip speed ratio, therefore has a critical influence on torque and power figures, and on the coefficients that apply to them.

The tip speed ratio $\lambda = v_u/v_1$ gives the ratio between blade tip speed v_u and the wind speed v_1 upstream of the rotor. The blade tip speed, or rather its maximum value, determines the rotor blade loading. Tip speed usually lies between 50 and 150 m/s. However, due to the problem of noise, tip speeds greater than 75 m/s are not usually possible. The tip speed is critical for establishing blade dimensions in both small and large machines. Since it is defined as the product of the blade radius and the rotational speed, low rotational speeds result for large machines, and vice versa. Hence kilowatt machines reach speeds of around 3 rps, i.e. 180 rpm, while in megawatt machines we observe about one revolution every 3 to 6 seconds, or 10 to 20 rpm.

As shown in Figure 2.13, for high-speed rotors (Darrieus, or two-blade rotors) having aerodynamically formed blades, at tip speed ratios of approximately $\lambda = 4$ to 7, performance coefficients from 0.4 to 0.5 are attained. Slow-running machines (American multiblade or Dutch four-blade rotors) with nonaerodynamic wood and sheet metal blades ($\lambda = 1$ to 2.5) have significantly lower performance coefficients, i.e. $c_p = 0.15$ to 0.3.

For operation over a wide range of wind speeds, horizontal-axis machines with high tip speed ratios (Figures 1.2 to 1.27, etc.) have advantages over vertical-axis machines (Figures 1.30 and 1.31). They also offer advantages in start-up behavior and control options.

In addition to high aerodynamic quality, wind turbines must also exhibit sufficient rigidity and strength for the lowest capital outlay. Here the number of blades, the rotor speed or tip speed ratio, the constructional technique, the geometry of the blades and the pitch control arrangements, plus the hub construction, play a decisive role.

2.2.1 Hub and turbine design

In stall-regulated machines the rotor blades are fixed directly to the hub (Figure 2.16(a)). In variable pitch turbines, on the other hand, the blades must be flange-mounted such that they can be rotated about their longitudinal axes (see Figure 2.54). In both systems, all forces, moments and vibrations resulting from gusts, tower shadow, etc., are transmitted via the hub to the tower. Improving smooth running and minimizing blade loading can be achieved using a teetering hub design (Figure 2.16(b)). Independently moving blades with cone hinges allow a freely self-adjusting cone angle (Figure 2.16(c)). This corresponds with the direction of the resultant of wind thrust (due to lift on the blade) and centrifugal forces as a result of the rotating blade masses, such that the articulation is free of bending moment in the direction of the wind. However, the two last-mentioned systems require relatively complicated constructions and are thus susceptible to repairs. They have therefore been unable to establish themselves on the market.

As shown in Figure 2.12, the rotor (seen from upwind) can be run in front of the tower (the upwind model, as in historical windmills) or behind the tower (downwind model). Three-bladed rotors (Figure 2.17) are by far the most widely used horizontal-axis types in all power ranges. Two-bladed rotors – widespread in the 1980s at the beginning of the modern development of wind turbines – today represent the exception (Figure 2.18). Turbines with a single blade (Figure 2.19), which also appeared in this phase of development, currently occupy an absolutely exceptional position. They have not been able to establish themselves on the market. In the machine shown in Figure 2.17 the blades are rigidly connected to the hub, while in Figure 2.18 the blades are attached to the hub such that they can rotate about

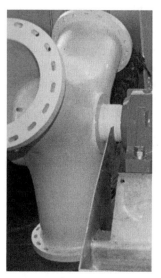  

(a) Rigid hub   (b) Teetering hub   (c) Cone-hinge hub

Figure 2.16 Hub types

Figure 2.17   Three-bladed turbine: Nordex N54, 1000 kW nominal output, 54 m rotor diameter

their longitudinal axes. The single-blade rotor (Figure 2.19), on the other hand, is fitted with a cone-hinge hub. Multibladed arrangements are commonly found in small machines in the kilowatt range and below.

2.2.2   Rotor blade geometry

When designing a machine to extract energy from the wind and to develop the resulting torque in the turbine, the airstream pattern shown in Figure 2.2 is the goal. It is then possible to achieve energy extraction using many blades at low speed or a few blades at high speed. Further, the optimal wind deceleration for the same speed of rotation can be attained using one very broad blade, or two or three blades of correspondingly smaller breadth. The optimal blade chord can be derived from the blade radius by the formula

$$t_B(r) = \frac{2\pi r}{z} \frac{8}{9} \frac{1}{c_a} \frac{v_1^2}{v_u(r) v_r(r)}. \tag{2.40}$$

Figure 2.18 Two-bladed type: Ned Wind, 1 MW nominal output, 52.6 m rotor diameter. Reproduced by kind permission of Nedwind

A rotor tip speed of

$$v_u(r) = \omega r = \lambda \frac{r}{R} v_1,$$

(2.41)

a target wind deceleration at the turbine of

$$v_2 = \frac{2}{3} v_1$$

(2.42)

and the resultant relative wind speed of

$$v_r(r) = \sqrt{v_u^2 + v_2^2} = v_1 \sqrt{\left(\frac{\lambda}{R} r\right)^2 + \frac{4}{9}}$$

(2.43)

yield a blade chord of

$$t_B(r) = \frac{16 \pi R}{9 c_a \lambda z} \frac{1}{\sqrt{\left(\frac{\lambda}{R} r\right)^2 + \frac{4}{9}}}.$$

(2.44)

Figure 2.19   Single-bladed type, Monopteros (MBB), 640 kW nominal output, 56 m rotor diameter

According to this, hyperbolic blade contours are to be expected. These, however, differ widely for different (design) tip speed ratios and numbers of rotor blades [2.13]. Figure 2.20 shows the relative blade contours for one-, two- and three-bladed machines having design tip speed ratios λ_A of 5, 8 and 12. This presentation makes it absolutely clear that single-bladed turbines are best for high-speed and multibladed turbines for low-speed machines.

When producing rotor blades, the optimal blade contour is approximated, mostly in trapezoidal or even in rectangular form (Figure 2.21). Trapezoidal blades, by far the most widely used, reach performance coefficients approaching those of optimally shaped blades. Rectangular outlines, in contrast, yield a markedly lower maximum performance coefficient at the design tip speed ratio. Outside the design area, however, extensive ranges of operation exist, some with even better performance ratios. In addition to blade outline, the position of the rotor blades relative to the wind direction or to the rotor plane is decisive in extracting energy from the wind.

The peripheral speed of the turbine is high at the blade tip and relatively low at the hub. This results, for the same airstream effect, in a low blade chord at the tip and a larger blade area near the hub. To obtain similar flow conditions over the entire length of the blade when

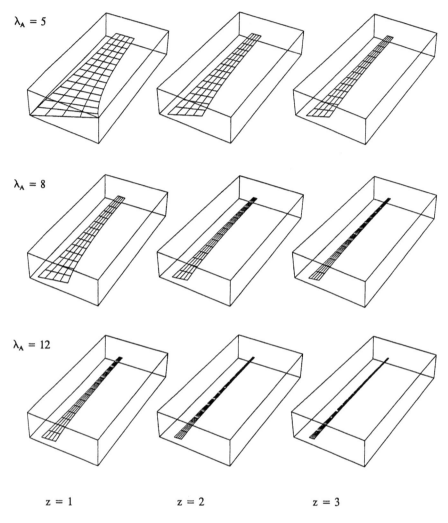

$\lambda_A = 5$

$\lambda_A = 8$

$\lambda_A = 12$

z = 1   z = 2   z = 3

Figure 2.20   Optimal rotor blade geometry for various numbers of blades and design tip speed ratios

running, similar flow directions relative to the aerofoil section must be achieved at all points between the tip and the hub. This is attained when the relative airspeed at design performance (i.e. nominal operation conditions) is the same at all radii (Figure 2.20).

Following Figures 2.4 and 2.22, the angle of attack α is the difference between the pitch angle ϑ and the resultant relative airflow direction. This is the vector sum of the wind deceleration $(v_2 = \frac{2}{3}v_w = \frac{2}{3}v_1)$ and the peripheral speed resulting from blade rotation. Hence the blade pitch angle

$$\vartheta\,(r) = \arctan \frac{2v_1}{3v_u} - \alpha \qquad (2.45)$$

(a) Optimal shape (Südwind). Reproduced by kind permission of Südwind GmbH

(b) Trapezoidal blade (Windmaster). Reproduced by kind permission of Windmaster GmbH

(c) Rectangular blade (Krogmann). Reproduced by kind permission of H. J. Krogmann

Figure 2.21   Produced rotor blade shapes

and its equivalent power complement

$$\beta(r) = \frac{\pi}{2} - \vartheta(r). \tag{2.46}$$

For electricity-generating machines, high-speed turbines are usually used. The rotor blades must exhibit the lowest possible moving mass and must be very strong. Further, glass-fibre-reinforced polyester and epoxy resin composite materials are mostly used to yield aerodynamically correct blade profiles and high strength. Even greater strength may be achieved using carbon fibre materials and the use of such materials is on the increase, in spite of the higher costs involved.

Figure 2.23 shows the manufacture of rotor blades in the megawatt range. In large turbines, in particular, the necessary distance from the tower can be achieved even in operation at high wind speeds with bent blades (Figure 2.24). Figure 2.25 shows the complete rotor being lifted for fitting to the rotor shaft.

Rotor blades are currently manufactured in two shells and put together into a single unit in the production halls for transporting and fitting. Figure 2.26 illustrates the dimensions of the rotor blades and the demands this places upon manufacture, logistics and fitting. Furthermore, it becomes evident that the limits of transportability can easily be exceeded in many regions. The importance of factories in harbour areas will therefore increase as the size of turbines – particularly those for offshore use – increases.

The construction and the fitting of rotor blades and turbines, which are discussed in more detail in references [2.3, 2.4] and [2.13], will not be considered further here.

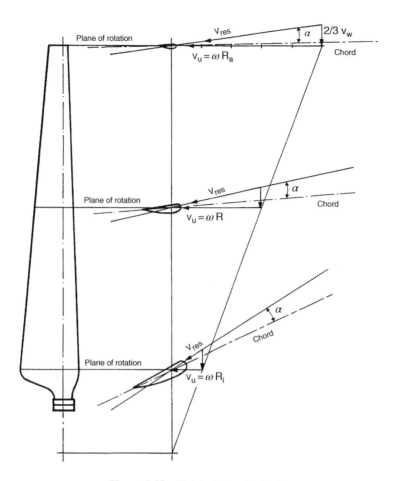

Figure 2.22 Twist of the rotor blade

Figure 2.23 Manufacturing a rotor blade. Reproduced by permission of Nordex

Figure 2.24 Bent rotor blade (NOI 37.5 / 77 m rotor diameter)

2.3 Power Control by Turbine Manipulation

In the middle of the last century, wind energy machines were protected from over-revving by manual intervention, e.g. by reefing the sails or by turning the plane of rotation to lie parallel to the wind. In American (slow-running) wind turbines it was even possible to control power take-up by yawing or tilting the rotor, depending on the structure. Since the middle of the twentieth century, principles founded on aerodynamics have come into use to limit power using blade stall or to modify it by varying blade pitch. In special designs, even in the 10 kW class, variable-speed generators allow an adjustable adaptation of turbine output to the prevailing wind conditions.

2.3.1 Turbine yawing

Power extraction in slow-running turbines can be limited by yawing or tilting the plane of rotation in the direction of wind pressure. In this way the effective flow cross-section of the

Figure 2.25   Installing the rotor. Reproduced by permission of DeWind

Figure 2.26   Transport of rotor blades (Nordex N80). Reproduced by permission of Nordex

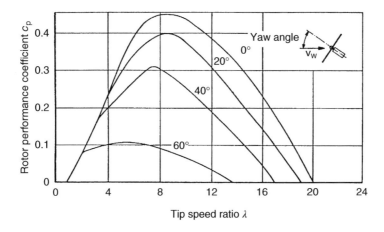

Figure 2.27 Rotor performance coefficients under skewed airflow [2.14]

rotor is reduced and the flow incident on each blade considerably modified. Figure 2.27 shows the drastic drop in performance coefficients resulting from turning wind turbines out of the wind, with consequent blade stall.

For multibladed symmetrical models that run at low speeds, this operation is particularly effective. The following shows why.

2.3.1.1 Control procedures and yaw accelerations

For high-speed wind turbines, turning the plane of rotation out of the wind by a predetermined amount provides a slow-acting means of protecting the machine. This procedure can be carried out by an active yaw-control system. Here, skewing the rotational plane with respect to the wind exerts moments on the rotating blades, the behavior of which must be formalized if their influence on the control system is to be determined. Further, the following considers the accelerations and moments occurring in a two-bladed model such as that shown in Figure 2.28.

The frame of reference x', y', z' is fixed with respect to the rotor. Then, for the centre of mass of a stiff blade, for the speed in this coordinate system

$$\frac{dr'}{dt} = v' = 0$$

holds, with the equivalent acceleration

$$\frac{dv'}{dt} = b' = 0.$$

For a yaw velocity of ω_A at the head pivot, a turbine rotating with a velocity of ω_R gives a resultant angular velocity of

$$\omega = \omega_A + \omega_R \tag{2.47}$$

and a rotor head yaw velocity of

$$v_0 = \omega \times r_0 = \omega_A \times r_0. \tag{2.48}$$

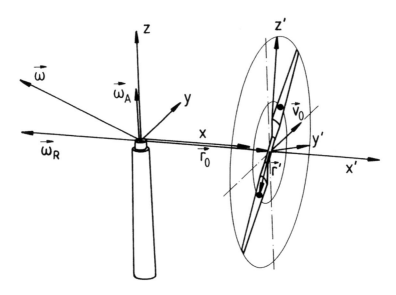

Figure 2.28   Relative motion of rotating blades for a wind turbine under azimuth angle yawing

The total velocity of the centre of mass [2.15] is then

$$v = v_0 + v' + \omega \times r' = \omega_A \times r_0 + \omega \times r' \tag{2.49}$$

and

$$b = b' + b_0 + 2\omega \times v' + \frac{d\omega}{dt} \times r' + \omega \times (\omega \times r'), \tag{2.50}$$

where

$$b' = 0 \qquad \text{and} \qquad 2\omega \times v' = 0.$$

The centripetal acceleration of the head

$$b_0 = \omega_A \times (\omega_A \times r_0)$$

and

$$\frac{d\omega}{dt} = \omega_A \times \omega = \omega_A \times (\omega_A + \omega_R) = \omega_A \times \omega_R$$

gives the acceleration of the centre of gravity

$$b = \omega_A \times (\omega_A \times r_0) + (\omega_A \times \omega_R) \times r' + (\omega_A + \omega_R) \times ((\omega_A + \omega_R) \times r')$$

or with components

$$b = \underbrace{\omega_A \times (\omega_A \times (r_0 + r'))}_{\substack{\text{centripetal acceleration} \\ \text{from } \omega_A}} + \underbrace{2\omega_A \times (\omega_A \times r')}_{\text{Coriolis acceleration}} + \underbrace{\omega_R \times (\omega_R \times r')}_{\substack{\text{centripetal acceleration} \\ \text{from } \omega_R}} \tag{2.51}$$

which we will now examine in detail, so as to be able to interpret the individual accelerations clearly.

Centripetal acceleration due to yawing
When the offset of the rotor head is

$$r_0 = |r_0| \, e_x = r_0 e_x$$

or that of the rotor blade centre of mass from the tower axis is

$$r' = y' e_y + z' e_z$$

then from Figure 2.29 the resultant is

$$r_0 + r' = r_0 e_x + y' e_y + z' e_z.$$

Allowing for a yaw control angular velocity of

$$\omega_A = \omega_A e_z$$

yields the determinant for yaw speed

$$\omega_A \times (r_0 + r') = \begin{vmatrix} e_x & e_y & e_z \\ 0 & 0 & \omega_A \\ r_0 & y' & z' \end{vmatrix} = -\omega_A y' e_x + \omega_A r_0 e_y$$

and thus the centripetal acceleration from ω_A

$$b_{\omega_A} = \omega_A \times (\omega_A \times (r_0 + r')) = \begin{vmatrix} e_x & e_y & e_z \\ 0 & 0 & \omega_A \\ -\omega_A y' & \omega_A r_0 & 0 \end{vmatrix}$$

or

$$b_{\omega_A} = -\omega_A^2 r_0 e_x - \omega_A^2 y' e_y. \tag{2.52}$$

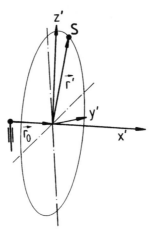

Figure 2.29   Centre of gravity of rotating blade

For rotors with two symmetrical blades, the component becomes

$$\omega_A^2 y' e_y = 0$$

and thus engenders an intrinsic moment that loads the blade structure and, in pitch-regulated machines, the blade bearings.

The residual acceleration

$$b_{\omega_A} = -\omega_A^2 r_0 e_x \tag{2.53}$$

only results in a moment along the x axis. Yaw control can generally attain only low angular velocities ω_A, so that b_{ω_A} and the resulting stresses play only a secondary role.

Centripetal acceleration due to blade rotation
From Figure 2.29, the radius of rotation of the centre of gravity is

$$r' = y' e_y + z' e_z$$

and its angular velocity is

$$\omega_R = -\omega_R e_x.$$

Therefore the velocity at S_p is

$$\omega_R \times r' = \begin{vmatrix} e_x & e_y & e_z \\ -\omega_R & 0 & 0 \\ 0 & y' & z' \end{vmatrix} = \omega_R z' e_y - \omega_R y' e_z$$

and the centripetal acceleration is

$$b_R = \omega_R \times (\omega_R \times r') = \begin{vmatrix} e_x & e_y & e_z \\ -\omega_R & 0 & 0 \\ 0 & \omega_R z' & -\omega_R y' \end{vmatrix}$$

Therefore

$$b_R = -\omega_R^2 y' e_y - \omega_R^2 z' e_z. \tag{2.54}$$

Although the forces or moments resulting from centripetal acceleration place loads on both blades and bearings, these components have no effect on yaw control in symmetrically arranged rotors.

Coriolis acceleration
From the velocity of the centre of mass

$$\omega_R \times r' = \omega_R z' e_y - \omega_R y' e_z$$

and the corresponding yaw angular velocity

$$\omega_A \times (\omega_R \times r') = \begin{vmatrix} e_x & e_y & e_z \\ 0 & 0 & \omega_A \\ 0 & \omega_R z' & -\omega_R y' \end{vmatrix} = -\omega_A \omega_R z' e_x$$

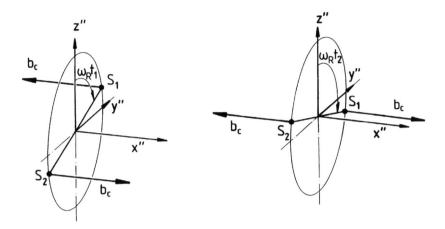

Figure 2.30   Coriolis acceleration of a rotor blade

the Coriolis acceleration may be derived:

$$b_c = 2\omega_A \times (\omega_R \times r') = -2\omega_A \omega_R z' e_x. \tag{2.55}$$

From this, we obtain, for an observer in a coordinate system x'', y'', z'' having a yaw velocity of ω_A but not a rotation of ω_R, a Coriolis acceleration b_C in the negative x'' direction when $z'' > 0$. For $z'' < 0$, b_C acts in the opposite direction (see Figure 2.30).

2.3.1.2   Yaw moments

Coriolis acceleration engenders moments that act on the rotor blades and on the yaw system in the tower head. When the nacelle is yawed the moment in the x direction acts not with a continuously damping effect, but either to accelerate or decelerate the movement: for a rotor blade of mass m_B the moment resulting from Coriolis acceleration with respect to the z'' or z axis is

$$M_{Cz} = -m_B b_C y''. \tag{2.56}$$

Where

$$z'' = r' \cos \omega_R t,$$
$$y'' = r' \sin \omega_R t$$

and

$$\sin \omega_R t \cos \omega_R t = \frac{1}{2} \sin 2\omega_R t$$

the moment in the x direction can be written as

$$M_{Cz} = -m_B r'^2 \omega_A \omega_R \sin 2\omega_R t \tag{2.57}$$

With a rotor blade moment of inertia of

$$J_B = m_B r'^2$$

this becomes

$$M_{Cz} = -J_B \omega_A \omega_R \sin 2\omega_R t. \tag{2.58}$$

In the angular ranges

$$0 \leq \omega_R t \leq \frac{\pi}{2} \qquad \text{and} \qquad \pi \leq \omega_R t \leq \frac{3}{2}\pi$$

the Coriolis acceleration for both blades, $2M_{Cz}$, works against the yaw movement.
In the ranges

$$\frac{\pi}{2} \leq \omega_R t \leq \pi \qquad \text{and} \qquad \frac{3}{2}\pi \leq \omega_R t \leq 2\pi$$

the $2M_{Cz}$ works with the yaw movement and tries to accelerate it. Thus damping or accelerating torques result from this overlaid rotation, depending upon the position of the rotor. This effect can be observed in two-bladed machines as jerkiness in the yaw movement. In three- and four-bladed machines, among others, these accelerating torques add up to zero and so have no effect on yawing.

Because of the moment along the y axis, which engenders a yawing moment on the rotor, swinging the turbine too fast can subject all arrangements of rotor blades to severe and even dangerous stresses. Such steering operations should therefore only be carried out very slowly.

2.3.1.3 Yaw control mechanisms

Very slow and controlled yawing can be achieved using active positioning mechanisms. Electric azimuth drives are most commonly used. Figure 2.31 shows all the main components of a typical control mechanism, with its motor–gearbox unit, ring gear and yaw control.

Hydraulic systems are usually of similar construction. They are, however, used only in large turbines (Figure 2.32).

In large turbines, internal ring gears (Figure 2.33(a)) are generally used instead of the external ring gear shown in Figure 2.31. Furthermore, several centrally mounted geared motors (Figure 2.33(b)) are established in the market. Their greater axial length, among other things, makes it easier to climb into the nacelle from the tower. Brake systems (Figure 2.33(c)) stop the yaw movement and thus protect the yaw gears and ring gears in particular.

Side-rotor yaw systems were widespread in earlier times, also being found in Dutch windmills (Figure 1.7), but for cost reasons they are hardly used today. They are mostly limited to wind energy machines that are not connected to the public grid supply (Figure 2.34).

Passive yaw systems are only used on small machines. The wind vane system (Figure 2.35(a)) was very popular, being used by the million on the American wind pumps known as the 'Western wheel' and similar machines (Figure 1.8). On the other hand, downwind turbines (Figure 2.35(b)), steered by wind pressure of the rotor blades, are not used so often. In these systems an azimuth drive is unnecessary, but Coriolis accelerations caused by changes in wind direction can exert considerable forces on the blades unless some damping or stabilizing system is provided.

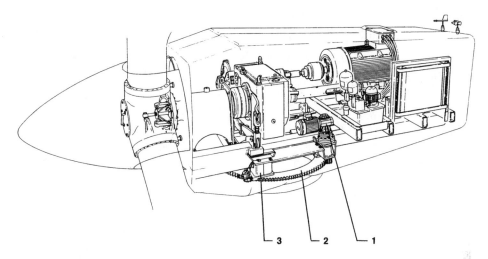

Figure 2.31 Electrically driven turbine yaw system (Vestas V27) with an external ring gear: 1, geared motor; 2, ring gear; 3, yaw control

Figure 2.32 Hydraulic turbine yaw system (Multibrid N5000, 5 MW). Reproduced by permission of Aerodyn Energiesystems GmbH

(a)   (b)

(c)

Figure 2.33 Electrically driven turbine yaw system for large turbines: (a) internal ring gear, (b) geared–motor pair and (c) yaw–brake pair

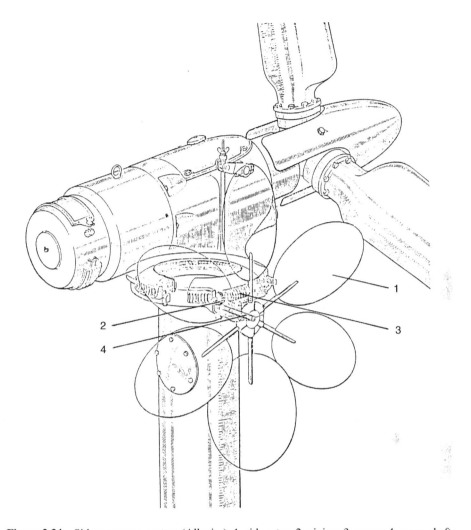

Figure 2.34   Side rotor yaw system (Allgaier): 1, side rotor; 2, pinion; 3, worm; 4, worm shaft

Because of continual changes in wind speed, wind turbines are subjected to ongoing and occasionally enormous available-energy gradients. To extract energy from the wind under such conditions, it must be possible to act swiftly on the turbine system. Rotor blade pitch control is particularly suitable for this purpose.

2.3.2   Rotor blade pitch variation

The rules developed in Section 2.1 have shown that the performance coefficient and therefore power production from the turbine is strongly influenced by variation of the blade pitch relative to the direction of the wind or the plane of rotation. As the example in Figure 2.6(b) clearly

(a) Upwind turbine with wind vane (Enercon)   (b) Downwind turbine with
                                                      wind-pressure yawing

Figure 2.35   Passive yaw systems

shows, at a tip speed ratio of $\lambda = 7$ the performance coefficient, and thus the output of the turbine, can be roughly halved by varying the blade pitch angle from 5 to 12°.

Rotating the blades about their longitudinal axes (see Figure 2.36) thus facilitates swift and active modification of the drive power on the rotor. To do this, a pitch variation mechanism must apply the necessary moment to the blade. In addition to the moment that the variation mechanism is designed and constructed to exert, moments arising from the following must also be taken into account:

- the inertia of the entire adjustment mechanism (motor, clutches, shafts, brake discs and lever mechanism or hydraulic cylinder, etc.);
- the springing and damping characteristics of the mechanism;
- adjustment springs (if present).

In lever-operated pitch adjustment mechanisms, these moments are also dependent upon the pitch angle. As a result, nonlinear relationships arise between drive power and blade pitch, which have an influence on transmission behavior and can, for example, cause variations in the amplification ratio.

In the following, the individual moments are explained, illustrated and examined in detail, with a view to establishing methods for handling them mathematically.

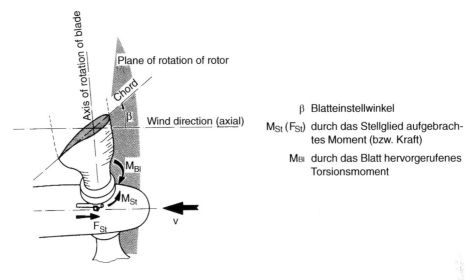

Figure 2.36   Blade pitch variation mechanism

2.3.2.1   Moments on the rotor blade

The following design characteristics of a rotor blade:

- lateral and longitudinal geometry,
- stiffness,
- mass distribution, dictated by the choice of constructional materials,
- possible degrees of freedom in the directions of teeter and wind thrust and in the rotational direction of the rotor and
- bearings,

can, depending on the operating conditions of a machine, give rise to the torsional moments described in what follows.

Propeller moments

Propeller moments arise as a result of the unequal mass distribution of the rotor blade (shown in Figure 2.37 as an aerofoil element) with respect to the axis of rotation of the blade due to the centrifugal force F_z acting on every partial centre of mass.

Breaking this down into its normal and transverse components F_N and F_Q yields the quantity F_{Pr}, which is largely dependent on the speed of rotation and blade pitch angle. Multiplying F_{Pr} by a_P, which is the offset from the blade axis at which F_{Pr} applies and integrating the product over the length of the blade yields the propeller moment M_{Pr}. For the usual operating blade angle of approximately 90° and the always-small blade angle with respect to the vertical axis ($\gamma < 10°$), a_P can be approximated for the moment-building offset, so that the propeller moment can be expressed as

$$dM_{Pr} \approx dF_{Pr}a_p. \tag{2.59}$$

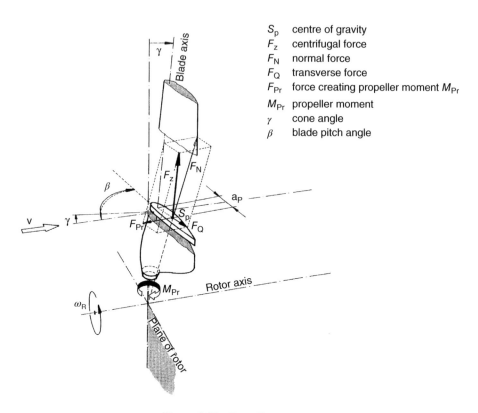

Figure 2.37 Propeller moment

The corresponding worked examples will be further detailed in Section 5.5. Equation (5.11) shows, for example, the entire propeller moment of a blade.

The lifting force F_A, acting to one side of the axis of the blade, develops moments M_{lift} as in Figure 2.38, which are largely dependent on the resultant wind velocity, the blade pitch angle, the blade profile and the offset at $t_B/4$ between the point of action of the lifting force and the axis of the blade. For a single profile element the following is approximately true:

$$dM_{lift} = c_a(\alpha) \, v_r^2 \frac{\rho}{2} a_a \cos \alpha dA_B. \tag{2.60}$$

The total moment resulting from lift forces on the blade is found in accordance with Equation (5.12). The restoring torques M_T cause, among other effects, a twisting of the profile into the direction of flow and are dependent on the resultant wind velocity v_r, the blade area A, the chord of the blade t_B and the airstream angle and blade profile, with the associated moment coefficient c_t.

For a single element, relative to $t_B/4$, the following statement is generally true:

$$dM_T = c_t(\alpha) \, v_r^2 \frac{\rho}{2} t_B \, dA_B. \tag{2.61}$$

The total restoring torque on the blade can be determined using Equation (5.13).

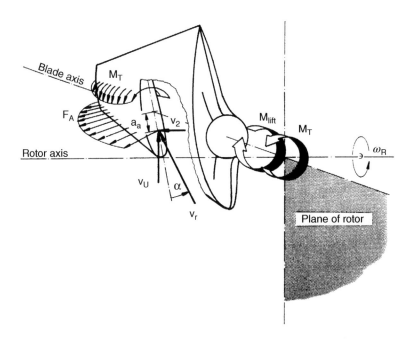

Figure 2.38 Moments arising from lift forces and aerodynamic restoration

Blade bending causes displacement of the lift and centre of mass. This generally results in an increase to the angular moment working against blade restoration, i.e. the moment M_{lift}, developed as a result of lifting forces, usually becomes larger due to aerofoil deflection. Due to mass shifts, extra propeller moments are generally developed. Furthermore, the moment of inertia of the deformed blade becomes significantly greater than it is in its undeformed state (see Figure 2.39).

Moments resulting from rotor teetering and the associated changes in pitch, which arise mainly in teetered hub models, are mainly dependent on blade angle and the amplitude of teeter during blade rotation (see Figure 2.40). In symmetrical blade arrangements, moments engendered (e.g. due to the acceleration of inert masses) cancel each other out with respect to the external drive. Propeller moments, on the other hand, can change considerably.

Because of the significant influences and the continually changing conditions during the rotation of a blade, these effects cannot be handled as they stand without unacceptable computing effort. The determination of extreme conditions is often sufficient, however, for dimensioning purposes.

Frictional moments in the blade bearings always work against blade movement (Figure 2.41). They depend upon the rotor position and speed of revolution, the speed of pitch variation and the wind speed, and have a quasi-damping character.

The frictional moment of a bearing can be expressed as the sum of a load-independent component and a load-dependent component. The load-independent component depends on hydrodynamic losses in the lubricant. This depends on lubricant viscosity and quantity, and also on rolling speed, and is dominant in swift-running lightly loaded bearings. The pitch angle varies only very slowly (maximum $\pi/6$ per second) and the load on the bearing is high, so this component can be neglected.

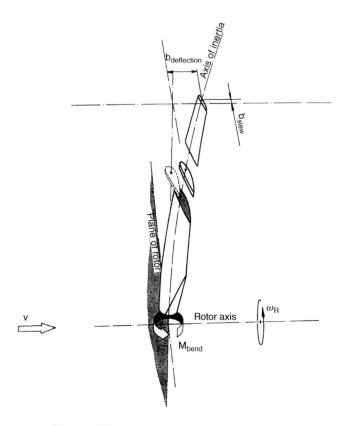

Figure 2.39 Moments arising from blade bending

The load-dependent frictional moment M_{RL} arises as a result of elastic deformation and partial local slippage of contact surfaces. This component predominates in bearings rotating slowly under load. It can be determined by the relation

$$M_{RL} = f_{L1} g_{L1} P_{L0} d_{m}. \qquad (2.62)$$

The quantities

f_{L1} coefficient depending on bearing type and loading
g_{L1} factor determined by load direction
P_{L0} equivalent static loading
d_{m} average bearing diameter

used here can be found in the manufacturer's catalogue (e.g. the SKF Main Catalogue).

The equivalent static loading is characterized by axial and radial bearing forces. These vary, however, in rotating rotors, depending on the blade angle of rotation ψ_{Bl}, so that

$$P_{L0} = f(\psi_{Bl}), \qquad (2.63)$$

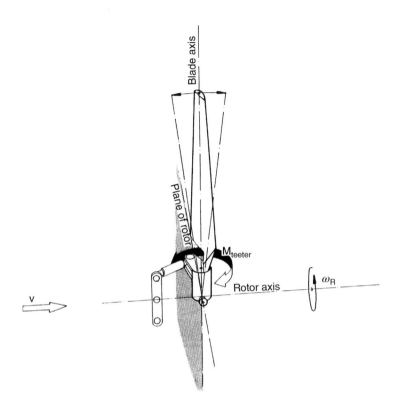

Figure 2.40   Moments arising as a result of teetering

e.g. for axial bearings

$$g_{L1}P_{L0} = F_a(\psi_{Bl})$$

and is therefore equal to the axial force. The simplified assumption for small deviations from the mean values

$$P_{L0} \approx \text{constant} \tag{2.64}$$

must, however, be investigated for each individual case.

Frictional moments are also dependent on the speed of rotation. As an approximation, when starting a movement the load-dependent component can usually be doubled.

The sum of frictional moments over all bearings active during pitch variation yields the total frictional moment

$$M_{\text{Frict}} = \sum_{k=1}^{n} M_{\text{RLk}}, \tag{2.65}$$

where n is the number of bearings involved. When the pitch of a rotating blade is altered, further moments act on the system, originating from the acceleration of air masses around the profile and from air damping (see Figure 2.42). A description of these quantities, which hamper movement, is only possible under predefined conditions.

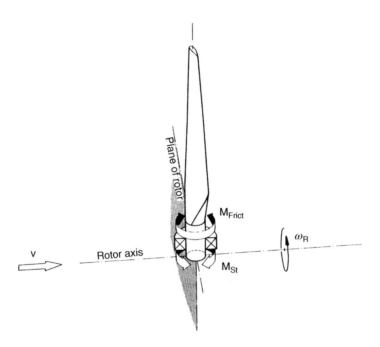

Figure 2.41   Frictional moments in blade bearings

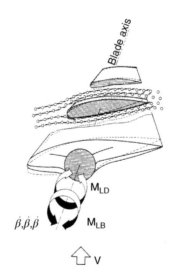

Figure 2.42   Moments arising from the acceleration of air masses and from air damping during blade positioning

For the limited case of

- blades that are stiff under bending and twisting,
- with an axis of rotation at a quarter of the chord $(t_B/4)$,
- when the blade is turned about the pitch variation axis only (i.e. no blade deflection),

these moments can be approximated per span-width element by the relation [2.16]

$$\frac{dM_L}{dr} = \pi \rho d^4 \left\{ \underbrace{-\left(\frac{1}{8} + a^2\right)\ddot{\beta}}_{\substack{\text{inertia of} \\ \text{accelerating air} \\ \text{masses}}} + \underbrace{\left[a - \frac{1}{2} + 2\left(\frac{1}{4} - a^2\right)C(k)\right]\frac{v_r}{d}\dot{\beta}}_{\text{damping of the moved air}} \right\}$$

where

β is the angle of rotation (pitch variation angle),

d

     is half the profile chord,

ad

     is the offset from the centre of the profile to the point of rotation,

$C(k)$

     is the Theordorsen function (quasi-stationary $C(k) = 1$) and

v_r

     is the flow speed.

The geometrical relations that underlie further considerations are shown in Figure 2.43 for the most common aerofoil sections.

In a quasi-stationary state the total moment resulting from air forces is found by integrating over the blade span from the inner to the outer radii $(R_i$ to $R_a)$ where $a = -\frac{1}{2} = $ constant and

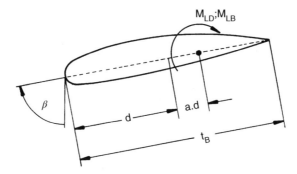

Figure 2.43 Definition of geometrical profile quantities for calculating moments arising from the acceleration of air masses and from air damping [2.16]

$d = t_B/2 = f(r)$ substituted. Therefore

$$M_L = \int_{R_i}^{R_a} dM_L = -\frac{3}{128}\pi\rho\ddot{\beta}\int_{R_i}^{R_a} t_B^4 \, dr - \frac{1}{8}\pi\rho\dot{\beta}\int_{R_i}^{R_a} v_r t_B^3 \, dr \qquad (2.66)$$

or

$$M_L = M_{LB} + M_{LD} \qquad (2.67)$$

where

$$M_{LB} = J_{LB}\ddot{\beta} \qquad \text{and} \qquad M_{LD} = k_{LD}\dot{\beta}.$$

For approximations of this kind, it is usually sufficient to divide the blade into five to ten sections, to derive the partial moments using the average chord and flow speed values, and to sum these to yield the total moment.

The results of such rough estimates show that components arising from accelerated air masses play a completely subordinate role in comparison with the large inert masses of the blades and pitch-setting mechanism, and can thus be disregarded. Moments arising from air damping are in general smaller than the associated frictional values, but achieve similar orders of magnitude and should therefore be taken into account.

The following section uses the equations of motion of the system to develop a model for rotor blade positioning.

2.3.2.2 Blade positioning model

In addition to the dependencies mentioned, the moments depicted above are also influenced by the momentary state of the rotor, i.e. by the rotor blades' vibration characteristics. If the wind-speed profile and tower-shadow effects are also taken into account, the position of the rotor during one rotation plays a decisive role. The block diagram in Figure 2.44 shows the quantities acting and the combined effects of all moments acting on the rotor blades.

In the following pages the quantities influencing the rotor blade will be more closely detailed. Starting from the general equation of motion of the blade turning about its longitudinal axis, the following differential equation may be obtained:

$$\frac{d\left((J_{LB} + J_{Bl})\frac{d\beta}{dt}\right)}{dt} + \frac{d\left((k_{DB} + k_{RL})\beta\right)}{dt} = M_{St} - M_{Bl}. \qquad (2.68)$$

After differentiating and separating out the values we can write

$$\left(J_{LB} + J_{Bl}\right)\frac{d^2\beta}{dt^2} + \left(\frac{dJ_{Bl}}{dt} + \frac{dJ_{LB}}{dt} + k_{DB} + k_{RL}\right)\frac{d\beta}{dt}$$

$$+ \left(\frac{dk_{DB}}{dt} + \frac{dk_{RL}}{dt}\right)\beta + M_{Pr} + M_{lift} + M_T + M_{bend} + M_{teeter} = M_{St} \qquad (2.69)$$

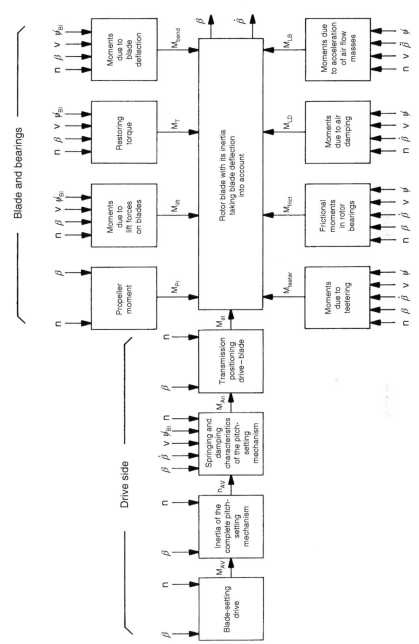

Figure 2.44   Structure and moments for rotor blade pitch adjustment

where

J_{LB} is the value equivalent to the inertia of masses due to accelerated air
           $(M_{LB} = J_{LB}\, d^2\beta/dt^2 -$ forming),

J_{Bl} is the mass moment of inertia of the rotor blade about its longitudinal axis, where
           J_{Bl} is subject to the influence of the centre of mass displacement due to blade
           bending,

k_{DB} is the damping coefficient for the mostly negligible structural component (k_{DS}) and
           the generally dominating aerodynamic damping component (k_{LD}),

k_{RL} is the coefficient of friction for bearings,

M_{St} is the drive moment of pitch regulation taking into account inertia, springing and
           damping characteristics of the regulation mechanism and the gearing ratio to
           theblade,

M_{Bl} is the load torque during blade positioning about the longitudinal axis, including
           propeller moments, aerodynamic components and moments due to blade
           deflection and teetering of the rotor.

Essential dependencies can be given for:

Moment of inertia of air mass	$J_{LB} = f(v, \psi_{Bl}, n, b_s, t)$
Moment of inertia of mass	$J_{Bl} = f(\beta, v, \psi_{Bl}, a_s, t)$
Damping coefficient	$k_{DB} = f(v, \psi_{Bl}, n, t)$
Coefficient of friction	$k_{RL} = f(\beta, v, \psi_{Bl}, n, t)$
Propeller moment	$M_{Pr} = f(\beta, n, t)$
Moment due to lift on blade	$M_{lift} = f(\beta, v, \psi_{Bl}, n, t)$
Moment due to blade bending	$M_{bend} = f(\beta, v, \psi_{Bl}, n, a_s, t)$
Moment due to teetering	$M_{teeter} = f(\beta, \dot{\beta}, v, \psi_{Bl}, n, t)$
Torsional effect due to aerodynamic restoring torques in the direction of flow	$M_T = f(\beta, v, \psi_{Bl}, n, t)$

Here the factors

- blade pitch angle β,
- wind speed v,
- rotor blade position relative to the tower, ψ_{Bl},
- rotor speed n,
- deflection of the blade a_s in the thrust and slew directions,
- acceleration of the blade b_s in the thrust and slew directions,
- observation time t during which the current status has remained unchanged,
- derivation of the named quantities

must be taken into account.

    The great number of components and the many effects preclude (without unreasonable computing effort) a complete description of the movement processes of rotor blades with a treatment of the corresponding moments. A simplified mathematical representation of events therefore makes good sense, even if the processes being examined are subject to certain limitations. This will be shown in the following pages.

2.3.2.3   Simplified rotor blade positioning model

In order to dimension a rotor blade adjustment drive and determine its governing parameters, the size and effect of the individual moments must be recorded as precisely as possible. Even calculations based on quasi-steady states, with the derivation of corresponding characteristic curves, allow the estimation of extreme situations, and are sufficient for dimensioning purposes. To this end, all the moments already mentioned must be taken into account under various operating conditions, e.g. even in the event of changes in surface roughness and icing up of the rotor blades, which particularly affect lift forces, propeller effects and blade inertia. Such variable operating conditions can best be mastered by self-optimizing regulation procedures based on parameters adapted to the locality. It is, however, possible to obtain good results by simply adjusting the most important parameters (see Section 5.5.4).

To examine the operational characteristics of wind power units, it was sometimes possible to use characteristic curves (see references [2.17, 2.18] and [2.19]) to determine load torque during blade positioning. In this manner, restoring torques, moments due to lifting forces, blade bending and propeller effects dependent on wind speed, speed of rotor rotation and blade pitch angle could be taken into account. Extensive calculations and simulations (e.g. in reference [2.20]) have shown that in a reliable design for a blade pitch drive and regulator mechanism different air damping and bearing frictional processes have little influence on general behavior. Considerable simplifications can thus be achieved.

In slim plastic blades, torsional moments resulting from blade bending predominate. The behavior of very stiff blades of heavy construction is, on the other hand, mostly determined by propeller moments. An estimate of individual moments based on constructional parameters allows recognition of the most important contributions and points the way towards further simplifications.

When the less relevant components are left out, the dominant moments that remain can be catered for by adding a safety margin to arrive at a total torsional moment M_{Bl}. When the transmission of moments between the pitch regulator drive and rotor blade is sufficiently rigid, springing and damping effects can also be left out of the reckoning and the inertia of the regulation mechanism can be attributed to the blades.

This simplified model of pitch variation moments, whether the pitch control system be electrical or hydraulic, is shown in Figure 2.45. Here I_{St} indicates the magnitude of electric current or hydraulic flow from the regulator required to position the blades. Air flow changes due to the tower as a result of compression effects in upwind turbines or shadow effects in downwind turbines can also be included in the calculations. Changes in transmission, e.g. between linear drive and radial blade adjustment, are also taken into account. Working under these last assumptions thus brings the effort involved in designing turbine pitch regulation and power control mechanisms within reasonable bounds.

2.3.2.4   Blade pitch control mechanisms and safety systems

The multiplicity of effects detailed in Section 2.3.2.1 demands a very good match between the rotor blade and the pitch regulation mechanism. Here the mechanical construction and the arrangement play an important role. Therefore, it is not possible to simply swap blades between different models and makes.

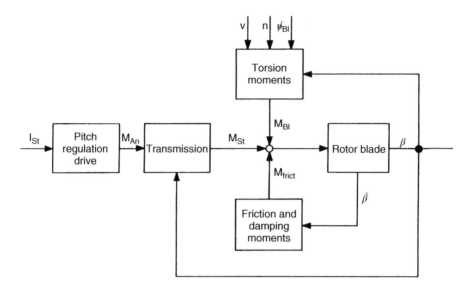

Figure 2.45   Simplified model of blade pitch regulation

Adjusting the pitch of the rotor blades, as shown in Figure 2.6, provides an effective means of regulating or limiting turbine performance in high wind speeds or storm conditions, or changing the speed of rotation. To put the blades into the necessary position, various mechanisms are employed. These may allow repositioning of the entire blade or just its tip. The necessary moment at the blade root can be applied by means of rotating masses, hydraulic lines or electric motors. A few concrete examples of pitch regulation mechanisms are presented in the following.

Mechanical systems of rotational speed control that exploit the principles of forces and moments in rotating masses have long been part of electrical engineering. Such mechanisms are usually simple and their solid construction usually makes for reliable operation. The energy they deploy for control processes is derived from the movement of rotating masses. To use such an arrangement for pitch regulation, however, the system must be accurately dimensioned, which in turn requires that the precise moments engendered in the blades during operation must be known. This is necessary, since modifying and adapting the parameters of mechanical devices usually involves considerable expense.

Figure 2.46 shows a pitch-regulation mechanism in which rotation of the turbine sets up a centrifugal force in a set of flyweights. Should the centre of mass lie outside the plane of rotation, the moments acting tend to bring it back into the plane. Load and restoring torques are applied by means of springs. By varying their length of travel the springs can be adjusted to suit machine and wind characteristics.

In the model shown in Figure 2.46, the pitch of each blade is regulated separately. In this way, every blade can be acted upon to protect the turbine from over-revving. Such in-built redundancy promotes reliability.

The three separate regulation mechanisms can, however, lead to different regulation procedures. These cause aerodynamic 'imbalance', which can place the rotor, hub and yaw gear under considerable loads.

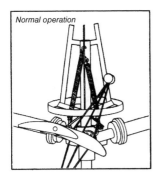
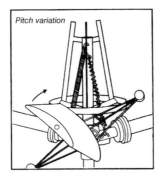
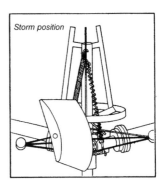

Figure 2.46 Pitch-regulation mechanism using rotating masses, with the possibility of varying the return spring moment (Brümmer). Reproduced by kind permission of Hermann Brümmer

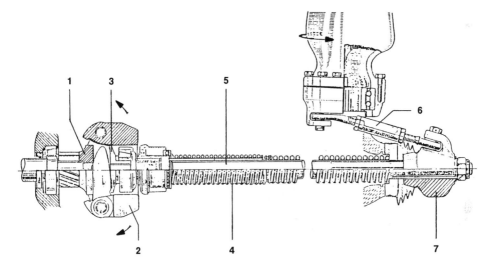

Figure 2.47 Centrifugal force regulator and control linkages (Allgaier): 1, regulator carrier; 2, centrifugal weight; 3, pressure plate; 4, return spring; 5, control shaft; 6, control rod; 7, spider

Figure 2.47 shows a centrifugal-force-based arrangement, used in the 1950s on Allgaier machines (see Figure 1.10). This mechanical regulation and positioning system was supposed to limit the speed of the turbine to its nominal speed. The regulator is built into the gearing and rotates at the same speed as the generator. The 2.5 kg centrifugal weights generate a centrifugal force of about 5000 N. They are therefore capable of working via the thrust rod and positioning crank, against the force of the return springs, to bring the rotor blades into a position appropriate for the current operating conditions. This regulation and positioning system thus makes it possible to control or limit the speed and power of the turbine.

In contrast to regulator systems that depend on rotating masses, hydraulic and electrical systems require an external energy source. The feeding of this energy to the rotating hub and blade positioning device considerably increases construction costs. In spite of this, hydraulic drives are in widespread use, principally in smaller and medium-sized machines, but are also found in large machines. Electrical adjusters, on the other hand, are generally found only in machines of 200 kW nominal output or above.

The power for hydraulic pitch-regulation mechanisms is usually supplied from units in the machine house. For turbines in the 10 to 100 kW class, the drive is usually fed directly via the rotor or generator shaft.

Figure 2.48(a) shows the pitch-setting and power supply systems for a turbine of the 30 or 40 kW class. The hydraulic pump is driven from the generator shaft. From the generator

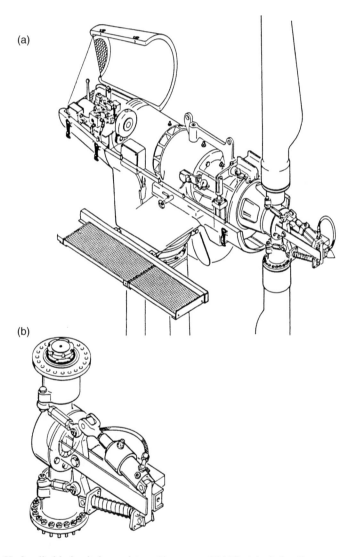

(a)

(b)

Figure 2.48   Hydraulic blade pitch regulators (Aeroman, MAN): (a) pitch adjustment and supply system in the nacelle and (b) setting cylinder and crank system

standpoint, the speed of rotation and power of the turbine are handled as electrical quantities and are compared with given control values. Should any deviation occur, electrohydraulic valves are opened or closed according to control characteristics and the positioning cylinder (see Figure 2.48(b)) is activated. When hydraulic power is applied to the cylinder, it turns the blades towards the plane of rotation. In the absence of hydraulic power, springs acting in the opposite direction return the blades to their original positions. This means that, if the hydraulic system fails or another emergency situation arises, the turbine is returned to a safe operating mode. Seen from the turbine, the control valves and the hydraulic power unit are located behind the generator.

The regulation and adjustment system shown here consists primarily of standard hydraulic drive products. In this way, in addition to the high reliability needed for wind turbines, a high level of solidity and cost-effectiveness can be attained.

Figure 2.49 shows the pitch-regulation system layout in the nacelle of a 4 MW wind power unit. Four positioning cylinders acting directly upon the blade positioning rods bring the 80 metre-diameter blades of the teetered-hub rotor into the positions dictated by the control system. This positioning system thus makes it possible to control power in both directions.

The hydraulic pressure supply with its pumps, fluid cooler and control computer are located at the rear of the nacelle. To provide a redundant backup supply for use during power outages, a hydraulic reservoir is located in front of the rotor hub.

A similar idea for pitch setting is to be found in the WKA 60 1.2 MW turbine shown in Figure 2.50. Here, too, the hydraulics unit and the cabinet with the electronics for control

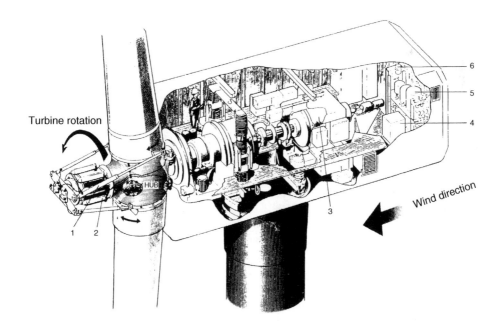

Figure 2.49 Hydraulic blade pitch-regulation system with direct cylinder transmission (WTS 4, Hamilton Standard): 1, hydraulic cylinder; 2, hydraulic reservoir; 3, generator; 4, control computer; 5, hydraulic pumps; 6, hydraulic fluid cooler

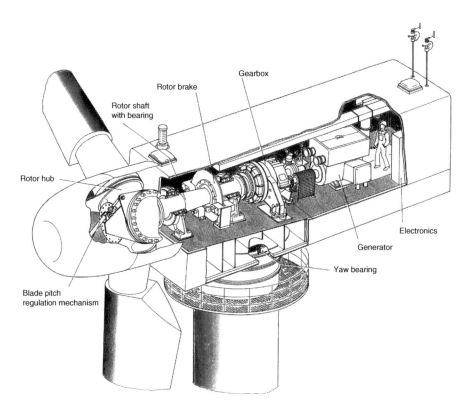

Figure 2.50   Hydraulic blade pitch-regulation system with positioning cylinder and lever transmission (WKA 60, MAN)

and regulation are located at the rear of the nacelle. Each of the three blades has its own direct-acting positioning cylinder. Double redundancy is thus afforded.

Manufacturers of wind turbines prefer mass produced, reliable components for the hydraulic pitch adjustment systems, which are currently in use. The buckling load on the positioning cylinder can be kept low if the mechanism is designed such that positioning routes are kept short and direct (Figure 2.51).

The pitch-setting systems shown above have the pressure supply in the nacelle and the pitch adjustment mechanism arranged around the rotor. This means that the connection between the supply and mechanism must be by rotary lead-through. A rotating hydraulic system, as used in the turbine shown in Figure 1, makes this unnecessary. Such a system is described briefly below (Figure 2.52).

In the pitch adjustment systems shown above, the blade is rotated over its entire length, e.g. into the wind, to reduce rotor power. Similar power changes can be attained by moving the blade tips only (Figure 2.52(a). This requires much smaller positioning forces. This means that the positioning cylinders (Figure 2.52(b)) and the rotating hydraulic power supply, with its pumps, motor, fluid reservoir and pressure reservoir (Figure 2.52(c)), can be correspondingly smaller. However, high turbulence levels must be expected in the area of transition between fixed and movable parts of the rotor blade.

Figure 2.51   Positioning system arrangement with cardanic bearing and short positioning routes

As mentioned earlier, electrically driven pitch-regulation systems are used only in large machines. Figure 2.53(a) shows the entire blade pitch-regulation mechanism of the largest turbine built in the twentieth century (100 metres in diameter), which operated in the mid-1980s (Figure 1.13). The pitch-adjustment mechanism, with its single motor (Figure 2.53(b)), is mounted to the fore of the rotor head and acts on its two blades via axial gearing and connecting levers. The drive used for positioning is an inverter-powered asynchronous motor. A hydraulic emergency shut-down system can, when necessary, force the entire electromechanical drive through a fast positioning procedure and bring the turbine into a safer operation mode.

The Enercon system shown in Figure 2.54 dispenses with the need for a redundant safety system. Three mutually independent electric servomotors drive one blade each. A precise pitch angle measurement system takes care of synchronous blade positioning. Figure 2.54(a) shows the electrically driven single-blade positioning system of a 200 kW Enercon E30. Figure 2.54(b) shows two positioning drives for a single blade of the 500 kW Enercon E40 turbine. Figure 2.54(c) shows the arrangement in the 1.5 or 1.8 MW E66 turbine. The motors, rotating with the turbine, always act with the same gear ratio during positioning.

Electric blade pitch adjustment drives predominate for turbines in the megawatt class. However, in contrast to the systems shown in Figure 2.54(a) and (b), these are usually fitted within the hub (Figure 1.2). The blade pitch adjustment system shown in Figure 2.55 is based upon a fundamentally different functional principle.

In the 1.2 MW VENSYS turbine with a permanently excited synchronous generator a safety and blade pitch adjustment system is used that requires no batteries, pressure reservoir or other energy storage devices for redundant operation and makes slip ring transmission superfluous. This is possible because the electromechanical safety system uses the rotor energy to bring

(a) Blade tip positioning   (b) Positioning cylinder

(c) Rotating hydraulic power unit

Figure 2.52   Hydraulic blade tip positioning mechanism with a rotary hydraulic supply (MOD2, Boeing)

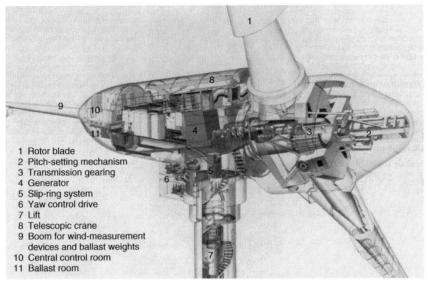

1 Rotor blade
2 Pitch-setting mechanism
3 Transmission gearing
4 Generator
5 Slip-ring system
6 Yaw control drive
7 Lift
8 Telescopic crane
9 Boom for wind-measurement
   devices and ballast weights
10 Central control room
11 Ballast room

(a) Complete system. Reproduced by kind permission of MAN Technologie/MDE

(b) Pitch adjustment mechanism. Reproduced by kind permission of LM Glasfiber A/S

Figure 2.53 Electrically driven blade positioning system with a hydraulic emergency positioning system (GROWIAN, MAN)

(a) Enercon E 30. Reproduced by kind permission of Enercon

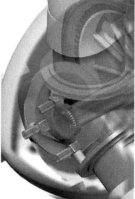

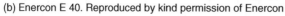

(b) Enercon E 40. Reproduced by kind permission of Enercon   (c) Enercon E 66/E70.

Figure 2.54   Electrically driven blade positioning systems

the rotor blades into feathered pitch, e.g. in the event of an emergency stop [2.20]. In stationary normal operation the turbine rotor and servomotor run synchronously so there are no differences in speed and angle of rotation in the distributor gearbox and the blade pitch angle thus remains the same. If the adjustment shaft is driven more quickly than the turbine, the rotor blades move in the direction of their working position. A slower adjustment shaft, on the other hand, adjusts the blade pitch angle in the direction of the blade's feathered pitch. The brakes, which in normal operation are actively released, drive the blades in the direction of their feathered pitch in the event of partial or total failure of the power supply and safety systems.

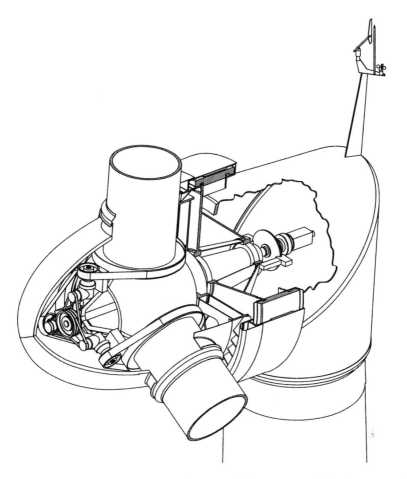

Figure 2.55   Electromechanical blade pitch adjustment with power transmission by means of toothed belts

In contrast to direct motor drives, nonlinear transmission ratio differences always occur in the above-mentioned systems, due to translation of the linear positioning cylinder or drive-rod movements into changes of the blade angle. These must be taken into account when dimensioning the system. The above paragraphs have looked at a few of the most important blade positioning systems out of the wide spectrum available, some of which also bring the blades into active stall (see the systems shown in Figures 1.23(c) and (d) and Figure 1.26(g)).

Fixed-speed machines without pitch variation, which will be described in Section 2.3.3, do not allow the manipulations described here. Such machines must be protected against over-speeding as a result of power outages (during which the generator develops no load torque and so cannot dictate the speed of rotation) by a so-called 'passive control' system. This may take the form of braking flaps in the blade profile (see Figure 2.56(a)) or the blade tips (Figure 2.56(b)). In addition, rotor brakes, usually in the form of disc brakes (see Figures 1.2 and 2.48 to 2.50), afford the possibility of bringing the rotor to a standstill in most machines.

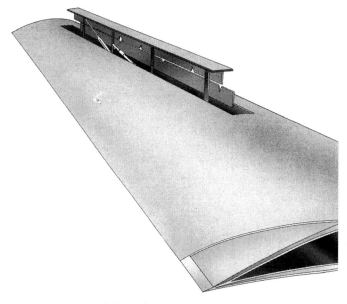

(a) Brake flaps in the blade profile

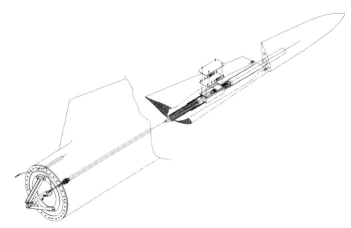

(b) Brake flaps at the blade tips. Reproduced by kind permission of LM Glasfiber A/S

Figure 2.56   Aerodynamic brakes

2.3.2.5   Blade positioning system design

Blade positioning drives – as control and safety systems – must guarantee the controlled operation of wind power units in all conditions. They must therefore ensure that in critical conditions the rotor blades can reduce the energy extracted by the turbine as quickly as possible to
protect the machine from possible damage.

Stationary operating states can usually be used in estimating positioning forces or for
dimensioning the required drive power. Dynamic processes, however, must also be taken

into account, considering the normal speed of blade swivelling in the direction of the operating position against the greatest possible opposing torque and the maximum return speed in the feathering direction against the highest blade torsional moment. In addition, the rotor-accelerating component and the entire positioning mechanism cannot be neglected when determining the size of the drive. In addition to the positioning moments, the forces required to keep the blades at the desired angle must also be considered in choosing the drive.

Hydraulic blade positioning systems that remain under steady pressure in active systems are able to exert this holding torque without extra energy expenditure. In these systems the moved masses, being limited to relatively small components such as fluid and positioning pistons, can normally be neglected in comparison to the inertia of the rotor blades.

Electric drives, on the other hand, can usually exert the necessary holding torque only when there is a continuous current flow. This can be eliminated when self-limiting reduction gears (e.g. axial or worm drives) are used to transmit motor torque to the rotor blades. Such systems, however, demand an extra redundant system in order to guarantee secure operation in case the drive motor goes down. Furthermore, during the operation of electric motors, internal decelerations due to the acceleration of rotating masses must be considered. The moments of inertia of motor rotors, gears, etc., taken in comparison with those of the rotating blades, may well be small but they work through large reduction gears and thus play a significant role in the design.

Wind turbines, particularly in turbulent gusty wind conditions, are subject to tremendous dynamic loads. These arise due to high winds and the resulting power gradients. To allow the turbine to continue to function safely, even under extreme conditions, changes to the turbine's torque must follow as swiftly as possible after the initiation of the pitch-adjustment sequence. This can be attained if the pitch-adjustment drive has a very short start-up time. Under constant acceleration torque and moment of inertia, this is determined by the following relation. For a blade moved directly by the drive,

$$t_{\text{APDd}} = \frac{\omega_{\text{BV}} J_{\text{tot}}}{M_{\text{St}} - M_{\text{Bl}} - M_{\text{Frict}}}. \tag{2.70}$$

If the drive acts upon a number z_a of blades then

$$t_{\text{APDz}} = \frac{\omega_{\text{BV}} J_{\text{tot}}}{M_{\text{St}} - z_a \left(M_{\text{Bl}} - M_{\text{Frict}} \right)}. \tag{2.71}$$

An inverter-powered drive with a current regulator allows such constant-torque procedures to be achieved.

The variables in the above are as follows:

ω_{BV}	design value for the angular velocity of the pitch-setting system
J_{tot}	moment of inertia of the entire pitch-setting system
M_{St}	positioning drive torque
z_a	number of rotor blades driven
M_{Bl}	load torque during blade pitch adjustment
M_{Frict}	moments of friction for all blade bearings in the pitch-adjustment system

The quantities given can relate to either the drive or to the blade(s).

For electrically driven positioning systems, the total moment of inertia is the sum of the motor component, the transmission elements and the rotor blades. Thus

$$J_{tot(A)} = J_{Mot} + J_{trans} + z_a J_{Bl(A)} \tag{2.72}$$

where

J_{Mot} is the moment of inertia of the drive motor,
J_{trans} is the moment of inertia of transmission gearing, clutches, etc., and
$J_{Bl(A)}$ is the moment of inertia of a rotor blade relative to the drive.

The moment of inertia of a blade relative to the drive is derived as

$$J_{Bl(A)} = \frac{J_{Bl}}{i^2_{M\,Bl}}, \tag{2.73}$$

where

J_{Bl} is the moment of inertia of a single blade and
$i_{M\,Bl}$ is the transmission ratio between the servomotor and blade swivelling, i.e.

$$i_{M\,Bl} = \frac{\omega_{Mot}}{\omega_{Bl}}. \tag{2.74}$$

For direct motor drives, i.e. where only rotational motion is involved, when

$$i_{M\,Bl,rot} = \text{constant} \tag{2.75}$$

the transmission ratio always remains the same. On the other hand, when the adjustment mechanism converts linear motion to rotary motion (e.g. in crank or lever transmission), the transmission ratio

$$i_{M\,Bl,lin-rot} = f(\beta) \neq \text{constant} \tag{2.76}$$

and is thus dependent on the pitch angle of the blade.

The use of electric motor drives opens up the possibility of selecting the motor speed (e.g. by setting the number of poles in three-phase motors) according to the transmission ratio. The aim should be to achieve the fastest possible blade positioning. This is attained by a short run-up time. It must be kept in mind here that, seen from the blade side, according to Figure 2.57, the moment of inertia of three-phase motors with greater numbers of poles increases less quickly than the moment of inertia of the motor increases in relation to the value relative to the rotor blade

$$J_{Mot(Bl)} = J_{Mot} i^2_{M\,Bl}. \tag{2.77}$$

Thus, in general, motors with higher pole counts, with correspondingly lower idling and maximum speeds,

$$n_{OM} = \frac{n_1}{p}, \tag{2.78}$$

manage shorter run-up times than high-speed motors with low pole counts. Furthermore, with high-pole-count motors, lower transmission ratios may be attained and so extra gearing levels may possibly be avoided.

In dimensioning blade positioning systems, both normal and fast positioning procedures must be considered. Both areas of operation are examined in the following.

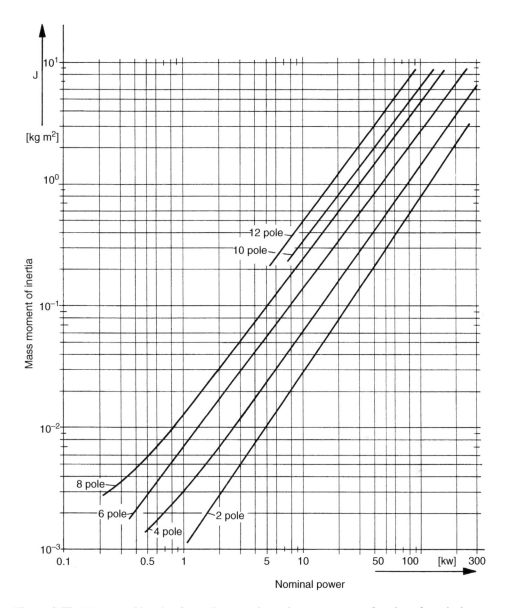

Figure 2.57 Moment of inertia of asynchronous three-phase motors as a function of nominal power with the number of poles ($2p$) as a parameter

Normal positioning procedures

During normal operation, blade pitch adjustments with rotational speeds of approximately

$$\dot{\beta}_n = 5 \quad \text{to} \quad 10°/s = 0.09 \text{ to } 0.17 \text{ rad/s}$$

are expected. This ensures that during normal regulation procedures the turbine will be protected from braking conditions, which place considerable loads on the drive train.

For normal positioning procedures the nominal torsional moments on the rotor blade $M_{\text{Bl n}}$ and the normal speed of displacement for blade positioning can be used. Under steady positioning conditions, ignoring the accelerating component, this yields a positioning torque of

$$M_{\text{St}} = \frac{z_a \left(M_{\text{Bl n}} + M_{\text{Frict}} \right)}{i_{\text{A Bl}} \eta_{\text{A Bl}}} \qquad (2.79)$$

for the drive side, where

$i_{\text{A Bl}}$ is the transmission ratio between positioning drive and blade rotation and
$\eta_{\text{A Bl}}$ is the efficiency of the transmission between the drive and blade.

An average torsional and frictional moment therefore acts against the positioning drive for the z_a blades driven. This figure is increased by the transmission efficiency and takes into account the transmission ratio between the drive and blade. The power required for positioning thus becomes

$$P_{\text{Stn}} = \frac{z_a \left(M_{\text{Bl n}} + M_{\text{Frict}} \right) \dot{\beta}_n}{\eta_{\text{A Bl}}}. \qquad (2.80)$$

For turbines in the 100 kW to MW class, positioning systems for normal operational ranges need between 1 and 10 kW. These figures should be understood as reference values for ongoing operation; as design values for a positioning system they are insufficient. In dimensioning the drive, the accelerating components and extreme situations must be allowed for.

Fast positioning procedures

In emergencies arising as a result of extreme wind conditions, sensor failure or similar states, fast positioning procedures with positioning speeds of

$$\dot{\beta}_s = 10 \ \text{to} \ 20°/\text{s} = 0.17 \ \text{to} \ 0.31 \ \text{rad/s}$$

are used to bring the turbine as swiftly as possible into safe operational ranges.

When dimensioning the drive needed, the greatest possible moment working on the rotor blade against rotation into the feathering position – arising, for example, due to gusting or over-revving – must be allowed for. In addition, the accelerating components of all parts of the system must be taken into account in the design calculations.

The power necessary to accelerate the inert masses of the positioning system is derived from the angular velocity and the accelerating torque of the drive, transmission and rotor blade components. Thus the power necessary for the positioning drive is

$$P_{\text{Sts}} = \frac{z_a \left(M_{\text{Bl max}} + M_{\text{Frict}} \right) \dot{\beta}_s}{\eta_{\text{A Bl}}} + \dot{\beta}_s J_{\text{tot(Bl)}} \ddot{\beta}_s. \qquad (2.81)$$

This quantity is generally decisive for the dimensioning of the positioning system. For turbines in the 100 kW to MW class, the drive rating usually lies between 5 and 50 kW. For the design of a drive, however, torques and the dynamic characteristics of the entire system must additionally be included, as in the following study.

Positioning drive design

In designing the drive, it must be ensured that the positioning system retains sufficient power reserves to overcome the highest possible torsional moments on the blades. Uncertainties in working out torsional moments and production variations in components (the rotor blades in particular) must be allowed for with a healthy safety factor. However, changes in the dynamic behavior of the entire system due to these power reserves must be taken into account. For example, a significant increase in the power of a two-pole motor will also considerably increase its inertia (see Figure 2.57), as a result of which blade positioning reaction time will be correspondingly longer. Such correlations show clearly that augmenting power reserves may be achieved at the expense of drive dynamics, which could possibly even compromise the safety of the entire system.

Therefore, during the design phase the transmission ratios and the drive power of technically feasible variants, such as, for example, those for electric motors, should be considered. In doing this, the time taken to attain the speed necessary for a fast positioning procedure

$$t_{Af} = \frac{J_{tot(M)}\dot{\beta}_s i_{A\,Bl}}{M_{St} - \frac{M_{Bl\,max} + M_R}{i_{A\,Bl}\eta_{A\,Bl}}z_a}$$

(2.82)

should first be determined. Thereafter the displacement processes, or the phases thereof, can be better described.

For blade positioning, the time t_0 required to turn a rotor blade through an angle of $\Delta\beta_0$ from a critical to a safe position can be determined. The process by which the blade is brought to this angle

$$\Delta\beta_0 = \frac{\ddot{\beta}_s}{2}t_{0b}^2$$

(2.83)

can, in high-inertia systems, be regarded as consisting only of acceleration over the acceleration time t_{0b}. In rapidly accelerating systems, on the other hand, after the run-up time t_{Af} the fast positioning speed is attained and movement progresses thereafter at the constant angular speed β_s, and can therefore be characterized by the relation

$$\Delta\beta_0 = \frac{\ddot{\beta}_s}{2}t_{Af}^2 + \dot{\beta}_s t_v.$$

(2.84)

Similarly, for pure accelerated motion until reaching the desired angle, the time

$$t_0 = t_{0b}$$

(2.85)

or, for overlying phases of acceleration and constant-speed displacement,

$$t_0 = t_{Af} + t_v.$$

(2.86)

Observations of differently designed positioning systems have shown that for motorized drives the dynamics of the mechanism are strongly influenced by the number of poles in the motor.

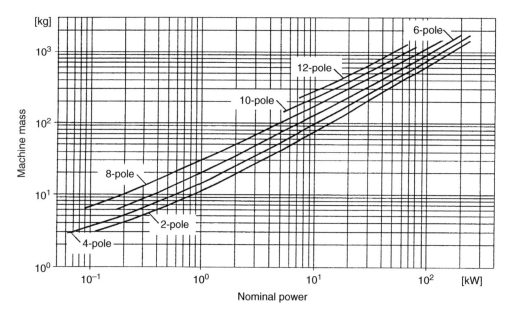

Figure 2.58 Mass of three-phase asynchronous motors as a function of the nominal power with the pole number $2p$ as a parameter

Two-pole, three-phase motors may well be lighter than multipole machines (see Figure 2.58), but as a blade positioning drive they take significantly more time to run up or to reach a safe blade angle $(\Delta\beta_0)$ than four- or six-pole motors. Furthermore, it is clear that increasing drive power above the level needed for fast positioning does not improve system dynamics. Generally, however, doubling the power of a two- or four-pole motor is sufficient for the run-up time to exceed the minimum. Further increasing power, particularly in two-pole motors, leads to a deterioration of drive dynamics. Six-pole motors, on the other hand, achieve even more favourable running-up behavior if the power is increased further.

When designing connection, control and protection devices for asynchronous machines (Figure 2.59), it helps to know the nominal current. If no current limiters are provided, high start-up currents (see Figure 3.65) should be expected. The higher current consumption of multipole motors is due mainly to the larger air gaps, which are necessary because of the greater mass and diameter.

If the behavior under acceleration of the positioning systems is given, then the safety margin during acceleration procedures can be derived in the form of a moment:

$$M_{res} = M_{St} - z_a \frac{M_{Bl} - M_{Frict}}{i_{A\,Bl}\eta_{A\,Bl}} - J_{tot(A)}\frac{d\omega_{St}}{dt}. \tag{2.87}$$

Calculations show that, here too, fast positioning procedures using two-pole motors can only be obtained with the help of blade torsional moments. Torque reserves are thus not available. Machines with greater numbers of poles, on the other hand, offer greater torque surpluses and thus higher safety margins because of more power and poles. Definite statements on the choice and behavior of blade positioning systems and their effects on the system as a whole can only be arrived at with the aid of simulations. The basic contexts for this are reproduced here.

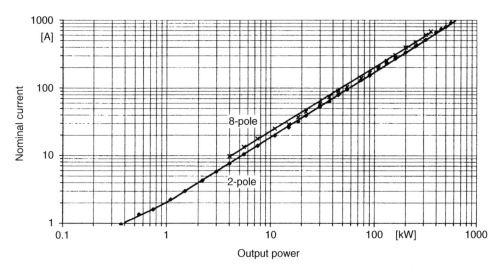

Figure 2.59 Nominal current of three-phase asynchronous motors as a function of nominal power with the pole count $2p$ as a parameter

2.3.3 Limiting power by stall control

In aviation, aerofoils, rotary wings and propellers are not generally run in stall for safety reasons. The power absorbed by wind turbines, on the other hand, can be limited by intentionally inducing stall conditions. In wind turbines with fixed rotor blades and asynchronous generators, coupled to the grid at a fixed speed, stall operation can be brought about passively by the altered profile flow that results from wind-speed increases. It can also be intentionally induced by varying the blade pitch angle or the turbine speed. In this active approach, however, a blade pitch adjustment system similar to that used in pitch-controlled machines must alter the blade position in the direction of stall or frequency converter systems (see Chapters 3 and 4) must alter the generator's (rotary) frequency, thereby changing the peripheral speed of the turbine and the flow over the blades. We will first discuss passive stall operation, which has predominated up until now.

2.3.3.1 Passive stall regulation

The vast majority of all installed wind power plants in the class up to approximately 1000 kW that supply power to the public grid are without blade positioning mechanisms, instead using so-called stall control to limit power extraction. Such plants are constructed up to an order of magnitude of 1.5 MW and are generally fitted with asynchronous generators (see Figure 1.22(d)).

In normal operation, laminar flow predominates at the rotor blades (Figure 2.60(a)). In these conditions, the lift values corresponding to the angle of attack are reached at low drag components (see Figure 2.60(c)). Thus in partial loading ranges a high degree of aerodynamic efficiency is attained. If, on the other hand, the wind speed approaches the value at which the

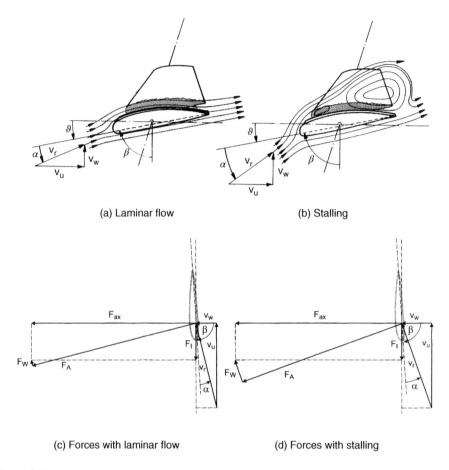

(a) Laminar flow   (b) Stalling

(c) Forces with laminar flow   (d) Forces with stalling

Figure 2.60   Airflow behavior and forces at the rotor blade as wind speed increases from (a) to (b)

generator reaches its maximum permanent load (usually the nominal power), further torque development at the rotor must be inhibited.

The largely rigid grid connection means that the generator (within the relatively narrow slip range of asynchronous machines) keeps the turbine at a near-constant speed; i.e. the peripheral speed v_p is approximately constant. Wind speeds exceeding nominal levels cause higher angles of attack and thus (in the appropriate design) stalling (Figure 2.60(b)), when the airflow 'unsticks' from all or part of the blade profile. Depending upon the angle of attack, therefore, as shown in Figure 2.5, the lift coefficient $c_a = f(\alpha)$ and the lift forces dF_A (see Figure 2.4) are reduced in certain ranges and the drag coefficients $c_w = f(c_a, \alpha)$ or the drag forces increase. As a result, the torque-creating tangential force F_t (the sum of all partial forces dF_t) does not significantly exceed its nominal values (Figure 2.60(d)). When the turbine is under full load and the wind speed climbs beyond the nominal range, this results – in spite of the greater levels of energy available – in lower rotor torque and lower performance coefficients.

The performance characteristics (Figure 2.61, top) of machines such as this are therefore largely dictated by their construction. In comparison with pitch-regulated turbines,

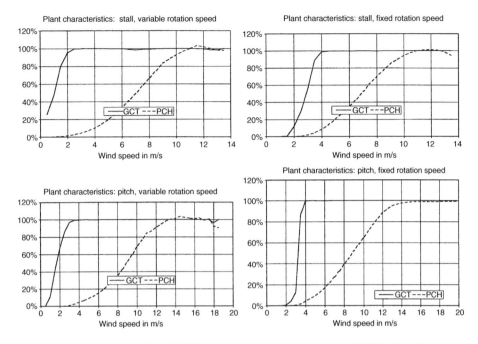

Figure 2.61 Grid connection time (GCT) and performance characteristics (PCH) of wind power plants (WMEP measurements ISET)

stall-regulated machines are often designed with asynchronous generators of higher nominal output. Rigid grid coupling – a basic requirement for safe operation – is thereby obtained.

The complex aerodynamic and vibrational processes involved in stall regulation will not be examined in this study. For stall-regulated machines with nominal outputs of 30 kW and above, variable-speed turbine systems with synchronous generators and frequency converters are gaining in importance. Such configurations allow control manipulations similar to those of active-stall or pitch-regulated units. Limited performance control must, however, be balanced against the advantages of smoothed output power.

Manoeuvres directed at regulating the flow of energy into the supply system (highly desirable in wind turbines running under partial load in isolated operation) are not possible with fixed-speed units as depicted in Figure 1.38(b). These call for active interventions to the stall operation.

2.3.3.2 Active stall regulation

By adjusting the rotor blades (usually in the opposite direction to that for pitch control) as shown in Figure 2.36 or by changing the turbine speed with the aid of the generator (see Figure 2.68), stall operation and thus turbine output can be actively influenced in accordance with Figure 1.38(a) and (b) and matched to the desired grid or consumer requirements. The term 'active stall regulation' is generally used to mean the former process.

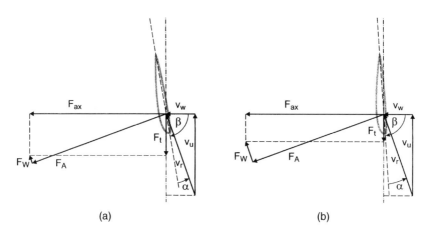

          (a)   (b)

Figure 2.62 Flow and forces on the rotor blade given the same wind speed for active-stall adjustment (a) Laminar flow (b) Stalling

In machines with blade pitch-adjustment systems as described in Section 2.3.2, the rotor blades can be moved in the direction of the feathering pitch in order to reduce the power absorbed by a turbine. In contrast, active-stall-controlled machines require that the blades are rotated in the direction of the plane of rotation in order to move into the stall range and thus reduce the power drawn from the airstream. In general, a blade pitch-adjustment range of a few degrees is sufficient to protect the machine from overload, for example, or to adjust output to the desired levels.

Figure 2.62(a) and (b) illustrates the change in the torque-producing tangential force F_t given the same wind conditions and only slightly changed blade pitch angle or angle of attack (β or α). On the other hand, stall-controlled machines with a synchronous generator and frequency controller or double-fed asynchronous generator permit variable-speed turbine operation. Such configurations mean that the peripheral speed of the blades and thus the angle of attack can be altered by adjusting the turbine speed (see Figure 2.63). Regulatory interventions similar to those of blade pitch-adjustable (active-stall or pitch-controlled) turbines are possible, such that the machine output can be matched to grid or consumer requirements by the adjustment of the generator speed. As a result, optimal output ranges can be approached and, as described in what follows, only a proportion of the available power drawn for the protection of machine components and energy consumers.

2.3.4 Power control using speed variation

In spite of the constant frequency of the grid, the speed of a wind turbine can be influenced

- mechanically by varying the transmission ratio if the generator speed is constant (cf. Figure 2.64(a) and (b)) or
- electronically by frequency converter if the speed of rotation is variable over the entire drive train (turbine, gearing and generator).

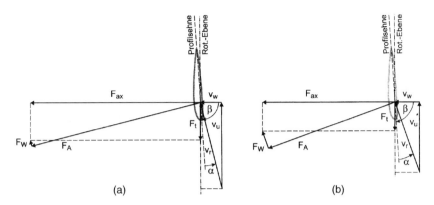

Figure 2.63 Flow conditions and forces at the rotor blade with (a) high and (b) low blade peripheral speed or rotor speed

The turbine speed can thus be adapted to meet grid and operational requirements (see Section 3.2). Varying the rotor speed in turn varies the power of the turbine, allowing it

- to be brought into the maximum range, depending on available wind, or
- to be reduced if required to meet grid or user requirements (see Figure 2.64(c) and (d)).

Such processes are currently used by machines in the 30 kW to MW range that employ power electronics (rectifiers, previously line-commutated and now self-commutated inverters, see Section 4.1) to convert the frequency to that required. Within their design limits, such models permit

- extensive running at optimal performance within the partial loading range without blade pitch adjustment, i.e. even in stall-regulated machines, and
- the reduction of power at turbine level under certain conditions.

In addition, when flexible grid coupling is employed, the effect of fluctuations in available wind can be reduced by making use of the rotating mass of the drive train (rotor, gearing, generator rotor) to:

- smooth out variations in rotor speed and
- reduce the dynamic load on the entire system, particularly in mechanical drive trains.

In this way, the power converter allows a very short-time intervention at the generator, which allows the requirements and desires to be achieved at the turbine via the drive train. Thorough explanations of different technical variants of

- possible aspects of design,
- control procedures and
- dynamic processes and behavior

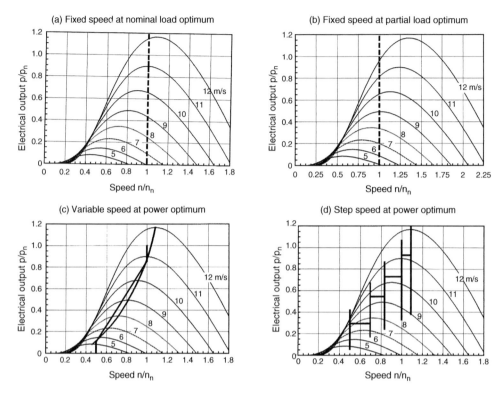

Figure 2.64 Output/rotational speed characteristics and dynamic operating ranges of wind turbines, with wind speed as a parameter

in systems of this type have already been given in reference [2.21] and, for subareas such as double-fed asynchronous generators, in references [2.22] and [2.23], and more detailed explanations will be given in the course of this book, particularly in Chapter 5.

2.4 Mechanical Drive Trains

The drive power of a wind turbine engenders torques in its mechanical drive train or generator that are subject to fluctuation as a result of both periodic and aperiodic processes, such as

- changes in wind speed,
- tower-shadow or tower-occasioned upwind overpressure,
- blade asymmetry,
- blade bending and skewing, and
- tower oscillation.

Figure 2.65 Mechanical drive train of a geared wind turbine

In addition, load moments in the generator and converter due to

- static,
- dynamic and
- electromechanical

behavior also act on the wind turbine via the drive train. The interaction of all torque effects works together with the flywheel-effect-dependent acceleration components to determine conditions in the mechanical drive train. A brief look at this is now given.

Figure 2.65 shows the structure of a typical drive train in a conventional geared turbine. From left to right, the rotor hub, the gearbox inclusive rotor bearing and the connecting flange of the generator may be discerned.

The considerably simpler construction of a gearless model is shown in Figure 2.66, which depicts the essential rotor–head components of a 200 kW Enercon E 30 turbine (see Figure 1.33(d)). The drive-train configuration is reduced to the rotor hub, the drive shaft and the disc generator. This system represents a further step in the development of medium-sized turbines, which are distinguished in having variable-speed rotor or generator operation and grid-compatible coupling to power utilities via frequency converter systems.

Turbine effects, which are not further quantified here, depend upon the wind conditions at the site, i.e. wind speed and wind gradients (due to gusting and turbulence), as well as the constructional characteristics of the wind power unit as a whole. Tower-shadow and tower-occasioned

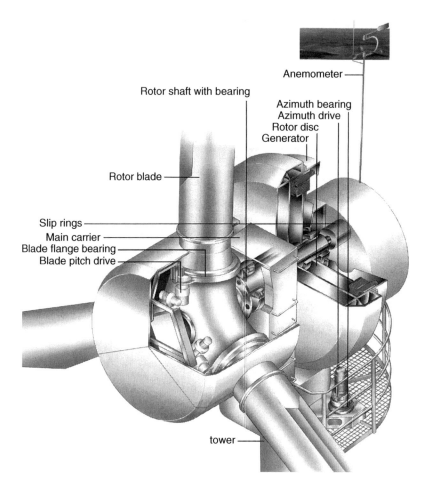

Anemometer

Rotor shaft with bearing

Azimuth bearing
Azimuth drive
Rotor disc
Generator

Rotor blade

Slip rings
Main carrier
Blade flange bearing
Blade pitch drive

tower

Figure 2.66 Schematic layout of an Enercon E 30 gearless wind turbine. Reproduced by kind permission of Enercon

upwind overpressure are especially important here as integer multiples of the rotor rotation frequency and blade count, occasioning torque oscillations in the hertz range that are generally more prominent in downwind than in upwind turbines. Load-alternation components arising due to

- rotor asymmetry (which acts according to the rotor frequency),
- blade bending and skewing, and
- tower oscillations,

which may be subject to influences caused by the respective resonant frequencies, are overlaid on the wind speed and thus show their effects additionally as wind gradients, in general play a subordinate role (from measurements, $\Delta M < 5\% \, M_N$) [2.24].

Torque oscillations at the meshing frequency of individual gear ratios are also possible, and measurements have shown that in unfavourable cases they can attain $\pm 10\%$ of nominal torque [2.4]. Particular influence is exerted by the design and type of the generator integrated with the grid and by the degree of elasticity of the entire system.

As shown in Figure 1.38(a) and (b), the drive torque M_A of the turbine acts on the mechanical drive train from one side and the load torque M_W of the generator acts on it from the other. Additionally, between these two main components there exists a speed-of-rotation or angle-of-rotation coupling via the mechanical elements connecting them. The rotating masses of the rotor blades, hub, gear train, brake disc, clutch, shafts and generator rotor are accelerated according to the difference in moments. The rigidity and damping characteristics of the individual components, and any play in the clutch and gears, exert a decisive influence on the behavior of the transmission.

Studies of designed and constructed machines have shown that in general the inertia of gears, brake discs, clutches and shafts play a subordinate role in comparison with the dominant components exerted by the rotor blades, the hub and the generator rotor. If necessary, the relevant inertial components can be allowed for by simply adding them to the slow-revving side (hub, rotor blades) or to the high-speed side (generator rotor) according to their effect or, depending on the coupling, they can be taken into account as multiple-mass systems.

Because of the shock load that can be expected and the possible reversal of the energy flow, play in gearing and clutch is to be avoided, or at least kept as low as possible. It will therefore be ignored in the following.

Calculations based on existing machines show that to make an approximate determination of the behavior of the transmission, the mechanical drive train, gears, clutches, etc., can usually be regarded as having zero mass according to the low moment of inertia within the system as a whole. This leaves only the two main components 'rotor' and 'generator', with an elastic, damped coupling existing between them (see Figure 2.67). Frictional components in the drive train can be ignored.

The torque at the generator coupling M_{Ku} can be represented by the simplified relation

$$M_{Ku} = k_{TS} \left(\varepsilon_W - \varepsilon_G \right) + k_{DK} \frac{d \left(\varepsilon_W - \varepsilon_G \right)}{dt} \tag{2.88}$$

or

$$M_{Ku} = M_{TT} + M_{TD} \tag{2.89}$$

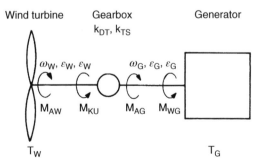

Figure 2.67   Simplified representation of a mechanical drive train

as the sum of moments of torsion–elastic (M_{TT}) and damping (M_{TD}) properties, where

M_{AW}	drive moment at the turbine rotor
M_{AG}	drive moment at the generator
M_{Ku}	coupling torque at the generator
M_{TD}	damping component of the drive train moment
M_{TT}	torsionally elastic component of the drive train moment
M_{WG}	electrical load torque in the generator
ε_W	angle of rotation of the rotor
ε_G	angle of rotation of the generator
$\Delta\varepsilon = \varepsilon_W - \varepsilon_G$	angle of torsion
k_{TS}	torsion resistance
k_{DK}	damping constant
T_G	run-up time constant of the generator
T_W	run-up time constant of the rotor

In stationary-state operation $d(\varepsilon_W - \varepsilon_G)/dt = 0$, so that the coupling torque reduces to its torsionally elastic component

$$M_{Ku} = M_N = M_{TT} = k_{TS} \left(\varepsilon_W - \varepsilon_G \right). \tag{2.90}$$

Thus the torsional stiffness of the mechanical drive train under stationary component loading

$$k_{TS} = \frac{M_{Ku}}{\varepsilon_W - \varepsilon_G} = \frac{M_{Ku}}{\Delta\varepsilon} \tag{2.91}$$

can be determined from the difference in angle between the transmission elements, which is relatively simple to measure.

Figure 2.68 shows a possible structure for the equation of moments as described above, including the rotating mass of the turbine rotor (J_W or T_W), on the one hand, and the inertia of the generator (J_G or T_G), on the other. In this figure,

M_{BG}	acceleration torque at the generator
M_{BW}	acceleration torque at the turbine rotor
$T_\varepsilon = p/\omega_0$	integrator time constants for determining the torsion angle (generator side)
ω_G	angular velocity of the generator
ω_W	angular velocity of the turbine rotor

In this way, torques and angular velocity values are generally related to the generator side, taking transmission ratios into account.

As predicted by calculation, existing machines show nontrivial angles of torsion (e.g. $\Delta\varepsilon = \pi/4$ at nominal torque). These are certainly of considerable significance for the design of the drive train, but this is not the object of the current study, which seeks an approximate determination of transmission behavior. In normal wind power plant designs a run-up time of around 10 seconds can be expected (see Section 2.5), while delays arising from the drive train are on the order of milliseconds.

Because of the large difference in time constants, in a good approximation a mechanical drive train can be regarded as constituting a perfectly rigid transmission, and can thus be described as having proportional behavior. Accordingly, the moment of inertia of mass of all rotating

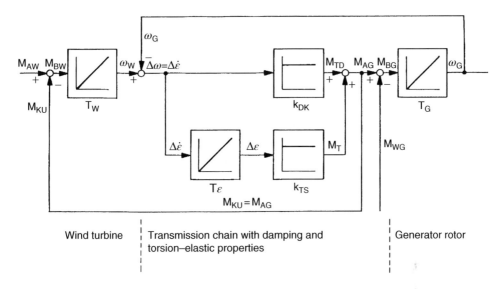

Figure 2.68 Structure of a mechanical drive train

parts can be taken as a whole, once any transmission ratios have been allowed for. In doing so, the generator side is the preferred frame of reference.

We thus obtain the simple equation of moments for the drive train

$$\frac{M_{AW}}{M_N} - \frac{M_{WG}}{M_N} = \frac{M_{BR}}{M_N} = T_R \frac{d\left(\omega/\omega_N\right)}{dt}, \tag{2.92}$$

where the integration or run-up time constant of the rotor system

$$T_R = \frac{J_R \omega_N}{M_N} \approx T_W + T_G, \tag{2.93}$$

which corresponds approximately to the sum of the time constants of the turbine and generator, where

J_R is the moment of inertia of all rotating masses,
M_{BR} is the acceleration torque of rotating components,
M_N is the nominal moment and
ω_N is the nominal angular velocity.

For cases of interest (e.g. dimensioning the drive train), the acceleration torque M_{BR} on the entire rotor system can be split into

$$M_{BW} = \left(1 - k_T\right) M_{BR} \tag{2.94}$$

on the turbine rotor side, and

$$M_{BT} = k_T M_{BR} \tag{2.95}$$

on the generator side, according to the inertia of their respective masses, as shown in Figure 2.16.

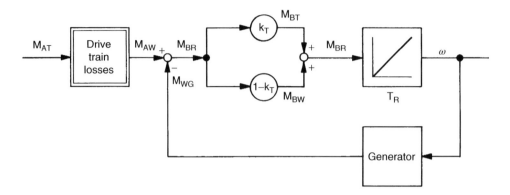

Figure 2.69   Highly simplified structure of a mechanical drive train

Transmission losses can be taken into account either by including them in the drive torque (Figure 2.11) or by introducing an extra block in the drive train structure, as in Figure 2.69. Here the input quantity M_{AT} stands for the drive torque of the drive train, which must be reduced by the losses therein.

Figure 2.69 illustrates the possibilities of controlling turbine speed by means of the generator, mentioned in Sections 2.3.3 and 2.3.4. This theme is complemented and continued in Chapter 3, which shows various ways of using generators to convert mechanical energy into electrical energy.

2.5   System Data of a Wind Power Plant

Standard system values are of great significance when planning machines, working out details and estimating everything from loads and expected behavior to transport and assembly problems. The following sets out essential construction data and currently available reference values for machine costs as a function of capacity.

The installation of turbines, and in particular screw connections, between highly loaded construction elements (rotor blades, tower flange, etc.) that facilitate the attainment of the defined prestressing forces with reproducible accuracy where there is no torsion or side loading (see end of Chapter 2) will not be discussed further here.

2.5.1   Turbine and drive train data

Within the normal ranges of wind speed (5 to 15 m/s), airstreams do not carry much energy. In consequence, wind energy converters need to have turbines of large diameter rotating at low speed, which means that construction costs are high. The following reference values for essential physical quantities of mechanical converters (turbine, drive train) permit estimates and calculations of the dynamic behavior of entire systems to be made. These values are given in Figure 2.70 in the form of a nomograph.

Reference values for nominal power, the necessary rotor radius, construction-dependent rotor speed and the resulting gear ratios, as well as run-up time constants for mechanical

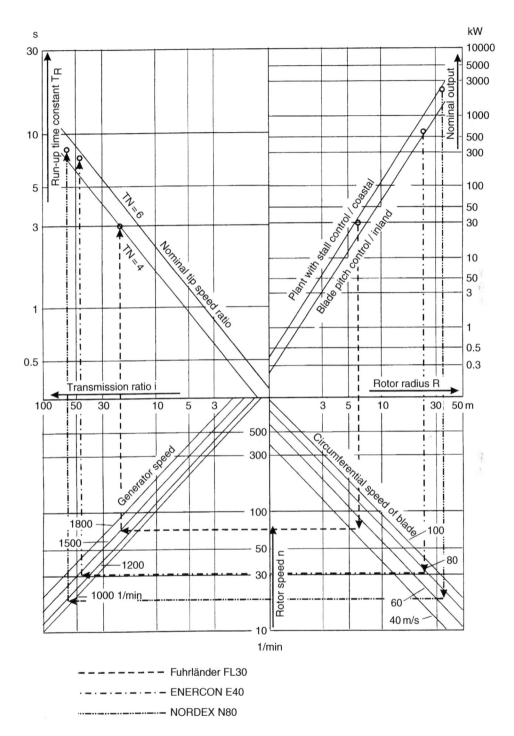

Figure 2.70 Standard values for turbine systems

drive trains (see Section 2.4, Equation (2.93)) for use in system studies, are summarized in Figure 2.70. This figure is based on data from over 100 different turbines, allowing for various design-specific parameters and constructional boundary conditions, such as differences in power or diameter in systems designed for inland as opposed to coastal areas, and the associated considerations of stall-versus pitch-regulated turbines.

In addition to the blade profile and tip speed ratio, blade tip velocity (which is limited by aerodynamics, rigidity and noise development) is decisive in determining the very low speed of turbines, particularly for large machines. The mechanical transfer of energy to the generator is usually kept at a constant speed by the use of gearing. In wind turbines, as in other types of power plant, four- and six-pole generators are usually used. Such generators, together with the associated gear trains, represent a low total mass for the converter system. This is why the gear ratios of these common generator variants, for use in 50 Hz and 60 Hz grids or for variable-speed generator variants, were taken as a basis for Figure 2.70. The data for wind turbines that have directly driven generators with a high number of poles, designed for gearless units, cannot be directly read off. Their 'virtual' speed, which corresponds with the electrical rotational frequency, is found from the product of the pole pair number and mechanical speed of rotation according to the general relation $n_{el} = pn_{mech}$.

In conclusion, we give standard values for rotor system run-up time constants as a function of the flywheel effect of all rotating masses. Here, in addition to nominal torque, the speed of rotation and tip speed ratios in nominal operating conditions are given for two characteristic turbine configurations, where $\lambda_N = 4$ and 6. In accordance with Section 2.3, a friction-locked torque transfer from the turbine rotor to generator was taken as a prerequisite in determining run-up time constants. The explanations above contain wind energy converter data that must be allowed for in calculating and designing electromechanical energy converters.

In the following, essential data for common wind power machines on the market are given in the form of standard values for further studies. Manufacturer's data for around one hundred of the most common standard models in the 20 W to 2500 kW class have been evaluated and are presented in Figures 2.71 to 2.80.

2.5.2   Machine and tower masses

We limit our study of the masses in a wind power machine to the transportable and dynamically loaded components. These mainly comprise those items found in the head of the tower (the nacelle and the rotor blades) and the tower itself. Much heavier items such as the foundations, control room and connecting cables are ignored. In Figures 2.71 to 2.73, specific values for machine nacelle mass including rotor blades, are plotted as functions of nominal power, swept rotor area and rotor torque. Comparing the figures shows that the mass per square metre of swept area climbs slowly with machine size, while the values related to machine power and turbine torque fall off distinctly. Furthermore, it can be clearly seen from the low degree of scattering of the turbine torque data that a reliable relation for nacelle mass exists.

Figures 2.74 and 2.75 show the difference in mass between steel tube, concrete and lattice towers as a function of nominal machine output and of the swept rotor area of the turbine. Because of the wide scattering of results, no standard rules can be immediately deduced. It can be seen, however, that in small plants steel tube masts have a lower mass than concrete or lattice towers. As swept area increases, however, they tend steadily to larger specific values.

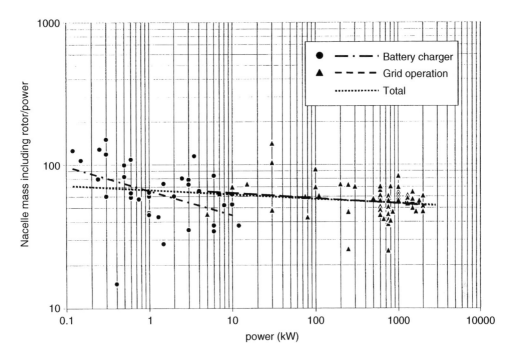

Figure 2.71   Specific mass per kW of nacelle plus rotor as a function of nominal power

The masses of concrete and lattice towers, on the other hand, decrease as the nominal power and swept area increase.

Figure 2.76 shows that turbines with outputs below 100 kW mainly have specific tower masses greater than 100 kg per kW, while turbines with outputs above 100 kW generally exhibit figures less than 100 kg per kW. High ratios of tower height to turbine diameter (e.g. $h/d = 4$) are generally selected for small turbines, while turbines in the megawatt class tend to exhibit constant height to diameter ratios ($h = d$). Although the tower mass per kW tends towards smaller values as the turbine size increases, at the same height to diameter ratio there is a clear increase in mass.

A completely different trend is seen in the relationship between the tower mass and tower height. Figure 2.77 illustrates the sharp increase of tower mass per metre mast height as the turbine output increases. This figure rises from approximately 20 kg/m for 2.5 kW plants to around 2000 kg/m for those in the 1.5 MW class. Furthermore, it is clear that small plants exhibit the same mass per metre at different tower heights, while large plants tend to have a significantly greater mass per metre as the mast height increases.

2.5.3   Machine costs

Figure 2.78 shows machine costs as a function of nominal turbine power. In spite of differing design variants, which are allowed for in this presentation, costs deviate only slightly from the average values. However, clear differences can be seen (in particular in small machines)

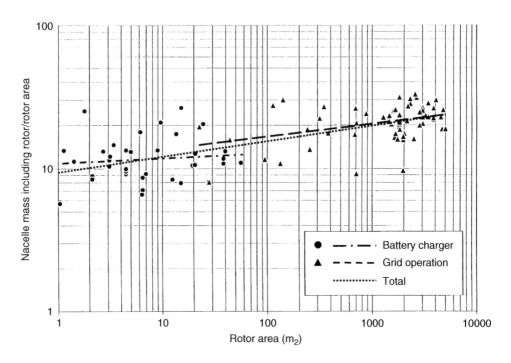

Figure 2.72   Specific mass per m² of nacelle and rotor blade as a function of swept rotor area

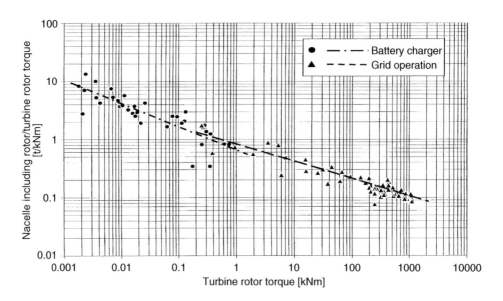

Figure 2.73   Specific mass per kNm of nacelle and rotor blade as a function of turbine rotor torque

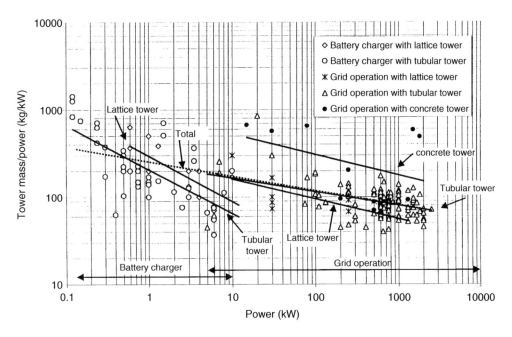

Figure 2.74 Specific mass per kW of tower as a function of turbine power

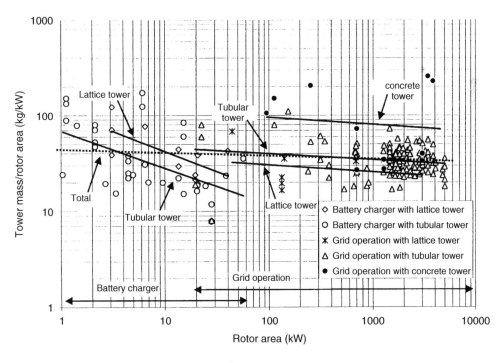

Figure 2.75 Specific mass per m² of a tower as a function of swept rotor area

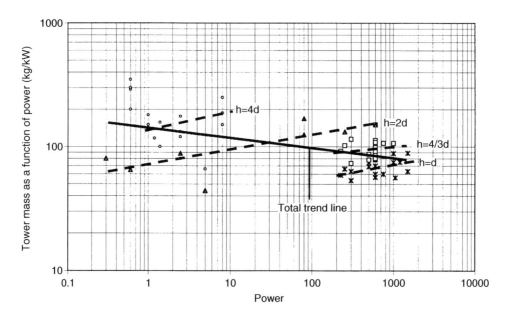

Figure 2.76 Specific mass per kW of a wind turbine tower as a function of the turbine output for various tower height to turbine diameter ratios

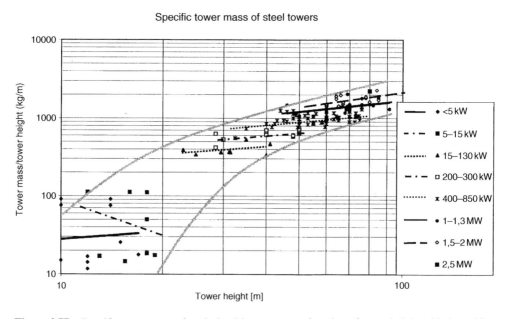

Figure 2.77 Specific mass per m of a wind turbine tower as a function of tower height with the turbine output as a parameter

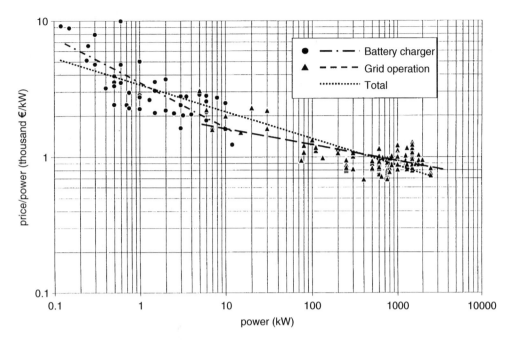

Figure 2.78   Specific machine cost per kW as a function of nominal output power

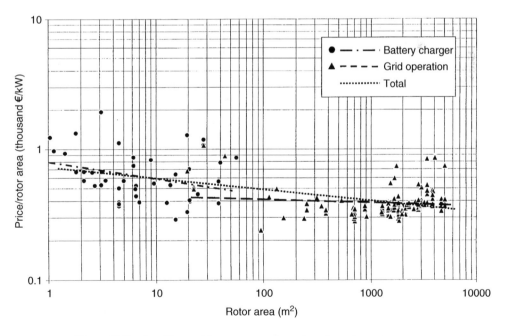

Figure 2.79   Specific machine cost per m² as a function of swept rotor area

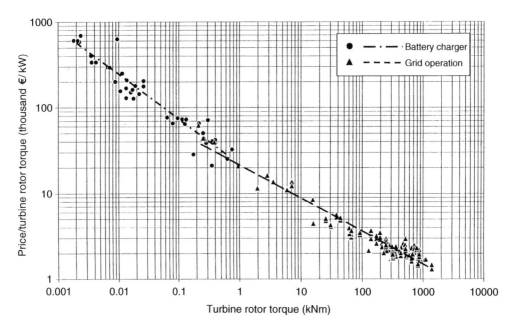

Figure 2.80   Specific machine cost per N m as a function of rotor torque

between battery chargers and grid-compatible machines. Furthermore, newly developed large-scale machines in the market introduction phase tend to have higher specific purchase costs. However, this trend could be discerned even in the previous developmental phases, meaning that cost digressions can be expected here too.

Better predictions, particularly as regards expected energy levels and economic questions, may be obtained with the help of a plot of cost against swept rotor area, as in Figure 2.79. A further variant is offered in the plot of machine costs against rotor torque as shown in Figure 2.80. In this, not only the nominal output power but also the turbine speed dependencies are included in the reckoning. The low scattering of the data shows that here too a clear relationship between cost and torque is to be observed.

In contrast to nacelle mass, it may be seen that with increasing machine size the area-related machine cost slowly decreases, while for nominal power and torque, cost falls off rapidly.

References

[2.1] Schlichting, H. and Truckenbrodt, E., *Aerodynamik des Flugzeuges*, Vol. 2, Springer, Berlin, 1960.
[2.2] Betz, A., *Wind-Energie und ihre Ausnutzung durch Windmühlen*, Vandenhoeck und Ruprecht, Göttingen, 1926.
[2.3] Gasch, R., *Windkraftanlagen. Grundlagen und Entwurf*, 3rd Edn, Teubner, Stuttgart, 1996.
[2.4] Molly, J.P., *Windenergie. Theorie, Anwendung, Messung*, 2nd Edn, Müller, Karlsruhe, 1990.
[2.5] Rijs, R.P.P. and Smulders, P.T., Blade Element Theory for Performance Analysis of Slow Running Wind Turbines, *Wind Engineering*, 1990, 14(2), 62–79.

[2.6]   Wilson, R. and Lissaman, P., *Applied Aerodynamics of Wind Power Machines*, Oregon State University, 1974.

[2.7]   Hernandez, J. and Crespo, A., Aerodynamic Calculation of the Performance of Horizontal Axis Wind Turbines and Comparison with Experimental Results, *Wind Engineering*, 1987, 11(4), 177ff.

[2.8]   Kraft, M., *Verfahren zur Leistungsbestimung für Windturbinen*, Thesis, Kassel University, 1990.

[2.9]   Man, D. T., Sullivan, J. P. and Wasynczuk, O., Dynamic Behavior of a Class of Wind Turbine Generators During Random Wind Fluctuations, *IEEE Transactions on Power Apparatus and Systems*, June 1981, PAS-100(6).

[2.10] Anderson, P. M. and Bose, A., Stability Simulation of Wind Turbine Systems, *IEEE Transactions on Power Apparatus and Systems*, December 1983, PAS-102(12).

[2.11] Amlang, B., Arsurdis, D., Leonhard, W., Vollstedt, W. and Wefelmeier, K., *Elektrische Energieversorgung mit Windkraftanlagen*, Abschlußbericht BMFT-Forschungsvorhaben 032-8265-B, Braunschweig, 1992.

[2.12] Moretti, P. and Divone, L., Moderne Windkraftanlagen, *Spektrum der Wissenschaft*, August 1986.

[2.13] Hau, E., *Windkraftanlagen*, 2nd Edn, Springer, Berlin, 1996.

[2.14] Divalentin, E., The Application of Broad Range Variable Speed for Wind Turbine Entrancement, in EWEA Conference, Rome, Italy, 1986.

[2.15] Eppler, R., *Vorlesungsmanuskript Technische Mechanik*, Stuttgart University, 1968.

[2.16] Försching, H.W., *Grundlagen der Aeroelastik*, Springer, Berlin, 1974.

[2.17] Heier, S. and Kleinkauf, W., *Regelung einer großen Windenergie-anlage (GROWIAN)*, Final Report, Part III: *Wirkleistungs-Drehzahlregelung*, Kassel University, Kassel, 1978.

[2.18] Cramer, G., Drews, P., Heier, S., Kleinkauf, W. and Wettlaufer, R., *Mitarbeit bei der Entwicklung der Windenergieanlage Aeroman. Regelung, Betriebsarten und Betriebsverhalten. Abschlußbericht*, Kassel University, Kassel, 1979.

[2.19] Cramer, G., Drews, P., Heier, S. and Kleinkauf, W., *Regelung und Simulation des Betriebsverhaltens der Windenergieanlage Näsudden/Schweden. Abschlußbericht*, Kassel University, Kassel, 1980.

[2.20] Albrecht, P., Cramer, G., Drews, P., Grawunder, M., Heier, S., Kleinkauf, W., Leonhard, W., Speckheuer, W., Thür, J., Vollstedt, W. and Wettlaufer, R. *Betriebsverhalten von Windenergieanlagen. Abschlußbericht zum BMFT-Forschungsvorhaben*, O3E-4362-A, BMFT-FB-T 84-154, Part I, Department of Electrical Energy Supply Systems, Kassel University, and Institute for Control Technology Braunschweig Technical University, Karlsruhe, 1984.

[2.21] Vollstedt, W., Variable Speed Turbine Generator with Low Line Interactions, in International Energy Agency (IEA) Expert Meeting, Göteborg, Sweden, October 1991.

[2.22] Vollstedt, W., Braussemann, H., Hanne, R. and Horstmann, M., *Kostengünstiges Generatorsystem für drehzahlvariable Windturbinen mit guter Netzeinspeisequalität. BMFT-Abschlußbericht F 0812.00*, 1995.

[2.23] Kleinkauf, W., *et al.*, *Windenergieanlagen im Verbundbetrieb, Abschlußbericht zum BMFT-Forschungs-vorhaben 03E8160-A*, Part 1, Department of Electrical Energy supply, Kassel University, Kassel, 1986.

[2.24] Hackenberg, G., *Interner Bericht zur Meßkampagne Aeroman 12,5/33*, Department of Electrical Energy Supply Systems, Kassel University, 1985.

3

Generating Electrical Energy from Mechanical Energy

In examining the functionality and behavior of wind power plants, particular significance must be accorded the generator as a source of load torque, depending on its situation and on how it is coupled to the system (see Figure 1.37). Seen as a whole, wind power stations, like other electricity supply sources of comparable output (e.g. hydropower, gas and diesel generators), should

- be simple to use,
- have a long useful lifetime,
- have a low maintenance outlay and
- have as low as possible initial cost.

To meet these requirements a suitable generator must be chosen. Because of their robust construction, the generators used in wind power plants for the conversion of mechanical energy to electrical energy should be almost exclusively synchronous or asynchronous. In making the choice, particular attention must be paid to the constraints imposed by the environment and the demands made of the electrical machines in consequence, as well as to the torques engendered in the generator.

3.1 Constraints and Demands on the Generator

On the one hand, drive torque characteristics are dictated by the properties of the rotor and the amount of wind energy delivered in the airstrip; on the other, the demands and, as far as possible, the wishes of the electricity supply company and the users must be catered for at the interface between mechanical and electrical interactions – the generator. Furthermore, the peculiarities of wind power machines must be particularly considered.

In horizontal-axis machines, schemes attempting to accommodate the generator at the foot of the tower (see Figure 3.1) were beset with difficulties, e.g. those resulting from torsional

Figure 3.1 Turbine with a generator in the tower foot (Voith). Reproduced by kind permission of J.M. Voith AG

oscillations in the drive shafts. Because of this, the generator is also put in the head of the tower (Figure 3.2) or more usually in the nacelle (Figure 3.3). Keeping costs and possible tower oscillation in mind, the lightest possible generator should be sought. As mentioned in Section 2.4, four- and six-pole generators used with gear trains yield good results. However, the gearbox mass associated with such designs is eliminated in gearless machines.

Periodic effects resulting from disturbed flow around the tower, asymmetries in the system, etc., and aperiodic effects due to wind-speed changes (e.g. gusts) make high demands on components. These should be eliminated as far as possible, since service life is essentially determined by component loading.

Shock moments to the drive train, the mechanical effects of which can propagate as far as the machine anchorings and the electrical effects of which can propagate across the windings into

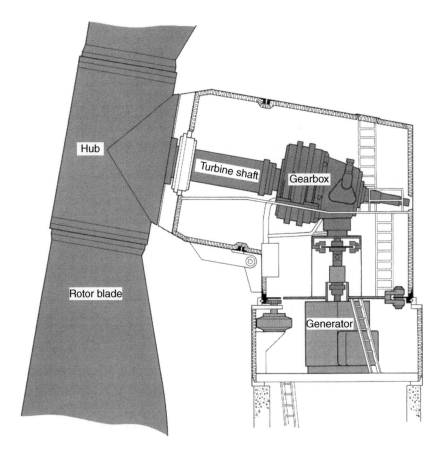

Figure 3.2   Turbine with a generator in the tower head (Aeolus). Reproduced by kind permission of Magnet Motor GmbH

the grid, are essentially dependent on the degree of elasticity, i.e. the torque/rotational speed gradient of the generator. Therefore, mechanical–electrical converters that are flexible both in their speed of rotation and in the way in which they are linked to the drive train have an advantage for use in wind turbines. By completely disconnecting the turbine or rotor speed from the grid, a greater degree of efficiency can be obtained under partial loading (see Figure 2.62) or a lesser amount of power can be drawn from the wind when the amount of energy required is lower.

Availability is of critical importance in guaranteeing security of supply and in the dimensioning and rating of storage and back-up plants. It has a great influence on capital investment and on the profitability of entire systems. Wind conditions at the turbine site play a significant role with regard to this aspect. Delivery of energy by wind power plants should start at low wind speeds and should be capable of maintaining small base loads. To accomplish this, standstill and run-up torques must be kept low. These can arise due to static friction in any bearings and brushes, etc. Also possible are dwell torques, arising mainly in permanent-magnet generators

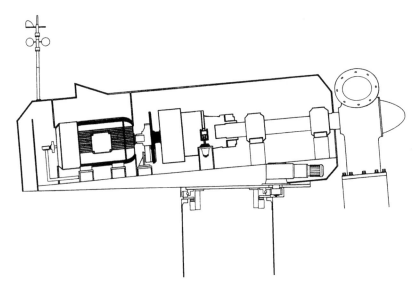

Figure 3.3 Turbine with a generator in the nacelle (Seewind). Reproduced by kind permission of Seewind

due to the reluctance effect. Further, the generator should show low no-load and exciter losses and high efficiency, even under low loading.

Systems for converting mechanical turbine power into electricity, which are suitable for incorporation into electricity supply schemes while fulfilling the above conditions and requirements, will be presented in the following. Based on personal studies and many years of experience in the design and optimization of such systems, guide values for dimensioning will be put forward, along with proposals for a forward-looking utilization of wind power.

3.2 Energy Converter Systems

Rotating field generators are used almost exclusively in the conversion of mechanical energy to electrical energy. Here, principally asynchronous and synchronous generators with direct grid coupling or with full or partial (rotor) inverter coupling are of significance. Three-phase generators employ a rotating magnetic field, known as a rotary field. This may be obtained by the use of rotating permanent magnets or by the rotation of excitation windings with the aid of current fed via brushes and slip rings. Such rotary fields excite an electric voltage in stationary conductors – the stator windings of the generator – the frequency of which is synchronized by mechanical rotation of the machine. In these synchronous generators, three (or an integer multiple thereof) coils are spatially offset by 120°. Such machines therefore produce three-phase voltages, which are displaced by a 120° vector in relation to one another – so-called three-phase alternating voltage. The voltage is dependent on the generator type, the speed of the rotor, the excitation and the load characteristics; in isolated and standalone operation this value can be regulated by varying the excitation. When connected to the public supply, both voltage and frequency are dictated by the grid.

If the three-phase alternating current stator of a generator is supplied with alternating current from the grid, this also sets up a rotary field. This field excites currents in the rotor windings of the generator, the frequency of which corresponds to the difference between the field rotation frequency and the mechanical speed of the rotor. These currents produce torques in the rotor, which, in synchronous machines, have a damping effect.

Asynchronous motors, on the other hand, cannot follow the rotary field. In their rotors a torque is set up that is proportional to the frequency difference (or the so-called slip) and acts in the direction of the rotary field. Asynchronous machines supplied at almost constant frequency and voltage from the grid go over to the generating mode when they are driven (e.g. by a wind turbine) past the synchronous rotation frequency. They then deliver power to the grid. Asynchronous motors and generators require inductive reactive power to build up their magnetic rotary fields, i.e. their electromagnetic excitation. This may be supplied by the grid or, for example, by capacitor banks. Rotary fields therefore set up torques in the rotors of grid-driven synchronous and asynchronous devices, which in turn enables them to extract energy from the driving turbine and convert it into electricity.

It is therefore important to differentiate between generator systems that draw their excitation or reactive power from the grid, and can thus serve only as grid support machines, and grid primary supply machines, which possess their own possibilities for voltage and reactive power control. All the most important configurations for converting mechanical energy into electrical energy are shown in Figure 3.4, which will serve as a reference for subsequent sections.

The energy user (e.g. the grid) usually sets low tolerance limits on frequency, voltage, harmonics, etc. According to the design, these can be met in various ways. Designs (a) and (g) in Figure 3.4 show extremely rigid grid coupling. In all the other systems shown, the design allows the mechanical speed of rotation to be dissociated in varying degrees from the electrical frequency or voltage in alternating or direct current systems by the incorporation of power electronics and the associated regulation procedures. In addition, these systems provide short-term energy smoothing during gusting, with the consequent reduction of loads on mechanical components. Variants (f) and (g), in addition to the requisite performance control, also allow a controlled delivery of reactive power. These systems can therefore be used as a primary grid supply or as grid support for alternating current (a.c.) grids, as can variant (h) for direct current supplies. Synchronous generators delivering power via grid-commutated frequency converters (models (i), (j) and (k)), which were used up until a decade ago, must draw their reactive power from the grid. By contrast, the self-commutated frequency converters in use today, such as designs (f) and (g), are themselves able to provide the reactive power necessary and to control the voltage in grid branches.

Direct converters in double-fed asynchronous generators, as shown in design (f), were used as far back as in the GROWIAN, for example. Further applications such as designs (c) and (l) are possible in both asynchronous and synchronous machines. Grid coupling by means of direct converters has not yet been able to establish itself on the market due to the high number of power semiconductor components that they incorporate. For the gearless systems (j), (k) and (l), only synchronous generators are used, excited either electrically (j) or by permanent magnets, (k) and (l).

The following section examines the main possible operational ranges for the rotary field machines mentioned above and the layout of the most important machines.

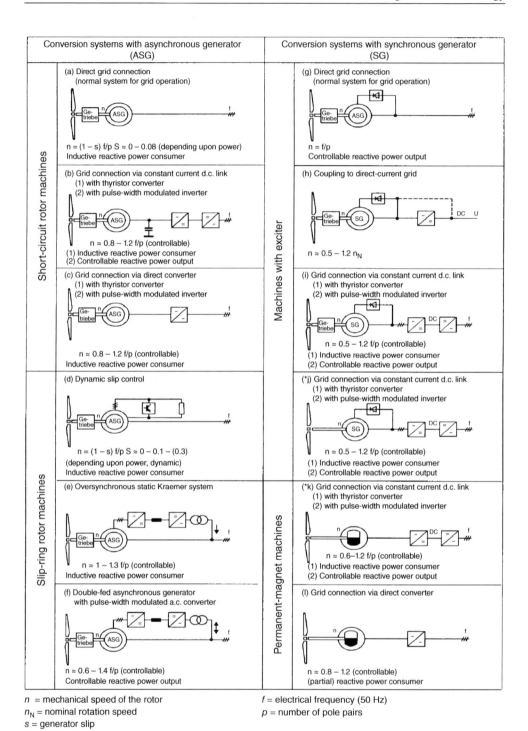

Figure 3.4 Systems for converting mechanical energy to electrical energy

3.2.1 Asynchronous generator construction

With a view to the exclusion of dust and damp and to prevent accidental contact, asynchronous generators (Figure 3.5) for wind turbines are usually manufactured as closed units. The essential components are the copper stator windings: three windings spatially offset by 120° in two-pole designs and six windings offset by 60° in four-pole designs. These windings are insulated from one another. The so-called three-phase alternating current windings are laid into slots in the laminated core. Since the core must conduct the magnetic flux of the rotary field around it, it is built up of laminations, which are insulated from each other to reduce losses.

In slip-ring rotor machines, equivalent alternating current windings are also laid into slots in the rotor core. Current is fed to or collected from them via rotating slip rings running against carbon brushes mounted on the generator chassis.

By contrast, cage rotors, as shown in Figure 3.5, have an aluminium diecast or bar winding laid into the slots of the rotor core. These conductors are connected to one another on the end faces of the rotor; in other words, they are short-circuited. Cooling vanes on these end rings improve heat dissipation.

The rotor and the fan are carried on a common shaft. This runs on two roller bearings mounted concentrically with the stator on two end plates, one on the shaft side and one on the fan side. The stator core with its windings and connectors is enclosed by the stator frame, the outside of which is fitted with cooling fins. The base of the generator provides mounting lugs and also serves as a support against transmitted torque. The grid connection box provides a well-protected electrical connection to the grid.

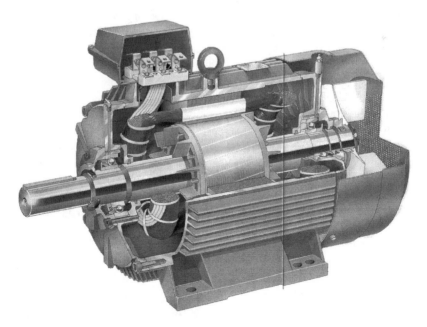

Figure 3.5 Cutaway of an asynchronous generator with a cage rotor (ABB). Reproduced by kind permission of ABB Motors

Cage generators are remarkable for their extremely simple layout, which permits robust construction and high operational reliability, even in the event of rough handling. Above all, these machines can be turned out in large numbers, making them correspondingly cheaper.

3.2.2 Synchronous generator construction

The synchronous machine is much more complicated than its asynchronous counterpart. Figure 3.6(a) shows a cutaway of the relatively complex layout of a brushless self-regulating generator. Following the sequence of the symbolic circuit shown in Figure 3.6(b) from left to right, the cutaway shows the main a.c. generator, the exciter and the pilot exciter. The stator with its a.c. windings, the shaft carrying the fan, the bearings and the housing with its base and connection box are as on the asynchronous machine.

Large generators in the 100 MW and GW range are usually of synchronous nonsalient pole construction (turbogenerators). Their rotor slots carry exciter and damper windings. Generators suitable for wind turbines, however, lie in the range between a few kW and around 5 MW. As shown in Figure 3.6(a), conventionally constructed synchronous machines of this size use salient-pole rotors. The rotor comprises the pole shoes, the poles lying beneath and the exciter windings. The stator consists of the stator core and a.c. windings. The rotor and stator together make up the generator.

In the brushless model shown here, the rotor is supplied by the exciter, which consists of exciter poles in the stator, rotary field windings and a rectifier bridge, to be seen at the end of the shaft (far right). The exciter or the exciter coils in the stator are supplied from the pilot exciter. Rotating an outer permanent magnet (right) generates current in the pilot exciter coils, which is fed via a voltage controller to the main exciter coils. In this way, the necessary magnetic field is set up in the poles of the exciter and crosses the air gap to the a.c. windings of the exciter. This sort of construction is preferred for grid-independent power supplies.

If a grid is available to power the exciter windings, the pilot exciter becomes superfluous. In brushless machines the exciter current is fed from the stator winding, across the air gap of the exciter and via the rotating rectifier bridge to the rotor. Its rotating magnetic field (rotary field) is transferred to the stator. Voltage regulation procedures must therefore be carried out over all systems of the windings, and are thus subject to delay times of some 100 ms. Much faster regulation procedures can be achieved if the exciter current is fed directly to the rotor of the main generator via slip rings. In this arrangement, delays of about 20 ms can be expected. The much simpler construction of the machine and the gain in dynamic characteristics must, however, be set against the severe disadvantage of feeding electricity through brushes and slip rings. Higher frictional losses, brush and slip-ring erosion and higher maintenance costs are the consequences.

In the following discussions, the operating ranges and the static and dynamic characteristics of these machines will be more closely examined.

3.3   Operational Ranges of Asynchronous and Synchronous Machines

Working from the structure, function and voltage equations of synchronous and asynchronous machines, the largely familiar equivalent circuits shown in Figure 3.7(a) and

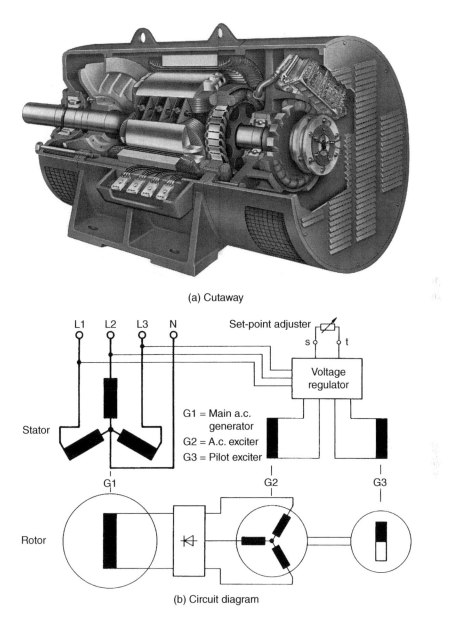

(a) Cutaway

(b) Circuit diagram

Figure 3.6 Synchronous generator structure (self-regulating brushless model, AvK). Reproduced by kind permission of AvK

(b) may be deduced for one machine phase. In the interests of clarity, a nonsalient pole synchronous machine with d.c.-fed exciter winding has been chosen by way of approximation. Permanent-magnet generators exhibit a similar equivalent circuit, but no interventions by means of exciter current are possible.

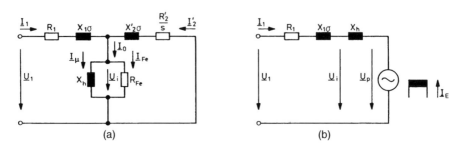

Figure 3.7 Equivalent circuits for one phase of (a) asynchronous and (b) synchronous machines

In the equivalent circuit diagrams, the variables with the subscript 1 characterize the stator values while the variables with the subscript 2 characterize the rotor values, i.e.:

R_1	represents the resistance of the stator winding,
R'_2	represents the rotor resistance in relation to the stator side,
R'_2/s	represents the slip-dependent value,
$X_{1\sigma}$	represents the leakage reactance of the stator winding,
$X'_{2\sigma}$	represents the leakage reactance of the rotor winding in relation to the stator side,
X_h	represents the air-gap reactance and
R_fe	represents the iron losses of the machine.

Furthermore,

U_1	represents the stator voltage,
I_1	represents the stator current,
I'_2	represents the rotor current in relation to the stator side,
I_μ	represents the excitation current,
I_fe	represents the current equivalent to iron losses,
I_0	represents the open-circuit current (of the asynchronous machine),
I_E	represents the excitation current of the rotating rotor of the synchronous machine, transformed on the stator side,
U_i	represents the induced voltage and
U_p	represents the synchronous generated voltage of the machine.

In addition,

s	represents the slip as a relative variable between the speed deviation from synchronism to the synchronous speed in accordance with

$$s = \frac{n_1 - n}{n_1} \tag{3.1}$$

In wind turbines there is a trend towards large systems in the kilowatt and megawatt range. For machines of this category, certain simplifications can be undertaken, while still obtaining a very good approximation for the following considerations. By ignoring

- iron ($R_{\text{fe}} = 0$) and frictional losses in asynchronous machines and
- resistance losses ($R_1 = 0$) in synchronous machines,

the simplified equivalent circuits shown in Figure 3.8 may be derived.
  By grouping the reactances

$$X_1 = X_{\text{h}} + X_{1\sigma} \quad \text{and} \quad X'_2 = X_{\text{h}} + X'_{2\sigma} \tag{3.2a}$$

or

$$X_{\text{d}} = X_{\text{h}} + X_{1\sigma} \tag{3.2b}$$

respectively, the stator current for asynchronous and synchronous machines

$$\underline{I}_1 = \frac{R'_2/s + jX'_2}{(R_1 + jX_1)(R'_2/s + jX'_2) + X_h^2} \underline{U}_1 \quad \text{or} \quad \underline{I}_1 = -j\frac{\underline{U}_1}{X_{\text{d}}} + j\frac{\underline{U}_{\text{p}}}{X_{\text{d}}} \tag{3.3}$$

may be derived in the usual form. From this, the simplified current circle diagrams as in
Figure 3.9 may be obtained. From these, the relationships between the mechanical input values
for torque and speed of rotation (see Figure 3.13 and Equations (3.9), (3.11), (3.12) and (3.15))
and the output values for electric current, voltage and power output may be recognized.
  In Figure 3.9 P_0 is the no-load point, P_N the nominal point, P_K the transition point and P_1
the short-circuit point, while M_N, M_K and M_{AM} are the corresponding torques when running
as a motor. The quantities shown in the figure represent the following:

$I_{1\text{M}}$ stator current in the motor mode
$I_{1\text{G}}$ stator current in the generator mode
M_{WG} generator load torque
P_{el} electric output power of the generator
P_{mech} mechanical power absorbed by the generator

  Particularly in view of the power station operation, regulation of the reactive power conditions is of great importance. From the current circle diagrams, it is clear that:

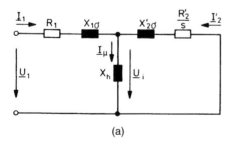

(a)

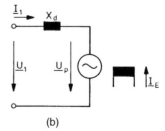

(b)

Figure 3.8 Simplified equivalent circuits for one phase of (a) asynchronous and (b) synchronous
machines

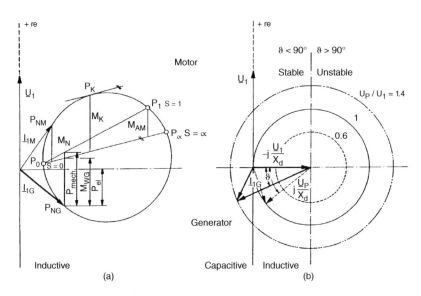

Figure 3.9 Current circle diagrams in simplified form for (a) asynchronous and (b) synchronous machines in grid operation (U_1, f = constant)

- For asynchronous machines, only motor and generator currents in the negative imaginary half-plane are possible, and these are always inductive.
- For synchronous machines, on the other hand, currents are possible in all four quadrants, i.e. in the motor and generator, and in inductive and capacitive ranges.

For power generation, therefore, synchronous generators offer distinct advantages. In addition to the positions of the circles, the radii also have differing effects on the operating state of these machines.

The full and simplified vector diagrams for the excitation status of asynchronous machines running under almost full load and synchronous machines under partial load, as shown in Figure 3.10, show different constellations. Here, the analogy between underexcited synchronous generator operation, which is highly variable, and design-determined fixed asynchronous generator operation becomes clear when, based upon the torque-producing rotor displacement angle ϑ for synchronous and asynchronous machines, a corresponding load angle λ (but with lower values) is introduced. Comparison of the vector diagrams in Figure 3.10 shows that the voltage U_1 induced at the magnetizing inductance

- is specified as a fixed value in asynchronous machines (examined more closely in Section 3.6), but
- is adjustable by means of the excitation state (over- or underexcited) in synchronous machines.

If the real axis is set in the direction of the stator voltage, it becomes clear that both types of machine allow current variations up to their stability limits in the real part of the plane. In generators, these arise as a result of changes in drive torque or active power. On the other hand,

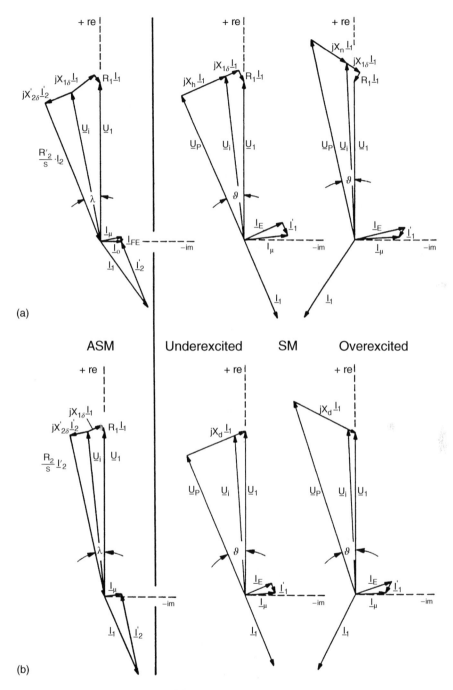

Figure 3.10 (a) Full and (b) simplified vector diagrams for asynchronous machines (ASM) and synchronous machines (SM) running as underexcited and overexcited generators

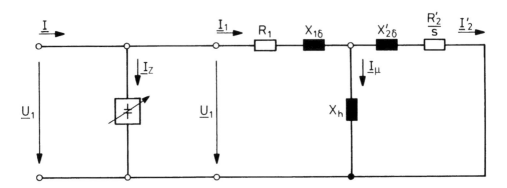

Figure 3.11   Equivalent circuit for an asynchronous machine with adjustable reactive power delivery

in asynchronous machines (in contrast to synchronous machines) the imaginary component of the currents, i.e. the reactive power alone, is not variable without resort to supplementary devices such as balancing capacitor banks.

If, however, reactive components can be pre-set by means of supplementary devices, e.g. capacitors, rotary phase shifters or static converters, then the same operational possibilities may be achieved with asynchronous as with synchronous machines. In asynchronous machines, the radius of the current circle is dictated essentially by the grid voltage. Deviations from circularity, which are ignored here, arise principally due to the saturation state and the construction of the machine (e.g. double-cage rotor). Fixed operating points can thus be explicitly assigned to predetermined load levels.

In synchronous machines, on the other hand, the radius of the current circle can be influenced by the excitation. This allows free selection of the operating conditions corresponding to a fixed load depending upon magnetization (over- or underexcited) or the reactive power demand (inductive or capacitive) as needed. In asynchronous machines, a supplementary device can deliver reactive current to the entire system. Looking at the asynchronous machine with adjustable reactive power delivery, shown in Figure 3.11, it is more or less possible to shift the axes to obtain, for an asynchronous machine, an operational range that covers all four quadrants, similar to that of a synchronous machine (see Figure 3.12).

Asynchronous generators working as shown here are usually coupled with cascade-switchable capacitors, and occasionally also static or rotary phase shifters, for standalone or isolated applications. Balancing capacitors in combination with machine and cable inductances can, however, set up grid resonance (see Section 4.3).

3.4   Static and Dynamic Torque

In the discussion of generators for use with wind turbines, when allowing for torque development, the distinction must be drawn between the fast processes of electrical balancing in the windings (and in the magnetic circuit), which are also influenced by the design of the mechanical drive train, and the much slower processes that have a direct effect on the dynamic behavior of the wind power unit. Knowledge of the torques involved in the complete system is important for deciding which measures are to be applied to protect the drive train.

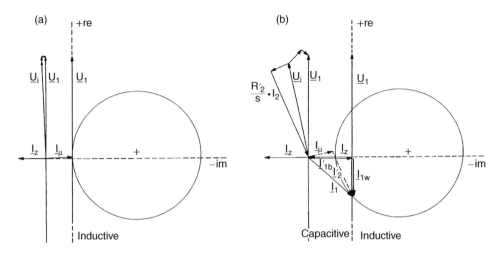

Figure 3.12 Operational ranges of asynchronous machines with reactive power delivery devices as entire systems with (a) no-load and (b) load compensation

Plots of steady-state torque against rotational speed are valid for speed or torque changes that take place much more slowly (quasi-steady-state) than dynamic processes (run-up and braking). Dynamic torques are decisive for fast processes, occurring within a short period immediately after any change of status. In wind turbines, their practical implications for dimensioning purposes are limited to special cases, such as switching-on procedures, line disturbances and generator short-circuits. The following examines parameters for estimating the order of magnitude and extreme conditions of static and dynamic torques.

3.4.1 Static torque

During quasi-static processes, the generator sets up a static load torque in opposition to the wind turbine. This load has a static and a dynamic component. The steady states of a rotary field machine are essentially determined by its fundamental-frequency torques. Synchronous and asynchronous harmonic components also exert their influence. In the generators mentioned, such harmonics can have considerable mechanical and electrical effects.

Due to the stepped magnetic waveform of successive phases, polyphase symmetrical stator windings produce rotary fields of order

$$v = \pm km + 1, \tag{3.4}$$

where $k = 0, 2, 4, 6, \ldots$ (even integer) and, usually, the number of phases $m = 3$. Apart from the fundamental frequency, harmonics arise, the ordinals of which are uneven integers indivisible by 3:

$$v = 1, -5, 7, -11, 13, -17, 19, \ldots.$$

Negative ordinals indicate rotation opposed to that of the fundamental field.

In assessing harmonic feedback in low-voltage networks, compatibility levels up to the 40th or 50th harmonic must be allowed for [3.1].

3.4.1.1   Asynchronous machines

The behavior of asynchronous generators is primarily determined by the steady-state torque/rotational speed characteristic of the fundamental-frequency field (see Figure 3.13). This is largely dictated by the design of the machine. In addition to the nominal torque, characteristic values include starting torque, pull-up torque and breakdown torque, plus the corresponding values for speed of rotation and slip. Nominal and breakdown values are of particular importance in the discussion that follows.

The ratio of breakdown to nominal torque M_K/M_N normally lies in the range of 1.8 to 3.5 and the ratio of start-up to nominal torque M_{AM}/M_N is usually between 1 and 3. Pull-up torque M_S/M_N, which usually lies between 1 and 2, generally gives rise to no dominant loads and so, like start-up torque, plays an important role only during the motorized run-up of a wind turbine.

Here, M stands for torque at the general slip value s, M_K is the breakdown torque, s_K the breakdown slip and n_1 the speed of rotation of the rotary field. Individually, the quantities represent the following:

Motor mode		*Generator mode*	
M_{AM}	start-up torque		
M_{SM}	pull-up torque	M_{SG}	pull-up torque
M_{KM}	breakdown torque	M_{KG}	breakdown torque
M_{NM}	nominal torque	M_{NG}	nominal torque
n_{KM}	breakdown-torque speed	n_{KG}	breakdown-torque speed
n_{NM}	nominal speed of rotation	n_{NG}	nominal speed of rotation

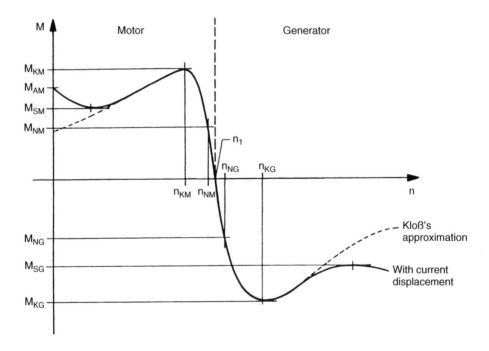

Figure 3.13   Torque/speed characteristic of an asynchronous machine

Ignoring stator resistance ($R_1 = 0$), the following is valid for the internal torque, which is transmitted by the air gap of the stator:

$$M_i = \frac{mU_1^2}{2\pi n_1 \sigma X_1} \cdot \frac{1 - \sigma}{s\frac{\sigma X_2'}{R_2'} + \frac{1}{s}\frac{R_2'}{\sigma X_2'}}. \tag{3.5}$$

The maximum torque is given by the derivative

$$\frac{dM_i}{ds} = 0$$

for the so-called breakdown slip

$$s_K = \frac{R_2'}{\sigma X_2'}. \tag{3.6}$$

Defining the primary leakage $\sigma_1 = X_{1\sigma}/X_h$ and the secondary leakage $\sigma_2 = X_{2\sigma}'/X_h$, the total leakage can be calculated:

$$\sigma = 1 - \frac{X_h^2}{X_1 X_2'} = 1 - \frac{1}{(1 + \sigma_1)(1 + \sigma_2)}.$$

Further, Equations (3.2a) define X_1 and X_2', where

R_2' represents the rotor resistance transformed to the stator side,
X_σ represents the reactance leakage from the stator and rotor ($X_\sigma = X_{1\sigma} + X_{2\sigma}'$),
m represents the number of phases (in three-phase systems $m = 3$) and
U_1 represents the grid voltage.

For low leakage values σ_1 and σ_2, which have very small values or tend to σ as they approach zero,

$$\sigma X_1 \approx X_{1\sigma} + X_{2\sigma}' = X_\sigma$$

and

$$\sigma X_2' \approx X_{1\sigma} + X_{2\sigma}' = X_\sigma.$$

Thus

$$s_K \approx \frac{R_2'}{X_\sigma} \tag{3.7}$$

and

$$M_K = \frac{mU_1^2}{2\pi n_1} \frac{1}{2X_\sigma} \tag{3.8}$$

Ignoring friction, the torque $M_i = M$. When the slip $s = (n_1 - n)/n_1$, the torque characteristic can be approximately determined according to Kloß's equation, referring to Equation (3.5) from the breakdown values by the relation

$$\frac{M}{M_K} \approx \frac{2}{\frac{s}{s_K} + \frac{s_K}{s}} \tag{3.9}$$

where $s_K \approx (5 \text{ to } 10) \, s_N$.

In normal operation between no-load ($n = n_1$) and motor and generator nominal load or permitted overload, a simplified approximation for torque development is given by the equation

$$\frac{M}{M_K} \approx \frac{2s}{s_K} \tag{3.10}$$

This means that, within the ranges mentioned, the machine characteristic is determined essentially by the breakdown moment and the breakdown slip. Allowing for stator resistance by applying the machine-specific parameters

$$s_K = \frac{R_2'}{X_2'} \sqrt{\frac{R_1^2 + X_1^2}{R_1^2 + \sigma^2 X_1^2}} \tag{3.11}$$

and

$$M_K = \frac{3U_1^2}{4\pi n_0} \frac{1 - \sigma}{R_1(1 - \sigma) + \sqrt{(R_1^2 + \sigma X_1^2)\left(1 + \frac{R_1^2}{X_1^2}\right)}}, \tag{3.12}$$

both quantities can be determined.

According to this, torque development in asynchronous machines is determined by the relation of ohmic to leakage components, particularly in the rotor windings. With slip-ring rotors, these can be altered over a relatively wide spectrum using supplementary resistors or power converters in the rotor circuit.

The size of the machine physically dictates the limits of slip. Large machines of this kind, in particular, are highly limited in elasticity in the event of torque fluctuations. Figure 3.14 reproduces average,size-dependent nominal slip values for given nominal power levels of production machines.

Figure 3.14 shows that machines of conventional design in the 100 kW range exhibit nominal slip values of around 1%. Six-pole generators show much greater elasticity than two-pole machines. In machines with such inflexible speeds of rotation, turbine torque fluctuations place heavy loads on the drive train because of the almost rigid coupling between the grid and generator. By raising design slip levels such systems may be made more flexible and thus offer greater protection to the drive train. However, this comes at the expense of greater volume, mass and losses.

This greater flexibility is desirable since the rotating blades of the turbine are subject to considerable airstrip turbulence, especially near to the tower. These result in fluctuations in turbine performance. Even when the blades react swiftly to limit turbine performance, the effects of tower-shadow fluctuations cannot be ironed out entirely. At low slip values, small variations in the speed of rotation therefore result in relatively large fluctuations in power.

Fluctuations in electrical output power can, as already mentioned, be considerably diminished by the use of asynchronous machines with higher nominal slip. Figure 3.15(a) and (b) shows comparative values for a 20 kW machine running under partial load, i.e. without application of blade pitch regulation. Considerable differences in output behavior may be observed.

While fluctuations in power of more than $0.2P_N$ can be seen for generators with 2% nominal slip (Figure 3.15(a)) due to tower-shadow effects, at 8% nominal slip (Figure 3.15(b)) these are reduced to some 5% nominal output ($0.05P_N$) [3.2]. The dominant variations in power shown here are measurable in seconds and are clearly due to variations in wind speed.

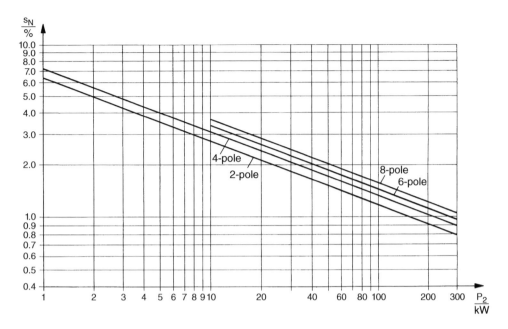

Figure 3.14 Nominal slip s_N for asynchronous machines as a function of nominal output power P_N for production machines (motors) with the pole count as a parameter

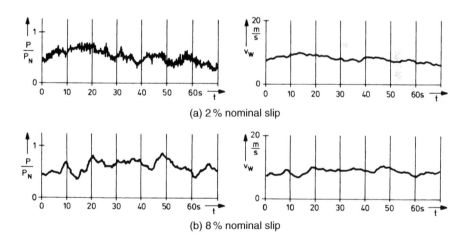

Figure 3.15 Power and associated wind speeds (measured values) for a wind power plant as a function of time for differently designed asynchronous generators

In accordance with Equations (3.6) and (3.11), high slip values can be achieved by designing the generator with

- high rotor resistance,
- low total leakage factor and
- lower rotor inductance,

e.g. by attention to such constructional details as

- slot form and
- choice of conductor materials and conductor cross-sections in the rotor.

Applying power smoothing in this way can also considerably reduce demands on turbine components.

In addition to setting higher fixed slip values by using extra resistors in the rotor circuit, slip-ring machines offer the possibility of dynamically adjusting or controlling slip to adapt to output power variations (Figure 3.16). The slip value and thus the proportional loss in power are thereby kept low and system efficiency is increased by varying the resistance of the rotor. It is also possible to limit the torque on the drive train to values around the nominal value [3.3] and to smooth the power output. This principle is employed in the so-called 'opti-slip control' in the Vesta turbines V 44 and V 60/63/66 and the USA export variant V 80.

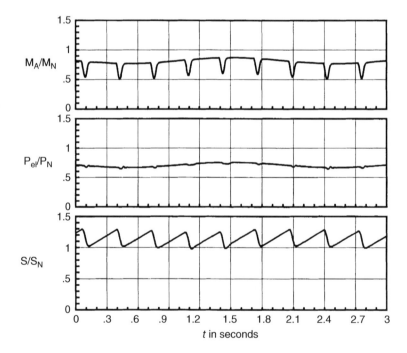

Figure 3.16 Dynamic slip control in slip-ring rotor asynchronous machines (simulation results)

Asynchronous harmonic rotary field torque
Harmonic fields induce a current in the rotor winding of frequency

$$f_{2\nu} = s_\nu f_1 \tag{3.13}$$

with slip

$$s_\nu = \frac{\Delta n_\nu}{n_\nu} = 1 - \nu(1 - s). \tag{3.14}$$

When the harmonics are synchronized the slip value $s_\nu = 0$ and the harmonic torque $M_\nu = 0$, i.e. the slip $s = (1 - 1/\nu)$. Outside these values, i.e. when running asynchronously ($s \neq 0$), these rotor currents set up a torque. All harmonic field torques are overlaid on that of the fundamental waveform (Figure 3.17).

Harmonic torques arise during run-up in the motor mode. They play a subordinate role if the windings are chosen appropriately and are not noticeable in normal generator operation.

Synchronous harmonic rotary field torque
Just as the stator current generates harmonic rotary fields, so too can the rotor current. If the stator harmonic and the rotor harmonic have the same frequency for a given operating point, they can generate torque together.

Such torques, however, are only significant during motor operation with high slip. During run-up in the motor mode, they can attain considerable levels relative to the start-up or nominal torque. In the generator mode, on the other hand, within the usual slip range of $s = 0$ to 10% synchronous harmonic torques are of little importance, so they will not be examined further below.

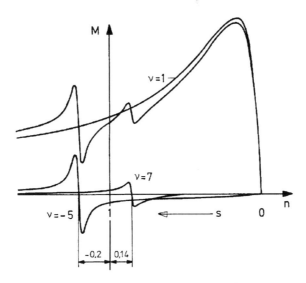

Figure 3.17 Fundamental (M_1) and harmonic torques (M_ν) and resultant torque (M) for an asynchronous machine

Parasitic torque

Skewed slots in a cage rotor build cross-currents from bar to bar across the iron, thereby generating torque, as do further eddy current losses due to slot harmonics. These torques are only of interest during start-up and will not be considered further.

3.4.1.2   Synchronous machines

For a utility connection, synchronous generators are used almost exclusively. For wind power plants in particular, they are used with and without gearing in isolated operation and via frequency converters in grid operation. Direct connection of synchronous machines to the grid is used in a few isolated cases in the megawatt class only, and these are mostly beset with problems (see Figures 1.11 and 1.12). While in conventional power stations nonsalient pole generators (turbogenerators) are predominant, wind turbines more often use multipole salient pole machines.

We must further differentiate between slip-ring fed rotors with small excitation time constants and brushless machines, which require much longer times for excitation changes (see Section 3.2.2). The torque in nonsalient pole machines is given by the general relation

$$M = -\frac{mU_1}{2\pi n_1}\frac{U_p}{X_d}\sin\vartheta \tag{3.15}$$

and for salient pole machines the following is valid:

$$M = -\frac{mU_1}{2\pi n_1}\left[\frac{U_p}{X_d}\sin\vartheta + \frac{U_1}{2}\left(\frac{1}{X_q} - \frac{1}{X_d}\right)\sin 2\vartheta\right], \tag{3.16}$$

where

m	number of phases (in three-phase systems, $m = 3$)
n_1	synchronous speed of rotation
U_1	grid voltage
U_p	rotary field voltage
X_d	synchronous direct-axis reactance
X_q	synchronous quadrature-axis reactance
ϑ	rotor displacement angle

According to this, the torque and the output power depend on the rotor displacement angle ($\sin\vartheta$ and $\sin 2\vartheta$). Further, by altering the rotary field voltage by means of the excitation, the torque curve can be influenced.

Directly coupled synchronous generators form a rigid system with the grid, which propagates sudden changes in torque more or less immediately in the form of energy flux changes. In wind power plants this leads to heavy loads being placed on the drive train, with correspondingly large fluctuations in output power.

Depending on the design and level of excitation, running at nominal rating, the following are common:

• for nonsalient pole generators,

- rotor displacement angle $\vartheta = 25$ to $30°$ and
- overload capacity $M_K/M_N \approx 2$;
- for salient pole generators,
  - rotor displacement angle $\vartheta = 20$ to $25°$ and
  - overload capacity $M_K/M_N \approx 2$ to 2.5;
  and, depending on the maximum exciter voltage,
  - rotary field voltage $U_p/U_N = 0$ to 2.

Thus in quasi-steady operation a maximum (pull-out) torque of

$$\frac{M_{max}}{M_N} = 2 \text{ to } 5$$

is to be expected.

To make matters worse, periodic torques arise (particularly due to periodic tower-shadow or upwind tower effects and their influences on the rotor blades) that are only weakly damped, which place considerable loads on the rotor, bed plate and tower. As a result, extra measures (oscillation dampers, etc.) are usually necessary (see Figure 3.18). Synchronous machines with network-coupled wind turbines are only found in combination with rotation-variable drives. Therefore, the only synchronous generators to make it into widespread use in grid-coupled wind turbines are those that use frequency converters to provide elastic grid coupling.

Drive train with rotation-variable transmission system
Figure 3.19 shows the structure of a 2 MW wind turbine drive train with synchronous generator, variable-speed WinDrive transmission, main transmission step and turbine drive shaft. The fixed-rotation synchronous generator coupled to the grid is uncoupled from the wind and the plant-conditioned turbine rotation gradients by the hydrodynamic variable-speed WinDrive transmission. Thus, the generator can be operated at the grid without vibrations and in a stable manner. In addition, the advantages of the grid-coupled synchronous generator can be used in a targeted manner with the grid-fed efficiency and reactive power, voltage or reactive power regulation and provision of short circuit performance, etc. These grid-forming properties thus offer similar grid services as conventional power stations.

3.4.1.3 Generators with power converters

Because of their extremely short reaction times (e.g. 2 to 20 ms), the use of power converters to couple generators to the grid allows very effective and dynamically efficient torque limiting values. This is the case in systems that feed their entire output power to the grid via rectifiers or inverters (see Figure 3.4, models (b), (c), (h), (i), (j) and (k)) and in configurations that feed only part of their output to the grid via power converters (see Figure 3.4, models (e) and (f)) or take off slip power via supplementary resistors (model (d)). If the regulation system is properly dimensioned, limiting values close to the nominal torque, such as, for example,

$$M_{max} = (1.2 \text{ to } 1.5)M_N,$$

(a) Gear layout

(b) Elastic suspension

Figure 3.18 Torsionally elastic gearbox suspension with oscillation damper (Hamilton Standard WTS 4 turbine)

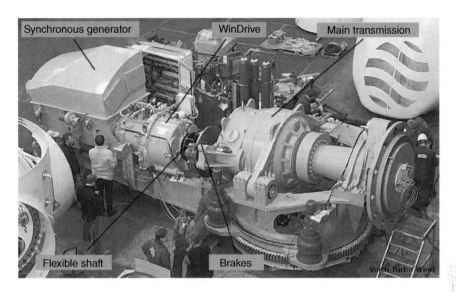

Figure 3.19   Drive train of a 2 MW turbine with a rotation-variable Voith transmission WinDrive [3.5]

can be attained. In doing so, the entire system must be designed such that fluctuations and instabilities resulting from internal and external disturbances are allowed for and eliminated by choosing component dimensions so as to protect the drive train.

3.4.1.4   Generator and turbine torque

The behavior of a wind power unit is determined by the interaction between the turbine drive torque and the load torque of the generator. The real drive of a machine can be approximated from its quasi-steady-state behavior. In this way, taking into account turbine and generator manipulations and using the torque–rotational speed curve, the operating and adaptive possibilities of using the characteristics of the energy converter system according to Sections 2.2 and 3.2 to control the power of the wind turbine can be determined. In doing so, the possibilities for the optimal exploitation of wind energy and for the adaptation of power to the grid will be considered [3.4].

From the performance coefficient (cf. Equation (2.8))

$$c_p = \frac{P_W}{A_R v_1^3 \frac{\rho}{2}}$$

and the characteristic given in Figure 2.6,

$$c_p = f(\lambda)$$

or from the equivalent torque parameter

$$c_m = \frac{M_{AW}}{R_a A_R v_1^2 \frac{\rho}{2}} \tag{3.17}$$

and the corresponding family of characteristics

$$c_{\mathrm{m}} = f(\lambda)$$

together with the tip speed ratio

$$\lambda = \frac{2\pi R_{\mathrm{a}} n}{v_1}$$

it is possible, for a fixed blade pitch angle (e.g. at around $\beta = 90$) at different wind speeds v_1 (as a parameter), to derive the power and/or the torque–rotational speed characteristics for wind turbines. These, together with the load torque–rotational speed curve of the generator being driven, determine the running mode and the stability of the entire drive, as shown in Figure 3.20.

If the wind turbine delivers its energy via a synchronous generator to a clock-pulse-generating grid, then at different wind speeds the turbine will be constrained by the generally constant grid frequency to a likewise constant rotation speed of $n/n_1 = 1$.

Similarly, if wind energy is delivered through an asynchronous generator then, when the wind speed increases, the turbine speed will also slowly increase (due to increasing slip). For normally designed machines, however, a change in rpm of only around 1% (see Figure 3.14) needs to be allowed for.

It should be pointed out here that the stable operating regime of the turbine may lie beyond the breakdown point of an asynchronous generator, should the turbine characteristic fall away more rapidly than that of the generator in question. In consequence, steady-state stability will always be attained when

$$\frac{\mathrm{d}M_{\mathrm{W}}}{\mathrm{d}\omega} > \frac{\mathrm{d}M_{\mathrm{A}}}{\mathrm{d}\omega} \tag{3.18}$$

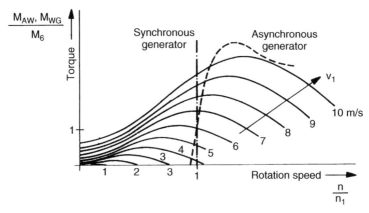

——— Turbine drive torque curve M_{AW} with wind speed v_1 as a parameter
—·—· Load torque M_{WG} of a synchronous generator
– – – – Load torque M_{WG} of an asynchronous generator

Figure 3.20 Torque–rotational speed relationship of a grid-connected wind turbine (for a turbine drive torque M_6 at a wind speed of 6 m/s)

This can be achieved by designing the asynchronous machine accordingly, particularly for the operating range that lies between breakdown and saddle points.

Figure 3.20 shows that sufficient turbine torque to drive the generator is only available at or above wind speeds of about 3.6 m/s for asynchronous machines and 3.8 m/s for synchronous machines. When transmission and generator losses are taken into account, an even higher wind speed must exist before any power at all is delivered to the grid. When the wind speed lies below nominal (minimum or no-load) levels, the machine acts as a motor and drives the turbine.

Under normal operation of wind power plants, continual changes in the wind speed occur. The tip speed ratio changes in consequence, so the coefficients and values for power and turbine torque are not constant either. If the generator is rigidly coupled to the grid, continually changing operating states result.

Under variable-frequency generator operation, the speed of rotation can be freely set within given limits. The turbine's utilization of available wind power may therefore be optimized within the possible range of adjustment. This can be seen from Figure 3.21. By plotting the torque of the turbine against speed of rotation, the power of the turbine can be given for every level of wind speed. The example in Figure 3.21 shows this for $v_1 = 5$ m/s. For every wind speed, therefore, an optimal performance may be determined; the corresponding values for the turbine drive torque and speed of rotation are shown in Figure 3.21 as a heavy dashed line. The load torque–rotational speed curve of a self-exciting synchronous generator under variable-frequency operation can be drawn in the turbine characteristic field for various excitation currents I_E. The level of excitation can be used to set the generator load torque according to the turbine drive torque at optimal turbine performance. This will always be attained at the

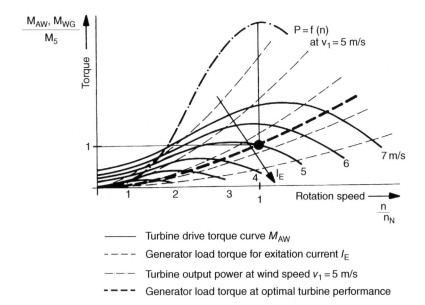

Figure 3.21 Variable-frequency adaptation of a self-exciting synchronous generator to a turbine (for a turbine drive torque at a wind speed of 5 m/s)

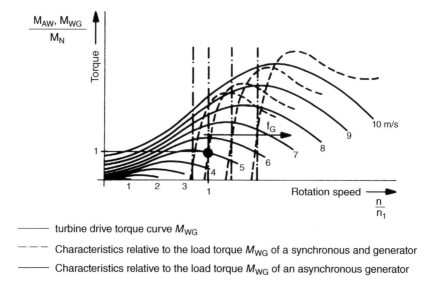

————— turbine drive torque curve M_{WG}

– – – – Characteristics relative to the load torque M_{WG} of a synchronous and generator

————— Characteristics relative to the load torque M_{WG} of an asynchronous generator

Figure 3.22 Torque–rotational speed relationship and characteristic adaptation by variation of the generator frequency f_G (for a turbine drive torque M_5 at a wind speed of 5 m/s)

speed of rotation corresponding to the intersection of the curve for the turbine torque at optimal turbine performance with that of the generator load torque for a given excitation current. This can be achieved by using frequency-converter systems with direct current intermediate circuits to decouple generator and grid frequencies.

If the generator frequency can be modified (e.g. by using an intermediate circuit converter or direct converter) then, by choosing the frequency, the speed of rotation may be set and the required torque level obtained. This is plotted in the turbine characteristic field in Figure 3.22 for different operational states of synchronous and asynchronous generators. Coupling this with Figures 3.20 and 3.21 allows a frequency, and thereby a speed of rotation, to be chosen that yields optimal turbine power. In Figure 3.22 this is shown, for example, as the operating state 1. The assumption is that the turbine output power for the chosen frequency is fed to the grid.

To adapt a machine to the optimum wind power, the operating points of the entire system must each intersect the maxima of the turbine performance–rotational speed curves, so that the turbine is running with constant tip speed ratios, as already explained. The speed of rotation of the turbine must therefore adapt to the prevailing wind speed. Controlling a machine for maximal exploitation of wind power therefore implies that, for a turbine of this design, electrical energy might be delivered to the grid, even at very low wind speeds. All the same, transmission and generator losses mean that a certain minimum wind speed is necessary to keep the turbine operating. If the wind speed drops below this so-called start-up speed, the turbine will stop.

In addition to producing the maximum possible performance figures in the range between cut-in and nominal wind speeds, generator frequency adaptation allows deliberate reductions in turbine power beyond this range. In this way,

- overload protection can be provided and
- mechanical effects due to pitch control, for example, anticipated.

This provides the possibility of reducing the mechanical interventions on the turbine system that must be made in response to changing wind speeds, thereby reducing component loading. As mentioned at the beginning of this section, in most wind turbine operating conditions, mainly static torques are determinative. Dynamic torques overlie these in marginal situations. The following considerations show that dynamic torques can far outweigh static torques and in consequence cause effects on the wind turbines that are relevant for design and dimensioning.

3.4.2   Dynamic torque

Immediately after a change in operating conditions due to

- cut-in and cut-out procedures,
- grid disturbances,
- rapid auto-reclosure in the grid or
- deliberate generator short-circuit,

short-time effects in the generator give rise to transient currents and torques. These can be estimated [3.6] and, as far as possible, can be attributed experimental values. Knowledge of such effects is important, especially when dimensioning mechanical (see Section 2.4) and electrical components and determining machine behavior.

3.4.2.1   Connection procedures

Connection of stator coils to the grid during

- motorized run-up from standstill or
- near synchronous operation

are of particular relevance. From the point of view of the grid, it is necessary to differentiate between the connection of the generator to

- a rigid combined grid and
- a weak isolated grid.

In addition, from the point of view of the generator, differences can be expected between

- unexcited and
- excited

motor state during connection.

   Motorized run-up from stationary or from very low rotation speeds is not normally an option for synchronous machines, and is only practised in exceptional cases with asynchronous generators. When calculating the maximum occurring moments in asynchronous machines that are connected to the grid with the rotor stationary, we need to differentiate between torque paths for

- running up and
- locked

machines (Figure 3.23(a) and (b)). The maximum values occur after a half-period. When the machine is locked

$$M_{\text{max}} \approx 2M_{\text{AM}},$$

i.e. the motorized start-up torque is doubled. When the generator is running up

$$M_{\text{max}} \approx 4M_{\text{AM}},$$

i.e. maximum torque is around four times the start-up torque.

At maximum start-up torque [3.7], similar values are obtained (depending upon machine layout) of

$$M_{\text{max}} \approx 2.6M_{\text{K}},$$

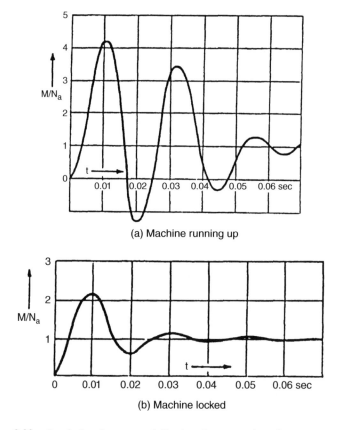

(a) Machine running up

(b) Machine locked

Figure 3.23   Graph showing torque following the connection of a cage motor [3.8]

i.e. 2.6 times the breakdown torque, with the expected torque peaks increasing with the size of the machine. Somewhat higher values are found using the equation [3.6]

$$M_{\text{max}} = M_{\text{K}} \left(1 + \frac{1}{\cos \varphi_{\text{K}}} \right),$$

where $\cos \varphi_{\text{K}} = 0.3$ to 0.5.

These torque peaks can reach three to four times the breakdown torque. However, maximum values only affect the shaft and drive train very briefly, i.e. for approximately one period with the rotor locked and for approximately two periods with the motor running up. A reduced component then occurs, which oscillates around the starting torque and dies away after a few further periods.

Synchronous generators coupled directly to the grid are (with a few exceptions) not important for use in wind power plants in practice, and for this reason connection procedures with the machine stationary will not be dealt with in detail here due to their complexity [3.9, 3.10].

If, however, the motorized run-up of synchronous generators needs to be considered, the field winding must be connected via an external resistor with approximately ten times the resistance of the exciter during run-up, as the load voltage would be too high for the field winding insulation with an open-field circuit. Currents that are generated in the starting/damping cage of the synchronous machine give rise to an asynchronous torque. The starting/damping cage, which is usually significantly weaker than in asynchronous machines, therefore develops much lower torques, which have a similar path for nonsalient pole machines as for asynchronous machines.

In salient pole machines the magnetic anisotropy of the rotor caused by the pole vacancy and the incomplete damping cage lead to further deviations. The effects of these are highly dependent upon the design, and cannot therefore be treated as generally valid. Maximum starting torques in relation to a nominal torque [3.9] of approximately

$$\frac{M_{\text{max}}}{M_{\text{N}}} = 1 \text{ to } 3.5$$

can be expected. As already mentioned in Section 3.4.1.2, the use of synchronous generators is basically limited to grid connection via frequency-converter systems. This permits the inrush current and starting torque to be limited to values close to the nominal value with the aid of the power converter, which means that in practice such systems are not usually associated with significantly increased moments in the drive train.

Grid connection at near-synchronous speed is possible in principle for excited or unexcited asynchronous machines, either directly or with the aid of a synchronization device of the type always used in synchronous generators. Low- and medium-output asynchronous generators can be connected to the grid either directly or via power converters, and large machines can be soft-connected using synchronization devices. Increases in the moment of a magnitude relevant to drive train loading can usually be avoided.

The maximum current and torque for direct grid connection near-synchronous speeds are highly dependent upon the degree of saturation of the main and leakage reactance of the asynchronous machine, as well as the timing of the connection (crossover or peak voltage) in all three phases, particularly when the generator is in an unexcited state.

In the case of asynchronous generators for wind turbines, which have lower saturation than motors due to their design, the nominal torque is only exceeded for a short period during connection in the unexcited state. During the first half-period they achieve similar torque peaks to those for stationary connection (see Figure 3.23). The torque drops rapidly, however, to the appropriate steady-state value for the load.

In order to determine reference values, asynchronous generators with outputs of 11 and 15 kW were selected for laboratory investigations. These exhibit behavior that is character-istic even for larger units. The generators were connected to a rigid public grid and to a weak isolated grid made up of a 28 kVA emergency generating set. The differences between the two configurations could thus be clarified. Due to voltage dips, lower initial current and torque peaks are observed for weak grids than for rigid public grids.

For the unexcited connection of an asynchronous generator at voltage zero, near-identical results are obtained for the cophasal and antiphasal connection states, which are charac-terized by their normal operation between the grid and machine. For the rigid public grid, the maximum current during the first half-period in the L_1 phase under consideration is as follows:

Cophasal $i_{max} \approx 8I_N$
Antiphasal $i_{max} \approx 7.5I_N$

and for the weak grid

Cophasal $i_{max} \approx 6I_N$
Antiphasal $i_{max} \approx 5.5I_N$

In the second half-period, however, the opposite behavior is observed. In the case of cophasal connection, the steady-state current level is approached after the first half-period, whereas in the antiphase connection a value close to that of the first half-period can be determined.

During the first half-period, these phenomena can be attributed to electrical transient processes, particularly those in the highly saturation-dependent magnetizing inductance of asynchronous machines. During the switching process in antiphase, a weakened electrical transient process is observed at the generator coil due to the voltage leap. During the second half-period, however, the mechanical transient process in particular comes into effect. This also terminates in this phase, so for connection in antiphase higher current and torque values should be expected during the second half-period. For the connection of asynchronous machines that have already been excited to no-load voltage, significant differences can be observed depending on the connecting angle between the machine and grid voltage.

In the case of cophasal connection, inrush currents are created that remain within the nominal range and disappear during the first three to four periods. Therefore no significant differences can be determined between rigid and weak grids. By contrast, excited machines that are con-nected to the rigid grid in antiphase produce approximately double the current peaks of those found in unexcited machines, e.g.

$$i_{max} \approx 15I_N,$$

which decay after three to four periods. The same trend can be observed for a weak grid, but with significantly reduced peaks of

$$i_{max} \approx 11I_N.$$

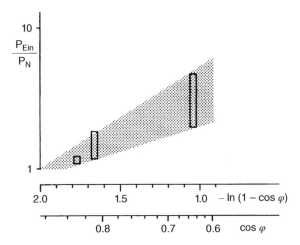

Figure 3.24   Effective inrush power range of asynchronous machines with different saturation levels (connection at synchronous rotational speed)

Figure 3.24 shows the results of investigations on the same asynchronous machines (11 and 15 kW) with the aid of a summary of current and voltage space vector variables plus a calculation of the effective output. The connecting angle between the grid and machine voltage pointer was varied between $0°$ and $360°$, using approximately 100 intermediate stages. The inrush power exhibits some dependence on the connecting angle; however, from the investigations to date this cannot be unambiguously defined and assigned individual machine parameters.

However, we can indicate the trend for inrush power ranges in relation to power factors during nominal operation on the basis of Figure 3.24. This figure shows that high-power-factor machines have low-making currents with little dispersion due to their low saturation. This is desirable for generator applications because of its beneficial outcome with regard to circuit elements, demands on the winding, foundation loading, etc.

Cophasal connection of excited synchronous machines at near-synchronous rotation speeds gives rise to relationships similar to those described for asynchronous machines. Good synchronization does not cause any significant increase in the moments.

Connection in antiphase in the transient range can be treated as double voltage, as for machine short-circuit, so that the maximum values of the moments described below are doubled. Saturation influences can, however, significantly increase inrush current. This type of loading can be avoided to a large degree by the use of suitable synchronization devices (see Section 3.4.2.4).

3.4.2.2   Generator short-circuit

If an excited generator is suddenly short-circuited, high current and torque peaks occur during the transient process, which die away after a few periods. In synchronous generators with the machine's excitation unchanged in relation to no-load excitation ($U_p = U_1$), the maximum moment of a nonsalient pole machine, as described in Section 3.4.1.2, is found to be

$$M_{K\,max} = -\frac{3U_1^2}{2\pi n_1 X_d} \qquad (3.19)$$

as a quasi-steady-state variable. For a sudden three-phase short-circuit of the machine the maximum value of the pulsating short-circuit moment, which only lasts for a few periods, is determined by the following equation [3.6]:

$$M'_{K\,max} = M_{K\,max}\frac{X_d}{X'_d} \qquad (3.20)$$

where the transient reactance is characterized by X'_d. Reference values for X_d, X'_d and X''_d are given in Section 3.7.3 for machines on the order of magnitude of 100 kW to a few megawatts.

For nonsalient pole machines it is found that [3.11]

$$\frac{X_d}{X'_d} \approx 7$$

and therefore an appropriate multiple of the pulsating maximum value of the quasi-steady-state pull-out torque is found, i.e.

$$M'_{K\,max} \approx 7M_{K\,max}.$$

If a subtransient current occurs due to the damping winding, an even higher maximum moment can be expected during the first period:

$$M''_{K\,max} = M_{K\,max}\frac{X_d}{X''_d} \qquad (3.21)$$

where X''_d represents the subtransient reactance of the synchronous machine. The relationship

$$10 \le \frac{X_d}{X''_d} \le 13$$

thus gives a peak value of around 13 times the pull-out torque.

For salient pole machines, transient peak values of an undamped moment can be expected that are approximately 30% higher than those for symmetrical rotors [3.6], i.e.

$$M'_{K\,max} \approx 10M_{K\,max}.$$

Assumed peak values of

$$M''_{K\,max} \approx 15M_{K\,max},$$

taking damping into account, lie on the side of safety. Due to the design-specific complexities of this process, it is not possible to conclusively derive a simple estimate. A detailed description will therefore not be provided.

No steady-state short-circuit current is created in a three-phase short-circuit with an asynchronous generator. The generator voltage collapses after the initial periods, so no further current flows and no load moment is maintained. During the first half-period, torques of

$$M_{K\,max} \approx 2M_K \quad \text{or} \quad (5\text{--}6)M_N$$

are reached. In two-phase short-circuit, on the other hand, peak values of

$$M_{K\,max} = (3{-}5)M_K \ \text{ or } \ (9{-}15)M_N$$

must be expected, due to the residual excitation.

Cautious estimates for synchronous generators with direct current intermediate circuits lead us to expect a maximum of four to five times the nominal moment during the first half-period for mains supply in two-phase short-circuit. Thereafter, no more electrical power is supplied to the grid and the torque drops to zero.

3.4.2.3 Grid abnormalities

Grid abnormalities are predominantly caused by grid short-circuits, rapid auto-reclosure, changes in grid voltage and grid frequency fluctuations. We can refer to the descriptions above when considering their consequences.

Single-phase, two-phase and three-phase grid short-circuits in the direct vicinity of the generator are almost identical to generator short-circuits. Grid short-circuits some distance from the generator cause currents to flow across the transformer and conductor impedances that lie between the short-circuit and generator. These currents and the resulting torques are less than the generator short-circuit values, in accordance with the limiting conduction data. The moments given in Section 3.4.2.2 are thus always based on the least favourable assumptions, thus producing safe designs, based on a simulation of extreme situations.

The rapid auto-reclosure familiar to electricity supply companies, in which grid sections become separated from the energy supply for approximately 100 to 500 ms, can, because of the connection between the wind power plant and consumers, lead to a brief, unintentional period in which the entire system operates in isolation. As a state of equilibrium does not usually exist between power supply and consumption, voltage and frequency will drift away from the grid values. In the worst case, the generator may be reconnected to the grid in antiphase after this interruption. Such phenomena are also possible if wind turbines are connected together but do not supply consumers. Because of the connection between direct-coupled asynchronous generators and variable-speed power converter supply systems, it is possible for the asynchronous generators to go over to the consumer or motor mode during the interruption. In this situation, component types and individual plant regulation systems greatly influence the behavior of turbines and the network as a whole (see Section 4.1.3).

Changes in grid voltage within the normal range do not cause any dominating torque levels. In grids supplied via power converters, rapid grid-frequency fluctuations can lead to significant loading of the drive train and generator of wind turbines. Although values remain well below maximum moments, such frequency variations can have a very negative effect on the lifetime of the drive train, particularly if they occur frequently.

3.4.2.4 Starting and synchronization devices

In order to connect motors and generators in a manner that will protect the drive train and winding, processes that reduce current, and therefore torque, are often used.

The star-delta connection is the simplest and most widespread measure for the reduction of inrush current peaks and the resulting torque shocks and grid reaction for motors in the 10 kW range, which start their run-up at no-load or at low-load moments. This is only possible if the coil is designed for connection voltage in the star-delta mode. Star connection start-up reduces phase voltage and current by a factor of $1/\sqrt{3}$. Electric power and torque drops to around a third of the value for direct connection. Moreover, due to the 42% reduction in phase voltage, significant saturation effects that would otherwise occur, significantly increasing current, are avoided. This minimizes current and torque peaks.

For the motorized running up of large turbines without blade pitch variation systems, these and similar measures are generally not sufficient. The starting current, which, as shown in Figure 2.55, amounts to around six to eight times the nominal value, will be reduced to around four times this value; however, the running-up time is around three times as long as that for direct connection and varies around values of 10 seconds or more. This means that the grid connection and the switch gear and protection devices must be designed to withstand these relatively short period currents. Therefore, for these procedures, so-called weak wind generators are fitted, either as a separate machine or in the form of a pole-changing coil within the main machine. These are designed for around 20% of the nominal output of the turbine and run-up at a low rotation speed. This means that the starting current amounts to around the nominal level for the main generator, or exceeds this value only slightly (see Figures 2.55 and 3.33). The available grid capacity can thus by fully utilized, both for running-up and full-load operation.

One further start-up method for wind turbines is the so-called soft start. In this method, gradual voltage and current increases are achieved and running-up currents limited to nominal operating values or similar, by using a thryristor actuator and drive angle alterations. During the starting-up phase, however, increased grid reactions must be expected due to harmonics. As these only occur for short periods and vary greatly due to the drive angle alteration, it is not feasible to compensate for them. Filtering devices are not generally used in this application.

Synchronization devices are common in power station operation for connecting synchronous generators to the grid and were/are used in the few exceptional cases in which wind turbines with synchronous machines are coupled directly to the grid. Synchronization is of much greater significance for wind turbines where large asynchronous generators are connected to the grid.

After running the turbine up and exciting the machine, the generator and grid voltage levels and positions in all three phases are compared with each other, synchronized by small speed and frequency changes and connected when almost synchronized. This keeps transient processes low and to a large extent prevents high drive-train loading and grid reactions.

3.5   Generator Simulation

Referring back to the above descriptions of generator-side steady-state and dynamic torques, simplified simulation models can be developed. These can be drawn upon for in-depth analysis of the interaction between wind turbines and the supply grid, and for the control of these systems. The normal five-coil model for synchronous machines [3.12] is taken into account in this context.

In addition to the previously common simulation procedures based upon PSpice, MATLAB/ SIMULINK, etc., new opportunities for the simulation of rotating and static power converter systems are opened with PV Simplorer [3.13]. It will be possible to simulate both conventional (diesel, battery) and renewable (photovoltaic, wind, etc.) power supply systems in graduated dynamic models. The simulation takes into account highly dynamic subtransient processes in the microsecond and millisecond ranges, transient processes in the high millisecond and second ranges and longer-lasting behavior (including that relating to energy and tariffs) in the ranges of several seconds, minutes and hours. Static and rotating loads are also considered. Furthermore, the use of implicit databases yields significant simplifications compared to previous simulation procedures.

3.5.1   Synchronous machines

Electrical energy supply plants (power stations, diesel generators, emergency generating sets) usually use synchronous generators. If rapid excitation processes are required, these are excited independently via a slip ring or, in the case of brushless machines (which are associated with much greater delays), they are designed to be self-exciting. These systems can be used for both the public grid and isolated operation, as well as standalone supply systems.

Up until now, wind turbines have been fitted with synchronous generators coupled directly to the grid only in exceptional cases. In such cases, measures for the control of mechanical vibrations (e.g. oscillation dampers on the generator foundation) were usually necessary (Section 3.4.1.2). Options for active damping by suitably designed coils are described in reference [3.12].

If short-term transient processes are of interest, the five-coil model [3.13, 3.14] can be used to describe the torque on the generator shaft in relation to the currents and voltages in the stator and rotor for grid operation of synchronous machines. The defined operating ranges can be accurately approximated using a system of differential equations, flux linkages and the mechanical torque–speed relationship or a corresponding logic block diagram. In this context, time constants for the electrical changes for stator and rotor currents lie in the millisecond range, and are thus significantly beneath the relevant times for mechanical processes in the seconds range. Furthermore, the following assumptions are normally made:

- losses are disregarded,
- the saturation of iron is assumed to be linear and
- only the effect of the fundamental wave is taken into account.

Transient, subtransient and saturation-related processes, the consequences of which are described in Section 3.4.2, cannot be adequately taken into account on the basis of these idealized assumptions. Therefore, simplified simulation models have been developed for synchronous machines, which simulate only those operating states and connection processes that are of interest. Thus relevant operating methods are approximated, or at least a trend shown, with justifiable calculation costs.

Results of very simplified simulations of the regulation of effective output for a turbogenerator [3.14] in grid and isolated operation, which only take account of the mechanical inertia of the system, the grid frequency, the turbine speed and the resulting rotor displacement angle ϑ (including power angle) and voltage, can indicate significant deviations from the actual rotor displacement angle. Wind turbines in particular tend to show such deviations due to the changes in excitation that are often necessary because of the varying dynamic loads. Furthermore, interesting special cases cannot be accurately represented.

However, we can use the above simplification to derive structures that allow us to simulate not only steady-state operation but also particularly relevant processes such as connection,

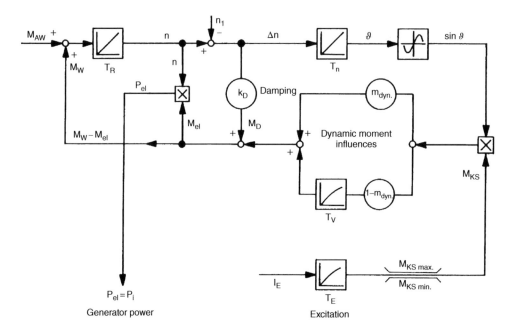

Figure 3.25 Simplified structure for representing the load moment at the synchronous generator

grid disturbances and generator short-circuit. Using the structure developed in accordance with Figure 3.25 we can approximate the load moment at the generator M_w (as a negative variable in the power consumption indication system) and the electrical output P_{el} from the difference in speed or frequency between mechanical and electrical rotation arising from the rotor displacement angle ϑ and its sinusoidal quantity (sin ϑ) in connection with the steady-state pull-out torque M_{KS}.

We can take account of the changes in moment according to the operating or extreme state under investigation by varying the excitation, while taking into account the time delay T_E and maintaining the minimum and maximum steady-state pull-out torques ($M_{KS\,min}$ and $M_{KS\,max}$). Furthermore, dynamic increases in the moment m_{dyn} acting in the short-time range (T_v) can be summarized in relation to the steady-state pull-out torque M_{KS} and damping effects (k_D) in accordance with the machine data in Section 3.4. Electrical and magnetic processes and the consumption by synchronous machines of their own power (e.g. due to excitation) are not taken into account here. The structure developed in Figure 3.25 can be used to obtain a good approximation of steady-state and dynamic processes, limited to electrical torque and mechanical effects, with the minimum of calculations.

Figure 3.26 shows a comparison of the simulation results between the five-coil model [3.14, 3.15] and the structure shown in Figure 3.25 for the neutral and overexcited state of the synchronous machine and for moment leaps of half and full nominal value. The reactions of the electrical generator moment and the rotor displacement angle shown display almost identical extreme values and paths for all four load states in both simulations. The structure represented in Figure 3.25 therefore gives the option of reflecting the load behavior of the synchronous

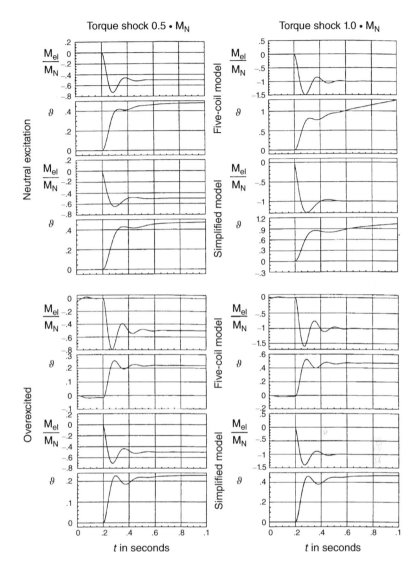

Figure 3.26 Comparison of simulation results (electrical generator moment and rotor displacement angle) between the five-coil model and the simplified model of a synchronous machine as shown in Figure 3.25 (input size of the torque leap from 0 to 0.5 or 1.0 M_N are not represented in this figure)

machine to a high level of accuracy, with significantly reduced calculation effort and machine data quantity.

Electrical torque is not determined on the basis of electric currents and magnetic flux: only their effects are – as far as possible – represented analytically or by structural blocks. The model represented here is only valid if the machine does not fall out of step, i.e. as long as the rotor displacement angle does not exceed 90° or the equivalent 1.57 radians. For the

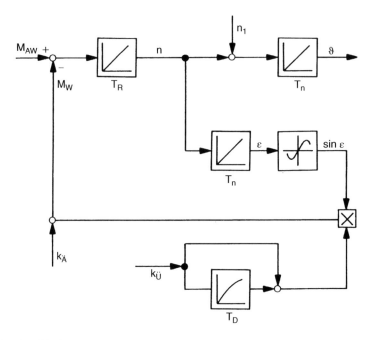

Figure 3.27   Simplified structure for the simulation of generator moment for three-phase connection processes and short-circuits in synchronous generators

simulation of three-phase short-circuit of the synchronous machine, the simplified structure shown in Figure 3.27 can be derived from the results of the five-coil model, where

- the time constant T_D determines the damping of torque oscillations,
- the factor k_U determines the maximum distortion of the generator moment, as well as damping, and
- the factor k_A determines the rate of change of the rotor displacement angle after the generator has fallen out of step.

At the instant of the short circuit, the values k_A and k_U of the dynamic process can be derived from the step-function response. Calculations based on this simplified model show only small differences when compared with the results from the five-coil model as shown in Figure 3.28. This presupposes that only the mechanical component of the system and the generation of electrical moment are considered.

In a similar manner, further representation options are also available for two-phase short-circuit, which will not be discussed further at this point. It is clear that, using this type of simplified model for specific investigations of torque behavior, results can be obtained that are a good match for those obtained by the five-coil model, which requires much more calculation effort. Furthermore, by making appropriate modifications to these models, we can also take into account approximations of transient and saturation-dependent processes.

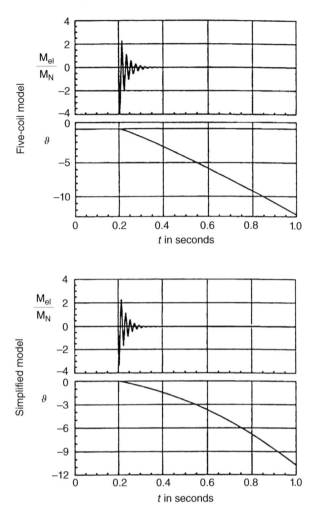

Figure 3.28 Comparison between simulation results (electrical generator moment and rotor displacement angle) between the five-coil model and the simplified model shown in Figure 3.27 for triphase short-circuit of a synchronous machine

The use of synchronous generators in wind power plants is almost always limited to independent (isolated) operation or corresponding conditions. In grid operation these are achieved by decoupling the generator speed from the rigid grid using power conversion devices (controlled or uncontrolled rectifier, direct current intermediate circuit, previously line-commutated or self-commutated inverter). Such configurations offer a great deal of potential for research and development, e.g. with respect to plant-specific control and operation strategies and with regard to the stability and controllability of synchronous machines when working alongside direct current circuits (see Chapter 4).

3.5.2 Asynchronous machines

The electrical transient processes for asynchronous machines, like those for synchronous machines, can be represented and summarized in a simplified form for the relevant time period that is decisive for mechanical loads. Further aspects specific to the use of asynchronous generators in wind turbines will be dealt with in Section 3.6.

Unlike synchronous machines, there is no completely fixed-speed connection to the grid in asynchronous operation. Furthermore, asynchronous generators generally have a damping effect upon exciting oscillations. Because of the variable-slip grid connection of asynchronous generators due to the large inertial mass of the wind wheel, dynamic moments take on a secondary role. Because of this, short-term increases in moment during dynamic processes can be disregarded in the first approximation or, if necessary, as in Figure 3.25, taken into account in the electrical moment M_{el} using the values given in Section 3.4.2.

The asynchronous load moment at the generator shaft can be simulated in a very simplified form using the steady-state torque–speed characteristic of the fundamental wave field. Therefore, in some cases large modifications of the operands (breakdown torque M_k and breakdown slip S_k), must be carried out here to the operands (breakdown torque M_K and breakdown slip s_K), so that Kloß's equation (3.5) can be used to represent realistically the slip-dependent torque of the most common current-displacement rotors in the form of the steady-state variable

$$M_s = \frac{2M_K}{\frac{s}{s_K} + \frac{s_K}{s}}.$$

Figure 3.29 shows a corresponding structure for the representation of the load moment for the shaft of an asynchronous generator (see Figure 3.13). Influences caused by changes of the grid frequency

$$f_1 = pn_1 \tag{3.22}$$

in connection with the number of pole pairs, p, can be taken into account here by the slip

$$s = \frac{\Delta n}{n_1} = \frac{n_1 - n}{n_1} \tag{3.23}$$

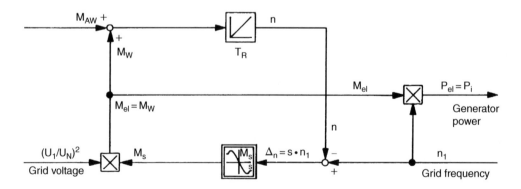

Figure 3.29   Creation of load moment at the asynchronous generator

as well as the static moment M_s and the electrical moment M_{el} in the generator output equation. Grid voltage fluctuations are taken into account by the approximate proportionality between the torque and the square of the voltage

$$M_{el} \approx f(U_1^2) \tag{3.24}$$

or in connection with the steady-state operand M_s in the creation of electrical moment

$$\frac{M_{el}}{M_N} \approx \left(\frac{U_1}{U_N}\right)^2 \left(\frac{M_s}{M_N}\right). \tag{3.25}$$

Under actual conditions, the additional dynamic moments mentioned above generally give rise to higher mechanical component loading and generator behavior that is less susceptible to overload. However, shaft loading as described in Section 3.4 is usually increased. The resulting electrical transient processes also have an effect on electrical energy transfer, as discussed below, and on the consumer.

Due to the slip-dependent speed variance of asynchronous machines, dynamic moments are of secondary importance in normal operation. Therefore relevant design options for generators are considered below.

3.6 Design Aspects

Normal design processes for electrical machines are based primarily on experimental values and the calculations in references [3.15–3.17]. Currently, these are usually used in a refined form with the aid of computer-aided programmes, and play a key role in the study of electrical machines.

The main dimensions of the machine, i.e. the diameter and length of the stator and rotor laminated core, can be roughly determined based on experimental values for the so-called output coefficient or tangential force figures, taking into account the rotation speed, output, cooling characteristics, slot and pole pitch, which have been predetermined or must be selected. In this context, the machine type, its size or output class, cooling (air, fluid, hydrogen) and the materials used, etc., have a significant influence on the dimensions of the generator. Calculations of the magnetic circuit with flux and field divisions and the determination of individual losses in the electrical, magnetic and mechanical components permit, with the aid of iterative processes, a near-optimal machine design. This process, which has been described in relatively simple terms, turns out in practice to be very costly and calculation-intensive, as the considerations contain a multitude of nonlinear relationships, and a great deal of design experience is required to obtain the desired optimization.

The full calculations upon which the design and construction of electrical machines are based will be omitted at this point, due to the wide-ranging prerequisites for the use of design methods, which lie far outside the scope of this book.

The representation that follows therefore aims to give some pointers and trends for design, based on the discussions in this chapter and on operating experience and simulation calculations for asynchronous and synchronous generators.

3.6.1 Asynchronous generators

Asynchronous generators with cage rotors were by far the most common type of generator for mechanical–electrical energy conversion in the wind turbines manufactured and installed in the past. Generators with slip-ring rotors have, up until now, been the exception. However, developments in power electronics over the last few years have meant that they are taking on increased importance for geared systems. In addition to slip-ring rotor machines with brushes, brushless machines (so-called cascade machines) are also used [3.18, 3.19]. The frequency converter is supplied via an additional auxiliary machine (see Figure 5.27(a)).

Different requirements are imposed on asynchronous generators for use in wind turbines than for those in diesel generators. In diesel units, high no-load losses in the generator can be very desirable from the point of view of operating characteristics. They bring about low motor fouling at only slightly increased fuel consumption. Furthermore, from the point of view of operating behavior, they bring significant advantages. Torque fluctuations are notably reduced, for example, when loads are decreased, resulting in lower frequency and voltage deviations. In wind turbines, however, no-load losses are generally undesirable, since they make it more difficult to convert low wind speeds into electrical output, and increase the standstill time of the plant.

In a generator based on a good mechanical design the saturation state of the magnetic circuit has a significant influence over no-load losses. High magnetic saturation usually results in high no-load losses. Load variations, however, bring about lower voltage changes with increasing saturation. Furthermore, the permissible current density of the generator (in particular for connection and shock load) is significantly increased due to the high magnetic energy content. The adjustibility and operating behavior of diesel units can therefore be favourably influenced by high generator saturation. However, operating problems can occur, particularly in low wind-speed ranges, when generators designed in this manner are used in wind turbines.

In wind turbines operated independently, generator excitation normally occurs after turbines have been run up. No-load losses caused in this manner cannot be covered at high saturation in the lower wind-speed range; thus the turbine speed can drop to the point where the generator is de-excited. The reduction of significant losses results in a subsequent increase in turbine acceleration. This pendulum process can repeat itself over and over again in a certain wind-speed range if no action is taken by the plant control system.

When wind turbines are operated as part of a grid, high inrush currents occur, which can trip fuses and significantly load windings (see Section 3.4.2.1), particularly if generator saturation is high. Furthermore, efficiency is sacrificed and generally poor power factors must be accepted. If the standard voltages were changed or adjusted, safety considerations [3.20] could increase in importance in addition to the aspects discussed here.

The behavior of the induced voltage U_i takes on particular importance in a detailed consideration of the processes in asynchronous generators, since it has a significant influence on the degree of saturation of the magnetic circuit. According to Figure 3.7(a), the induced voltage is found to be

$$\underline{U}_i = \underline{U}_1 - (R_1 + jX_{1\sigma})\underline{I}_1 \tag{3.26}$$

Based on the general current circle diagram, normally represented in simplified form as circle I_1, the circle U_i shown in Figure 3.30 can be derived for induced voltage. The U_i circle intersects the real axis at U_1 near the nominal voltage at no load and at approximately $U_1/2$ when stationary. Both circles are represented in Figure 3.30 for an asynchronous machine designed

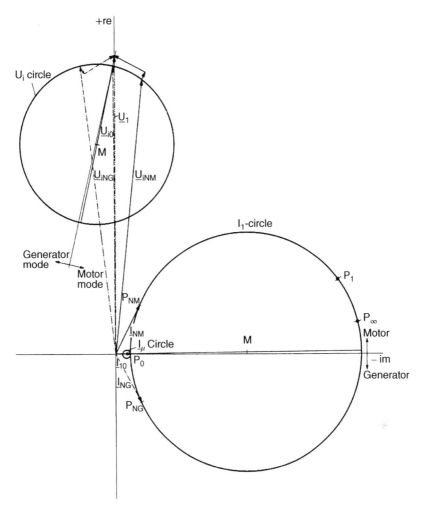

Figure 3.30 Circle diagram of a 22 kW asynchronous machine for stator current (I_1 circle), magnetization current (I_μ circle) and induced voltage (U_i circle)

as a generator with 22 kW nominal output, as an example for larger asynchronous machines operating at a nominal rating for the motor and generator mode. For the sake of completeness, the I_μ circle, which will not be dealt with in detail here, is also drawn. Saturation-dependent influences on the circles, which lead to significant deviations from the circular shape, will not be considered.

The circle diagram (the so-called U_i circle) traces the path of the pointer tip for induced voltage. We see a clear increase in U_i for a generator's normal load operation when compared with motor operation (compare U_{iNG} with U_{iNM}). Asynchronous machines therefore fall significantly further into the saturation state in generator operation than in motor operation. No-load losses increase and the power factor falls. Both variables therefore take on much poorer values. This is clearly illustrated by the following discussion.

3.6.1.1 Performance characteristics from measurements

In order to demonstrate the operational options and problem formulations, and the energetic effects that are dependent upon machine selection, we will show the results of metrological investigations on five different asynchronous machines of the same size, with

- 11 kW nominal output;
- motor- or generator-type design;
- output-class-dependent slip value and increased slip values due to
  - constructional changes in the case of cage rotor machines or
  - circuitry measures with the aid of additional resistance in the rotor circuit of slip-ring rotor machines;
- for stator voltages between 280 and 500 V or between 320 and 480 V;
- in the electrical output ranges of
  - generator no-load operation or
  - motor no-load operation respectively and
  - up to a maximum of 1.3 times nominal output.

To record the measured values in a stable-temperature, steady-state condition, a warm running phase of around two hours was necessary for each measurement. The rating of the machine was selected such that the characteristic attributes of the large unit class would be accurately represented.

To demonstrate the transient behavior of individual machines, and in order to specify guidelines for the design of generators specifically for use in wind turbines, there follows a small selection from the comprehensive programme of investigations based on some performance characteristics from a selection of machine types (generator or motor design, slip-ring rotor). The relevant performance characteristics for

- active power consumption,
- apparent power,
- reactive power,
- stator current,
- efficiency,
- power factor and
- generator slip

are represented in the following in relation to

- electrical output power and
- stator voltage.

The grid voltage was adapted to the selected stator voltage with the aid of a regulating transformer. Due to the fact that the power values were measured at various distances, they had to be interpolated for spatial representation in the form of computer graphics before they could be represented in the selected power grid (based upon 1 kW increments).

In Figure 3.31 the measured values are shown for a generator-type asynchronous machine, from the maximum voltage downwards. This particularly illustrates the trends that increase steadily with voltage, which is true for apparent power, reactive power and mechanical power input, and partly so for efficiency. Values that decrease with increasing voltage, such as the power factor and slip, remain difficult to recognize due to being partially obscured. These trends are, however, clearly shown in the subsequent figures, which show voltage increasing.

Figure 3.31(a) and (b) illustrates the increase in apparent and reactive power with increasing stator voltage and output power. The increases with voltage are caused by the effects of

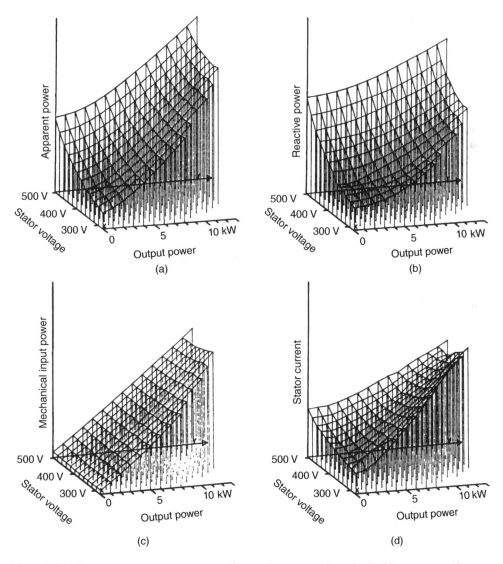

Figure 3.31   Parameters: (a) apparent power, (b) reactive power, (c) mechanical input power, (d) stator current, (e) efficiency, (f) power factor and (g) slip of a generator-type asynchronous machine (11 kW nominal output) as a function of electrical output and stator voltage (results from a series of measurements)

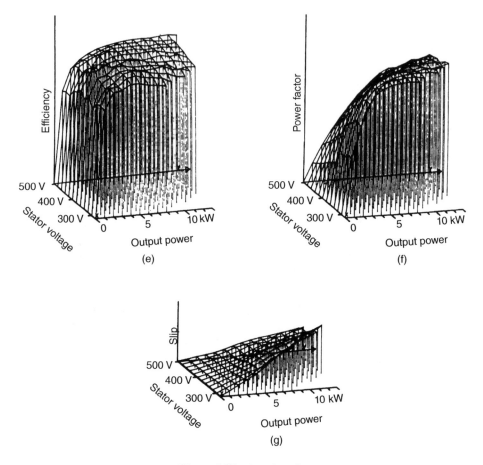

Figure 3.31 (*continued*)

saturation in the magnetic circuit and the associated increase in current with rising voltage, particularly at no-load and low outputs. The resulting increases in copper and iron losses in the machine require higher drive input power values, as illustrated by Figure 3.31(c).

As shown in Figure 3.31(d), the minimum current runs from the lowest voltage values (280 V) at no-load to maximum output at approximately 460 V. High electrical output at low-voltage values thus causes high stator currents due to output demands. At high voltages, high stator currents are caused by the high level of magnetization due to high saturation.

Figure 3.31(e) shows that efficiency increases progressively at high voltages, i.e. if the machine is designed in the heavily saturated range. Conversely, at low voltages, and thus dimensioning in the unsaturated range, maximum efficiency is achieved at approximately half the nominal output. Higher output values only bring about a small reduction in efficiency in such designs, a characteristic that can be very desirable in wind turbines, which are usually operated at partial load. Furthermore, Figure 3.31(f) shows that for highly saturated machines the desired power factor values typical of generators are only achieved in the overload range. Unsaturated design, on the other hand, gives the maximum power factor at only approximately two-thirds of nominal output. Such machine types also exhibit high slip values, as shown by the slip levels (in the lower-voltage range) in Figure 3.31(g), which, in combination with the torque fluctuations that are always present in the turbine, results in a high degree of elasticity and compliance in the transmission. Output fluctuations are thus reduced and highly fluctuating component loadings in the entire power transmission section are significantly reduced. Measurements on machines of differing designs have shown that, assuming that driving torque fluctuations are similar, generators with slip values four times higher than nominal slip transfer around a quarter of the output fluctuations to the grid compared to a normal design (see Figure 3.15).

Therefore, the generator parameters of a motor-type slip-ring rotor machine will be described below for different slip behaviors, which can be adjusted with the help of additional resistance in the rotor circuit. Figure 3.32 shows a comparison of the relevant parameters for different levels of slip. The machine in question was operated with a short-circuited rotor at approximately 3% nominal slip or with additional resistors in the rotor circuit, which are selected to achieve 7, 14 or 21% nominal slip at nominal load and voltage.

Figure 3.32 shows an increase in input power at higher slip values due to increased losses in the rotor circuit, and thus a worsening of the machine's efficiency, particularly in the full-load range. The mechanical input power increases with increasing slip. Stator current and power factor values, on the other hand, show only small differences. They are almost identical for all slip values at corresponding stator voltage and generator load.

Finally, the slip levels illustrate the very strong voltage dependence of these variables and their significant increase as voltage decreases. This trend is highlighted in a very striking manner by the levels at the highest slip values (21% operating at the nominal rating). It is thereby possible to achieve the desired degree of slip by the selection of

- rotor resistance in connection with
- rotor leakage (see Equations (3.7) and (3.11))

and the specification of

- stator voltage.

As was clearly shown at the beginning of this section with the help of the derived circle diagram for induced voltage, very different behaviors can be expected from the motor-type design compared with generator-type dimensioning. Figure 3.33 shows marked trends and constructive areas of influence, by direct comparison of the relevant parameters.

Mechanical input power and, in particular, stator currents are higher for the motor-type machine than for the generator-type machine across the entire generator operating range. The excessive increase in current at high voltages clarifies the high saturation dependence of the motor. Efficiency and, to an even greater degree, power factor values are much more favourable for machines of generator dimensions, particularly at higher voltages. At low voltages, however, only small differences are apparent. The generator-type machine with 3% slip compared with 4% nominal slip of the motor demonstrates significantly more rigid behavior, due to the more efficient design, resulting in a more rigid grid connection.

In summary, we can state that the asynchronous machines used in wind turbines should be designed to achieve a low degree of saturation. Figure 3.30 illustrates this point, since the

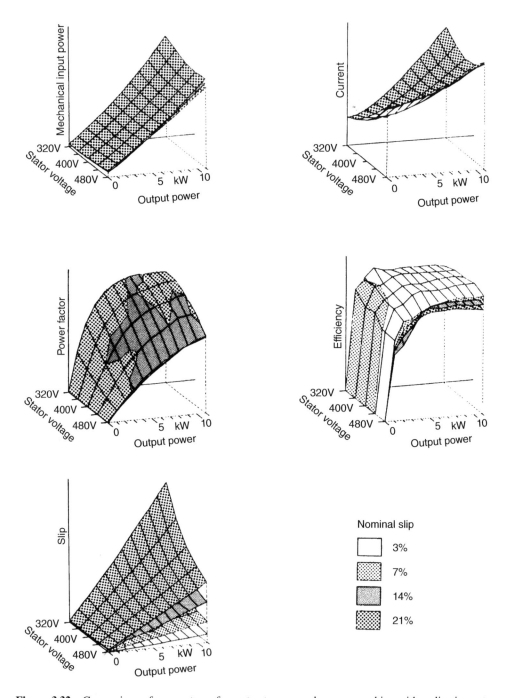

Figure 3.32 Comparison of parameters of a motor-type asynchronous machine with a slip-ring rotor (11 kW nominal output) at different nominal slips in relation to electrical generator output and stator voltage (results from a series of measurements)

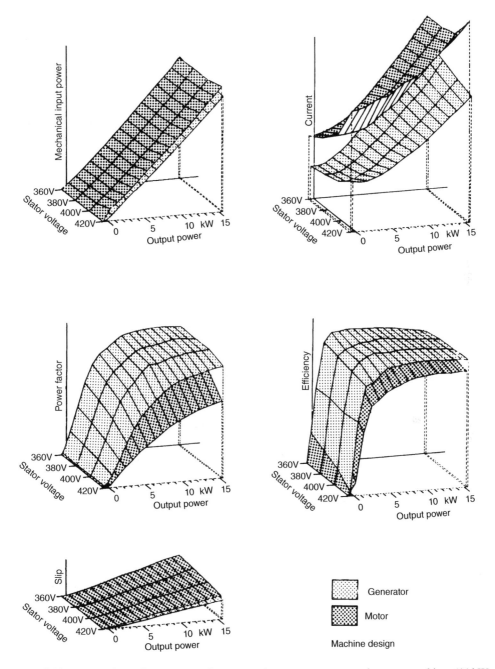

Figure 3.33 Comparison of parameters of motor- and generator-type asynchronous machines (11 kW nominal output) in relation to electrical generator output and stator voltage (results from a series of measurements)

generator always achieves higher induced voltage values and thus higher saturation states in normal operation than in motorized operation. This achieves

- a low mechanical input power,
- stator currents as low as possible for the prevailing load state and
- favourable efficiency and power factor values, both in the partial load range and at high slip.

Altogether, this type of design allows the following to be achieved in asynchronous generators:

- low reduction of efficiency,
- high rotation speed flexibility and
- correspondingly low drive-train loading

due to better flexibility in the turbine rotation system. In the following, we will describe specific designs for operating ranges. The grid connection is described in Chapter 4.

3.6.1.2   Weak wind generators

As mentioned in Section 3.4.2.4, weak wind generators are used for the motorized running up of wind turbines without blade pitch variation. They are designed to achieve 20% of turbine nominal output at 50 to 75% nominal speed, which means that the weak wind generators can operate in a favourable output range at low rotation speeds. This permits energy yields in the lower load region to be significantly increased (Figure 3.34). Thus – similarly

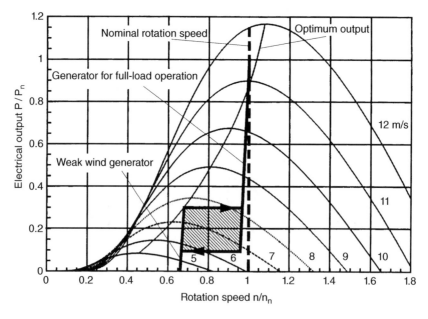

Figure 3.34   Group of turbine characteristics for output as a function of rotation speed, with wind speed being a parameter, and the working range of weak wind and a full-load generator

to the variable-speed operation of plants with frequency converters but with only two fixed speeds – the potential output of the turbine can be more fully exploited. A weak wind generator can either take the form of an additional machine that rotates with the turbine and has been specially designed for this application or it can be a pole-changing generator designed for weak wind operation at low speeds and for full load at higher speeds.

To avoid a high frequency of changeovers on the borderline between weak wind and full-load speeds, and the associated loads to the drive train and switch gear, an output overlap must be provided between the two operating conditions. The changeover hysteresis shown in Figure 3.34 must be adapted to the plant circumstances and the prevailing local conditions.

3.6.1.3 Drive-train junctions

Drive-train junctions provide further options for the improvement of starting procedures and partial-load operating ranges. In this approach, two to four generators G_1 to G_4 (see Figure 3.35(a) and (b)) are coupled to the gearbox. Transmission loading can thus be reduced and the design of the system significantly improved.

These configurations permit the linking of different features. For example, using a (modular) 250 kW generator, wind turbine outputs of 250 and 500 to 1000 kW can be achieved by fitting one to four machines. This permits significant cost reductions for development, storekeeping, etc. In the four-generator layout, moreover, a specific design is possible for the motorized run-up. Furthermore, by pole changing, the generator in the foreground of Figure 3.35(a), for example, can be operated at different rotation speeds. The weak wind characteristics of the plant can thus be improved. It is also possible to operate all generators in high-efficiency ranges by selecting appropriate individual changeover levels. Depending upon requirements and design goals for the entire system, the option of flexible grid connection can be included, as well as other features that we will not describe in detail here.

3.6.1.4 Variable-slip asynchronous generators

As already described in Section 3.4.1.1, power gradients and grid effects can be reduced at increased slip values by the special design of the machine in accordance with Equations (3.6) to (3.12). As shown in Figure 3.16, significantly better results can be achieved at a more favourable utilization ratio by slip regulation. In this approach, slip values less than 5 to 10% are generally used, which are controlled by a microcontroller through additional rotor resistors rotating along by using pulse-width-modulated insulated gate bipolar transistor (IGBT) power output elements. With regard to the control dynamics, similar reaction times can be achieved to those in systems with a frequency converter supply, despite the simple layout and low system cost. The operating range of slip-regulated asynchronous generators is not limited to variable-pitch wind turbines. According to reference [3.21], stall-regulated turbines also permit the use of such systems.

The reduction of the transmission and generator load as well as the resulting more favourable flicker values and so-called K factors (See Section 4.3) are achieved by design and construction measures. To this end:

- climatic influences due to temperature changes, air density, moisture, etc., must be compensated and

(a) Double generator configuration (Adler 25, 165 kW). Reproduced by kind permission
of Feus Peter Molley

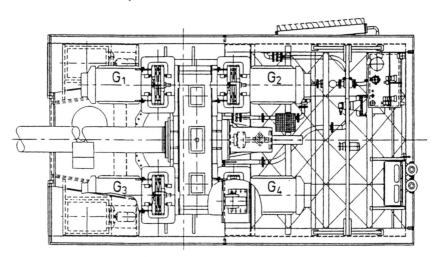

(b) Four-generator configuration (NEDWIND 50, 1000 kW)

Figure 3.35   Multiple generator systems

- high intrinsic safety,
- good control dynamics of the slip-control unit and
- rapid (target and actual value) data transfer guaranteed.

Furthermore, all important operating states of the generator and the power electronics of the higher management of the wind turbine must be made available (by a communication interface).

3.6.1.5 Shaft voltages and bearing currents of three-phase machines in a frequency converter operation

Parasitic effects in machines equipped with frequency converters give rise to currents that propagate through shafts, the running surfaces, rollers and housing of bearings, and which can (depending upon the intensity of the current) cause damage to bearings. Shaft voltages are particularly common in low-voltage, high-output machines, which are used, for example, in the generators of wind turbines [3.22–3.24].

Asymmetries in the magnetic working flux circuit, e.g. caused by ventilation openings, laminated core mountings, magnetic anisotropy of the laminations and asymmetries in the windings, result in a parasitic magnetic flux. This surrounds the rotor, increases with the level of flux per pole and induces voltages in the shaft of 50 and 150 Hz in particular, the so-called classic shaft voltage. Magnetic asymmetries must therefore be avoided in the design and manufacture of the machines. Furthermore, in large generators a bearing can be constructed in insulated form, in order to interrupt circulating currents [3.22]. According to reference [3.24] permissible limit values for shaft voltage in roller bearings lie at 0.5 V.

The use of high-speed IGBT frequency converters further promotes the generation of bearing currents [3.25–3.28]. Pulse-width-modulated a.c. converters are generally designed with a direct current intermediate circuit and generator-side inverter. This connects the rotor phases with positive and negative intermediate circuit potential on an alternating basis. As a result the total voltage at the output of the pulse-width-modulated a.c. converter (in contrast to symmetrical three-phase voltage systems) exhibits a pulse-shaped path that deviates from zero. This so-called common-mode voltage is dependent upon the frequency converter pulse pattern or the rate of rise of the common-mode voltage and can be measured at the neutral point of the generator winding. It is responsible for the bearing currents caused by the pulse-controlled a.c. converter.

Furthermore, in electrical machines capacitive couplings exist between the windings, the stator housing and in particular the rotor, which give rise to potential differences between the rotor and stator [3.29]. These arise in the same direction at both bearings, break down the common-mode voltage and are generally dependent upon the machine geometry. Since the capacitances are very small and lie in the nF range, these potential differences are limited by discharge routes or they lead to discharges that can break down the bearing lubrication film. This results in microscopic marks on the running surfaces of the bearing. These effects can be minimized by a common-mode voltage-minimized pulse pattern. In slip-ring rotors, earthing brushes connecting the stator and rotor are used. Furthermore, the stator housing is connected to the frequency converter via the cable shield in order to guarantee this as a good path for the return of common-mode current [3.24]. Since the impedance of these current routes is often too high, stray earth currents sometimes flow through the generator bearing or the bearings of components that are connected to the generator shaft (e.g. gearbox, sensors). These capacitive earth currents, which can reach levels of up to 10 A, put the bearings at risk and have an unfavourable effect upon electromagnetic compatibility (EMC). This situation can be remedied by the use of voltage-rise filters, known as dv/dt filters, which limit the gradient of voltage rises; these filters ensure the best possible return line for the high-frequency earth currents by providing a two-sided earth shield. Bearing damage to components connected to the generator shaft are avoided by the provision of an insulated coupling or the use of lower-inductance earthing than in the generator. Additional protection is provided by the earthing of the shaft at the non-isolated bearing.

Similarly to the classical shaft voltage mentioned at the start of this section, frequency-converter-dependent circulating currents arise as a result of common-mode current flow in the motor winding due to the occurrence of capacitance between the windings and the laminated core. Asymmetries in current distribution across the machine give rise to a flow that induces pulse-shaped voltages along the generator shaft. These are overlaid upon the classic shaft voltage and can give rise to low-frequency currents. For this phenomenon, too, the steepness of the common-mode voltage can be reduced by the use of a voltage-rise filter, and circulation currents can be weakened or prevented by good insulation of a bearing.

Shaft voltages and bearing currents thus give rise to various damaging effects, which are also overlaid upon one another. To prevent bearing damage, several measures are therefore necessary. To insulate generator bearings, isolated bearings can be used or, if standard bearings are used, an isolating layer can be inserted between the bearing and shaft or high insolated ceramic bearings can be used. The bridging of the non-isolated generator bearings represents an additional preventative measure. Connections that are as short as possible and exhibit low impedance, even at high frequencies, are necessary in conductors, cable shields and earth connections. Furthermore, the use of voltage-rise filters and the design of common-mode voltage-minimized frequency converters are of great importance for avoiding bearing damage.

The following considers synchronous machines, which in combination with frequency converter systems permit variable-speed turbine operation.

3.6.2   Synchronous generators for gearless plants

In addition to the simply constructed and very robust asynchronous machine with a cage motor and the increasingly common double-fed asynchronous generator, the conversion system with a frequency-converter-coupled synchronous machine has also achieved a large market share. For the conditioning of electrical energy and grid connection, load-side rectifiers for intermediate circuits (until the start of the 1990s with line-commutated inverters based upon thyristors and thereafter with line-side self-commutated converters based upon IGBT technology) have gained the greatest importance. Gearless plants dominate.

However, in synchronous generators the entire machine output must be fed through the frequency converter. In double-fed asynchronous generators, on the other hand, the frequency converter output is oriented towards the selected speed variation range or the slip power to be transmitted (e.g. 30 to 40% of the nominal value).

This type of generator and frequency converter system offers the advantage of decoupling rotor speed from grid frequency over a wide variation range (e.g. $(0.5-1.1)\ n_N$). Thus the potential for intermediate storage of kinetic energy in the rotating mass is utilized over a wide range for output smoothing and, in contrast to fixed-speed converter systems, by selecting optimal output operating points for better efficiency (Figure 3.36).

The normal type of synchronous generators for use in wind turbines are – unless more stringent requirements are imposed upon the control speed of the exciter – of a brushless design with an integral exciter. The auxiliary generator thus supplies the field winding of the main generator via a rotating rectifier (see Figure 3.6). In order to increase the subtransient reactance of the machine and to limit the intermediate circuit current in the case of a short-circuit, the generator rotors can be designed without damper windings (compare the equivalent circuit diagram in Figure 3.37(a) and (b)). This can reduce the maximum value of stator current by around 30% (see Figure 3.38).

Gearing represents a considerable cost factor in the drive train of a conventional wind turbine. It is also subject to dynamic load shocks, e.g. due to gusts. In addition, gears bring about losses, and require constant monitoring and maintenance during operation. Therefore, since the early days of wind energy technology [1.3], the option of installing plants with directly driven generators with no gears has been considered. Such configurations are in a period of rapid market growth and consolidation. More and more manufacturers are currently going over to developing gearless conversion systems. The electrically excited generators used in these

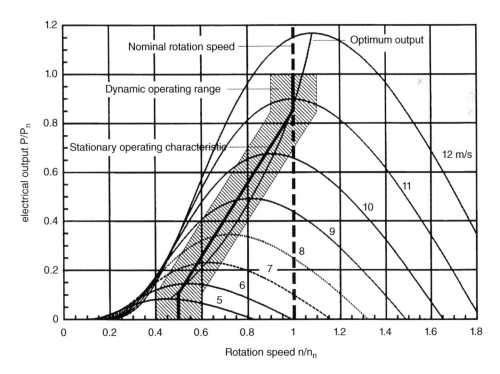

Figure 3.36 Power–speed diagram of a variable-speed wind turbine with steady-state and dynamic working ranges near to optimal output

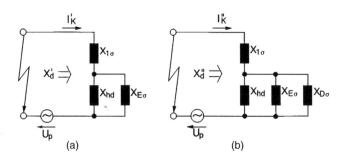

Figure 3.37 Equivalent circuit diagram for a synchronous machine in short-circuit (a) without damper winding and (b) with damper winding

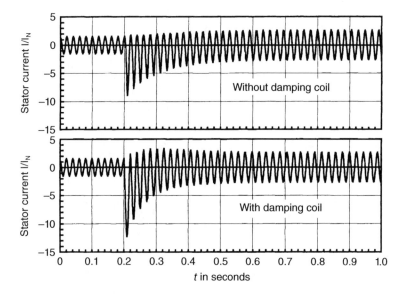

Figure 3.38 Stator current during the short-circuit of a synchronous machine with and without damper winding (simulation results)

systems have, up until now, not been fitted with an auxiliary generator. The field winding is fed via slip rings.

The new type of configuration must incorporate

- modular power components, in particular
  - a generator and
  - power converter;
- stepped output, with
  - 10 to 100 kW for hybrid plants and
  - 0.5 to 5 MW class for large plants;
- good transportability and

- simple installation.

Furthermore, the entire system must facilitate both

- excellent compatibility for the grid supply and
- the option of connection in weak grids and hybrid plants.

These requirements can usually only be achieved by conversion systems, which

- can be connected to form a grid,
- have voltage and/or frequency controlling characteristics, as well as
- energy optimization and supply-compatible tuning for the control and operation of the plant.

   Different system configurations are available that fulfil these requirements. Conventional generator types, which require geared transmissions, will not be considered further in the following discussion, which will concentrate on the new type of configuration.

   Gearless synchronous generators can basically be divided into the following lines of development. In particular,

- separately-excited salient pole machines or
- permanent-magnet designs, with
- axial,
- radial or
- transverse air gap

come into consideration. In addition,

- non-salient pole,
- unipolar and
- claw pole designs

are possible. Some of these systems are described in further detail below.

   Firstly, it should be mentioned at this point that for wind turbines in the 100 kW to MW class at a normal speed range between 60 and 15 rpm, 100 to 400 poles would be necessary to achieve a generator frequency of 50 Hz. To incorporate this large number of poles and the associated stator windings in the generator around the rotor and stator would require a relatively large machine diameter in the conventional layout, resulting in a correspondingly high component mass for construction elements.

   By decoupling the generator rotation speed and grid frequency, lower frequencies can be selected with the aid of frequency converters, thus reducing the required number of poles. Alternatively, a higher number of poles can be achieved using new layout types, e.g. in permanent-magnet machines. Lower generator frequencies, however, generally increase the requirement for smoothing in the direct current intermediate circuit. This can be significantly reduced by a trapezoidal generator voltage pattern (compare Figure 3.39(a) and (b)), which is achieved by the design and operation of the machine at high saturation. This, however, results in higher losses.

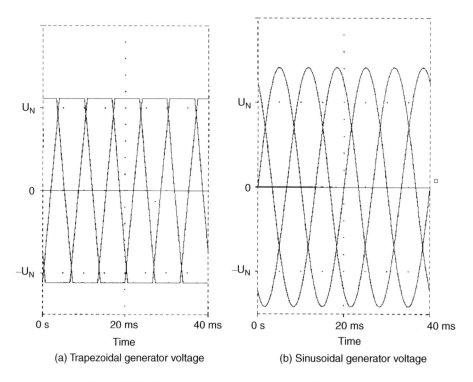

(a) Trapezoidal generator voltage   (b) Sinusoidal generator voltage

Figure 3.39   Voltage pattern for a six-pulse bridge connection

Stringent demands are imposed on wind power plants and their components with regard to reliability and service life. Plant availability above 98% is currently the norm. However, these values can only be achieved with repeatable manufacturing and installation procedures and precise execution. Components and subsystems are therefore manufactured and fitted, as far as possible, at the factory. After testing, the units are transported to the site and installed. Then, the entire system is commissioned. Thus, in the best case, the generator, as the main component of a gearless wind turbine, is fully assembled and tested at the place of manufacture and, where possible, transported to the site as a single unit incorporating the nacelle and hub and integrated into the tower head. The maximum dimensions for the transport route must be adhered to.

To facilitate road transport, generators with a housing diameter of around 4 m are desirable. If a good route with suitably dimensioned bridges, underpasses, tunnels, etc., is available significantly larger units can also be transported by road and rail. A generator/nacelle diameter of 5 to 6 metres should, however, not be exceeded.

Based on this preliminary discussion, the numerous design options and limiting factors become clear. The following description is limited to four different systems.

3.6.2.1   Separately excited salient pole machine with radial air gap field

The key components of the Enercon E30/33, E40/44, E48, E53, E66/70, E82/92, E101/115, and E112/126 gearles wind power plants are the synchronous generators driven direct from

Figure 3.40   Electrically excited synchronous generators for Enercon E40 gearless wind turbines; stator (standing) and salient pole rotor (lying)

the turbine. These are separately excited salient pole machines with a radial air gap field, of conventional design and using established materials technology. Two rotors with excitation windings on T-shaped pole cores are shown at the centre of Figure 3.40. Almost 80 poles are fastened on to a yoke ring. The stators, as shown in Figure 3.40, partly wound on the left and almost fully wound on the right, are made up of identical armature segments, which are joined together, overlapping in the direction of travel. The stator windings are located in the slots of the stator laminated core, as shown in Figure 3.40 in the right-hand foreground.

3.6.2.2   Permanent-magnet synchronous machine with radial air gap field

In the alternative form of vertical axis turbines, the so-called H-Darrieus system, gearless systems were initially introduced in the 30 and 300 kW classes. The ring generators conceived for these plants are designed with a radial air gap field and permanent-magnet excitation. They are connected to the grid via a frequency converter, thus permitting variable-speed operation. The plant was originally designed for the supply of electrical power for arctic regions, and is thus suitable for operation at very low temperatures.

The 300 kW prototype shown in Figure 1.31(a) is fitted with a ring generator, which was installed at ground level and had a diameter of 12 m (Figure 3.41(a)). Subsequent plant designs (Figure 1.31(b)), which were manufactured in a short production run, had the generator at the top of a three-legged mast (Figure 3.41(b)). As already mentioned in Section 1.2, however, this type of system has not yet managed to break through into the wind generator market.

(a)

(b)

Figure 3.41 H-Darrieus HM 300 plant with the generator at (a) the mast foot and (b) the mast head

Radial air gap generators with permanent-magnet excitation are particularly common in low-power horizontal-axis plants. With this design, it is possible, as in the layout shown in Figure 3.41, to fit the rotor with excitation poles of alternate polarity and the stator with salient poles. This makes it simple to dismantle the stator, e.g. for transport. Furthermore, the grain-orientated structures in the electrical sheet steel can make good use of the flow of magnetic flux to reduce magnetic losses. The insulation of differing phases can also be significantly reduced, since no intersections protrude from the end windings. This results in a large generator diameter and high material cost. Compact plant designs are thus not possible.

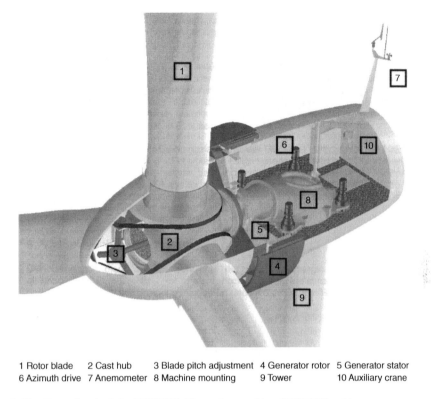

1 Rotor blade   2 Cast hub   3 Blade pitch adjustment 4 Generator rotor 5 Generator stator
6 Azimuth drive 7 Anemometer 8 Machine mounting   9 Tower   10 Auxiliary crane

Figure 3.42   Tower head of the VENSYS 62 gearless turbine (1200 kW) with a permanent-magnet external rotor generator [3.30]. Reproduced by permission of Vensys

Smaller pole divisions and diameter, and the associated higher generator voltage frequencies and better utilization of the active materials, can be achieved using split windings in the slots. However, this necessitates more insulation in the stator. A division of the stator can only be achieved by making sacrifices in utilization and increased use of coil segments. For this generator design, the stator diameter should be as small as possible. For machines in the megawatt class, external diameters between 4 and 6 m can be achieved. A range of single-ring generator modules, e.g. the 500 kW size, could also facilitate favourable layouts here, thus keeping the costs for development, manufacture, storage and spare parts service low.

A further possibility for reducing the generator diameter in comparison with standard machines was used by Prof. Dr F. Klinger (HTW Saarbrücken) in the development of the permanent-magnet generators for the GENESYS 600 and VENSYS 62 wind turbines (1200 kW) (Figure 3.42). In addition to the axial iron or laminated core length, the decisive quantity for the dimensioning of the mechanical–electrical energy converter is the air gap diameter of the generator, which when increased leads to a quadratic increase in power.

Normally, synchronous machines are designed with the rotor running inside a stator. The three-phase windings are, almost without exception, located in the slots of the stator laminated core. Thus, in addition to the air gap diameter that dominates in multipole machines, the slot height, the magnetic return (the so-called armature return) and the mechanical support structure

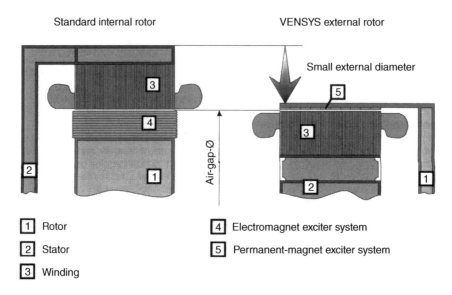

Standard internal rotor   VENSYS external rotor

Figure 3.43   Generator diameter comparison between external and internal rotor designs (page 181)

that encompasses all electromagnetically active components determine the external diameter of the generator.

Given the same air gap diameter or the same power, significantly smaller generator diameters can be designed by the use of permanent magnets that are just a few millimetres thick, using the external rotor variants as shown in Figure 3.43. Lower machine masses can thus be achieved, along with the associated advantages during manufacture, transport and installation.

3.6.2.3   Transverse flux machines

Motor and generator designs with magnetic circuits laid out across the direction of motion can mainly be attributed to the work of Prof. Dr Ing. Weh at the Institute for Electrical Machines, Drives and Rails at the Technical University of Braunschweig [3.31]. Figure 3.44 illustrates the difference in flux direction in a laminated core and current flow in one phase of a stator between conventional longitudinal-flux (a) and transverse-flux machines (b).

The principle layout of a two-sided transverse flux machine with permanent-magnet excitation as shown in Figure 3.45 illustrates the significant manufacturing cost of magnetic rotors and stator parts. The cross-section shown in Figure 3.46 shows the complete layout of the machine. Designs based on the reluctance principle, in which very cheap magnetically soft materials are used instead of expensive permanent magnets, are also possible. The advantage of this cheaper variant is traded off against the disadvantage of poorer utilization.

3.6.2.4   Permanent-magnet synchronous machine with axial air gap field

A further option is to design the generator as a permanent-magnet, multiple-pole synchronous machine with an axial air gap field [3.32, 3.33]. Permanent magnets of alternating polarity are

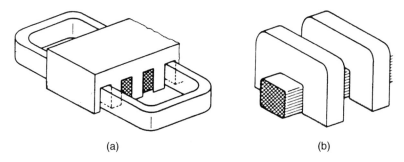

(a)   (b)

Figure 3.44   Stator laminated core and winding arrangement of (a) a longitudinal-flux machine and (b) a transverse-flux machine [3.34]

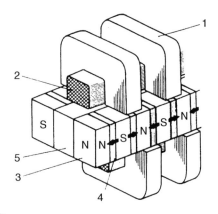

1 Stator core
2 Stator phase winding
3 Rotor
4 Permanent magnet
5 Non-magnetic material

Figure 3.45   Principle layout of a double-sided transverse-flux machine [3.34]

fastened to the rotor disc to create an excitation field, which can, as shown in Figure 3.47, comprise many single magnets or, for large quantities, one or a few individual parts. A large number of poles can be attained using this layout. This means that this type of machine can be designed for very low speeds (e.g. in the 100 kW range at 30 rpm with 100 pole pairs and a grid frequency of 50 Hz).

The ring-shaped stator (Figure 3.48) carries the armature coils. It can be made up of normal alternating current windings with projecting end windings that cross one another or, as shown in Figure 3.49, of several identical single-phase sectors. The winding can be constructed of parallel conduction layers, which may be made of copper or aluminium plates, or suitable preformed coils.

Axial-flux machines have a constant air gap. The electrically and magnetically active component of this type of generator can therefore be made of a few different components, which can be manufactured and assembled in large quantities. The layered winding, as well as having

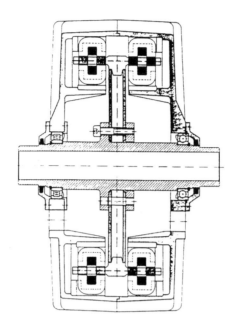

Figure 3.46   Cross-section through the stator and rotor of a transverse-flux machine [3.34]

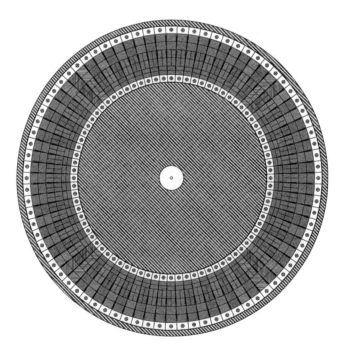

Figure 3.47   Rotor disc of a permanent-magnet multipole machine with an axial air gap field

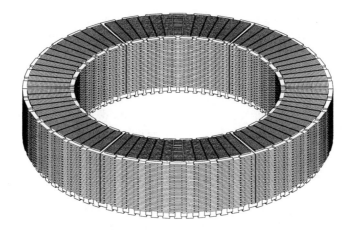

Figure 3.48 Stator with a laminated core and winding from a multipole machine with an axial air gap field

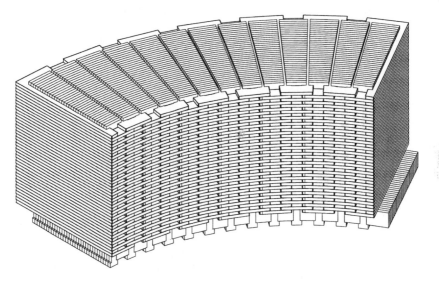

Figure 3.49 Single-phase stator segment (laminated core and stack of coils) of a multipole machine with an axial air gap field

a high slot space factor, gives compact end windings (Figure 3.48), since only small or zero bending radii have to be observed during manufacture, thus making short connecting routes between the grooves possible, despite the high conductor cross-section. Since, in principle, it is only the section of the conductor within the magnetic field that actively contributes to energy conversion, small end windings give high power density.

Due to its construction in sectors and the use of stacked windings, the cost for the insulation of adjoining conduction layers can be kept low. An insulation layer thickness of a few µm is generally adequate. The slot area can thus be fully exploited for the conduction of electricity.

Due to the short end windings, the proportion of the conductor that is inside the magnetic field is very high. These constructional characteristics can lead to a significant reduction in weight compared with other multipole generators. When machine-side pulse-controlled a.c. converters are used, however, the generator phase windings must be protected against partial overvoltages by voltage rise filters.

Due to the space-optimized layout of the windings in multipole generators, higher overall efficiencies can be achieved than is the case for conventional designs. Excitation losses are also eliminated by the use of permanent magnets. The generator therefore has a higher efficiency than electrically excited induction machines, particularly in the partial-load region. Furthermore, the need for maintenance is reduced in this type of generator, since no slip rings and brushes are required to supply the excitation current to the rotor.

The specific labour and material costs for highly permeable permanent magnets (Neodym) lies at around 100 to 150 euros per kilogram, i.e. around ten times higher than aluminium or copper conductors, at 5 to 8 euros per kilogram, or low-loss electric sheet steel and welded constructions, at 5 to 10 euros per kilogram [3.36]. Thus an economical generator design requires the use of magnetic materials to be minimized. This can be achieved, for example, if the generator is designed to be underexcited. However, step-up converters or the supply of exciter reactive current via the machine-side frequency converter are necessary to maintain the voltage in the intermediate circuit (see Section 4.1).

Permanent magnet generators can only be economically designed with a relatively small air gap due to the high cost of magnetic materials. System configurations with a one-sided air gap thus exhibit enormous axial forces (e.g. approximately 500 kN for a 100 kW generator). These require correspondingly rigid and expensive constructions.

Double air gap arrangements allow the forces to be largely compensated and thereby these disadvantages avoided and the design simplified. This variant (Figure 3.50(a) and (b)) can be designed with both axial and radial air gaps. In the magnetic circuit there are two air gaps connected in series. As a result, the normal components of magnetic force acting upon the rotor are enormously reduced. If approximately 25% of the normal air gap is maintained on both sides the axial force is reduced to a maximum of around 10% of the value for a one-sided

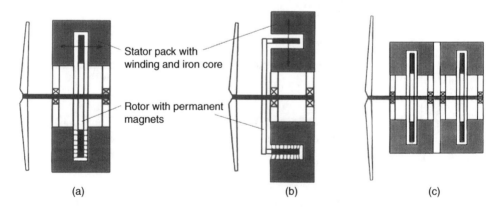

(a)   (b)   (c)

Figure 3.50   Schematic representation of double air gap designs for machines with (a) axial and (b) radial air gaps and (c) modular structure of system groups

air gap. Furthermore, due to the modular structure of system groups (Figure 3.50(c)) relatively wide power ranges can be covered by a few construction sizes.

To sum up:

- The 'Enercon concept' with a radial air gap flux and electrical excitation by salient poles has proved itself very well, and currently dominates the market. However, further improvements are still possible using a layout with a high number of poles, etc., which can give a more compact form, for example.
- Corresponding designs with permanent-magnet excitation necessitate high expenditure on the frequency converter and protection concept. Double salient pole machines facilitate simple division of the stator into several sectors. This results in larger models.
- Transverse-flux machines are expensive to construct due to their layout. For the permanent-magnet design, the magnet costs are high due to the double air gap. Cheap reluctance machines, however, have a high mass.
- Highly permeable permanent magnets for the excitation of the machines still lead to significantly higher generator costs. At an additional price in relation to electrically excited synchronous machines of 50 euros per kW, for example, and at an exciter power component of 2%, amortization times are around five to ten years.
- Permanent-magnet multipole generators with the stator divided into sectors, either with an axial or a radial air gap, have some advantages, particularly with regard to machine utilization and expected efficiency. In the case of the axial air gap field, a double stator layout is advantageous, to compensate for axial forces in the rotor. Modular construction lends itself particularly to this type of generator.
- Radial-flux machines with external rotors permit smaller generator external diameters than standard internal rotor variants.
- Encapsulating the electrically active generator components significantly reduces the risks of corrosion and insulation damage. However, correspondingly higher requirements are imposed upon the cooling systems. High-quality vacuum impregnation of the three-phase windings additionally increases operating reliability.

To round off this chapter, the next section briefly describes performance-related data of conventionally designed machine

3.6.3 Multi-generator concept (Dissertation A. Ezzahraoui)

The high average wind velocities at sea promise enormous energy potentials. However, the costs of construction, maintenance and repair of the plant at sea are very high. Yet, it can be assumed that a great increase of utilization of offshore wind energy will take place. For this purpose, gearless wind turbines with permanently excited synchronous generators (PMSG) offer a highly efficient solution with little maintenance effort. The costs and time spans for the development of wind turbines and their key components, in which the generator–converter system is the main part, is increasing far out of proportion to the increase in power. Thus, a module-based configuration that is developed and manufactured for a small unit, with its multiple use (two, three or even four modules of the generator–converter system) can achieve corresponding performance increases. In this way, very high development costs can be saved and production, transport, installation and storing, etc. can be substantially simplified. For this purpose, however, the overall operation of the mechanical and electrical components, as well as their vibration behavior in normal and malfunction operation cases must be extensively investigated.

Multi-generator systems can be designed in various ways. Systems can be directly driven by the turbine; for instance, two, three or four generators can be coupled together or driven by a common shaft also with supporting bearings, whereby several generators require a careful vibration-sensitive design. Single- or multiple drive models, in contrast, permit a direct drive of a generator or a drive train branching to two, three or four generators. The corresponding torque distribution to the individual generators has an important effect on the design of the drives, bearings, etc. and on the vibration behavior of the overall system.

Generators that are driven directly by the turbine require large rotor diameters in order to withstand the high torque of the wind turbine. Also, a large number of poles are required to obtain a corresponding frequency (10 to 100 Hz) at the low speeds. This results in a large generator volume and weight as well as high production costs. An important part of this is the high cost of the magnetic materials (such as neodym−iron−boron). They can easily be a factor of ten higher in comparison to conventional materials of electrical machine construction (iron and copper). The offshore use requires that the generator and rectifier systems are incorporated in the gondola. Thus, direct-driven multi-generator systems with fully rated converters in compact construction offer excellent prospectives. Figure 3.51 shows the principle structure of a wind turbine with a horizontal axle method of construction and with two directly turbine-driven generators and converters of identical type. An extension of the conception to three or four generator and converter systems seems possible but requires further investigation.

The implementation of the multi-generator concept for wind turbine of larger capacity in the 10 MW range is realized by two slow-turning PMSG. The example shown in Figure 3.51 produces for an overall power of almost 6 MVA per generator unit a conductor current of approximately 5000 A at a voltage of 690 V. As shown in Figure 3.52, the electrical energy is

Figure 3.51 Concept of a gearless wind turbine with multi-generator model (two generatorasa and two converters) in the machine room [3.37]

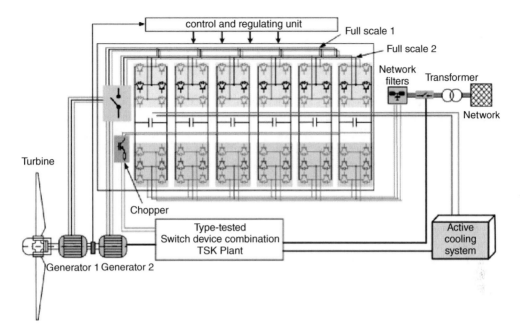

Figure 3.52   Electrical concept of a 10 MW wind turbine with multi-generator and cooling system

fed via a fully rated converter and medium-voltage transformer into the grid. A so-called grid or sine filter is used to limit the harmonic portions of the grid current according to DIN EN 61 558-2-20. The demands on the cooling management of the components increase with the increasing power of the plant. The whole of the heat built up by the generator and converter, etc. must be dissipated by a cooling system. This must be compact, light and powerful and permit being placed in the limited space available in the gondola. Moreover, cooling systems must be designed for stable and low-maintenance use, especially in offshore plants with the raw environmental conditions such as corrosion-promoting and saltwater-containing sea air as well as the strong temperature variations and this means that the metal surfaces must also be protected against the salty sea air.

The electrical concept in Figure 3.52 is made up of the following main components:

- Two generator units
- Machine switch for load separation
- Machine-side converter
- Intermediate voltage circuit with integrated chopper (dc link)
- Grid-side converter
- Control and regulating unit
- Grid filter
- Grid-side current switch
- Cooling system for converter

The utilization of this plant concept in the offshore region requires interruption-free and low-maintenance operation. The plant must have a interruption-free duration of at least

180 000 hours and reset times of less than 30 minutes. Furthermore, all safety standards in stationary as well as in dynamic cases must be fulfilled.

PMSG in direct-driven wind turbines are already in wide use for small wind turbine plants. The technical conditions for capacities of a few kilowatts are already available. The converter systems necessary for feeding the generated current separate the turbine from the grid. In this performance category, a cooling by means of good heat conductors (e.g. aluminium sleeve) can achieve good results [3.37]. For capacities in the megawatt region, however, powerful converter systems and corresponding cooling systems are necessary. In the operation of wind turbines and independently of the strength of the wind, energy is converted to heat and dissipated by means of cooling systems. In the case of air cooling, the outside temperature influences the efficiency of the whole plant as the failure rate of the electrical components and, especially the converter rises with the temperature. The IEC EN 61439 standard prescribes the proof of the heating limits for the components of low-voltage circuit combinations. The limiting values are used within the plant when the ambient temperature does not fall below $-5\,°C$ or does not exceed the maximum value of $+40\,°C$. In addition, it must be shown that the internal temperature of the switches is based on an average value for a time of 24 hours and does not exceed $+35\,°C$. For this, the relative humidity must be limited to 50% at the highest temperature of $+40\,°C$ or to 90% at $+20\,°C$ [3.30]. Because of these conditions, the utilization of normally available switchboxes that have a life of maximum 35 000 hours and require maintenance every three months is not possible in offshore wind turbines. As is seen in Figure 3.52, an active cooling is indispensable.

3.6.3.1 Drive train as a three-mass system

If the turbine and the two generators in Figure 3.51 are taken as individual masses, then the drive train can be described as a three-mass system. In this, the wind turbine is seen as a rigid body with the moment of inertia J_{WT}, generator 1 with J_{G1} and generator 2 with J_{G2}. At the wind turbine, the drive torque M_{WT} acts at generator $1 M_{G1}$ and generator $2 M_{G2}$. Further, the shaft between the wind turbine and generator 1 is considered as a torsion spring with a spring constant K_{01} and the shaft between generator 1 and generator 2 as a torsion spring with the spring constant K_{12}. The connection of the components is carried out by means of corresponding rotation dampers with the damping constants d_0 for the wind turbine, d_{01} between wind turbine and generator 1, d_1 for generator 1, d_{12} between generator 1 and generator 2 as well as d_2 for generator 2. Figure 3.53 shows the schematic view of the drive train as a mechanical three-mass system.

For describing the rotation or torsion vibration conditions, the angle of rotations $\varepsilon_0, \varepsilon_1, \varepsilon_2$ of the three masses with the associated derivations $\dot{\varepsilon}_0, \dot{\varepsilon}_1$ and $\dot{\varepsilon}_2$ as angular velocity and $\ddot{\varepsilon}_0, \ddot{\varepsilon}_1$ and $\ddot{\varepsilon}_2$ as angular accelerators are necessary. Equation (3.27) represents the differential simultaneous system of the second order in a matrix depiction.

$$\begin{bmatrix} J_0 & 0 & 0 \\ 0 & J_1 & 0 \\ 0 & 0 & J_2 \end{bmatrix} \cdot \begin{bmatrix} \ddot{\varepsilon}_0 \\ \ddot{\varepsilon}_1 \\ \ddot{\varepsilon}_2 \end{bmatrix} + \begin{bmatrix} d_0 + d_{01} & -d_{01} & 0 \\ -d_{01} & d_1 + d_{01} + d_{12} & -d_{12} \\ 0 & -d_{12} & d_2 + d_{12} \end{bmatrix} \cdot \begin{bmatrix} \dot{\varepsilon}_0 \\ \dot{\varepsilon}_1 \\ \dot{\varepsilon}_2 \end{bmatrix} + \begin{bmatrix} k_{01} & -k_{01} & 0 \\ -k_{01} & k_{01} + k_{12} & -k_{12} \\ 0 & -k_{12} & k_{12} \end{bmatrix} \cdot \begin{bmatrix} \varepsilon_0 \\ \varepsilon_1 \\ \varepsilon_2 \end{bmatrix}$$

(3.27)

With this simultaneous equation, the behavior of the plant in normal operation and for a change of load as well as for a malfunction case (e.g. grid short circuit; FRT, fault ride-through) and

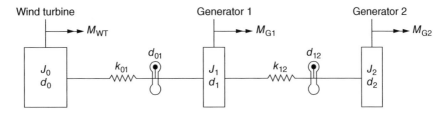

Figure 3.53   Schematic depiction of the drive train as a mechanical three-mass auxiliary model

other critical conditions can be investigated. The torsion vibration behavior and the resulting mechanical loadings can thus be determined, and with changes in the design of the components (e.g. shaft strength, inertia), harmonic vibrations can be avoided. In addition, with fast control access, torsion vibrations developing in the generators can be actively countered.

3.6.3.2   Validating the system

In order to ensure the continuous supply of mechanical power from the driving WT rotor, the drive torque of the turbine was kept constant. However, the counter-rotation torque at the PMSG was reduced for a duration of 430 ms to 100% for the simulation time of 698 ms. The simulation results of generator torques, rotation as well as torsion angle between the turbine and the first generator are shown in Figure 3.54. The impulse type of excitation of the generators leads to a new transient process of the whole drive train in which its revolutions increase from the time of occurrence of the malfunction and the torsion angle of the shaft change. The new torsion angle or the torsion torque is dependent on the height of the electrical counter-torques remaining at the bearings and the static damping. The larger the sum of the counter-torque and static damping torques, the greater is the angle.

A validation of the three-mass model by means of comparison tests at a small test drive train in the 10 KW range showed similar torques or speed sequences. As shown in Figure 3.55, a transfer of the load to the generator results in causing vibrations in the test drive train from the stationary operating condition. The counter-torques at the generator break up at the time point of the load transfer, and the drive train components start to accelerate. An increase of the rotation of the test drive train can be clearly seen in Figure 3.55.

The renewed impulse-type excitation of the drive train due to switching of the loading again leads to a transient process in the original stationary condition in which the twist angle as well as the rotation return again to their starting values. The counter-torques are again at their full strengths at both generators.

3.6.3.3   System control in normal and fault operation

In order to be able to utilize a wind turbine in all operating conditions, it is necessary, besides the mechanical components, also to consider the electrical part-system (generator, converter, filter and transformer). The wind turbine should achieve the biggest possible energy feed during the part-load operation, i.e. with wind velocities between switch-on and rated operation. This is normally achieved by operating the turbine revolutions in the highest possible power

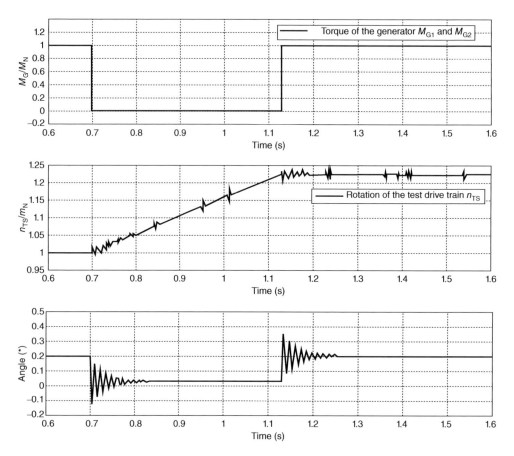

Figure 3.54 Simulation results of generator torques, rotation and twist angle of the shaft between the turbine and the first generator in dynamic operating condition

range (see Section 3.6.2). However, in the nominal load operation at a wind velocity between rated and switch-off velocity, it must be ensured that turbine, generator, converter, etc. are not overloaded. This is achieved by limiting the output power with the aid of pitch regulation of the turbine. Both methods of operation must then be adapted to the wind availability or the turbine side.

A grid fault, on the other hand, first decisively affects the electrical system of the wind turbine and thus has a substantial feedback on the mechanical part system.

For instance, in order to be able to safely master the voltage dips in the grid with the resulting drop in torque, it is necessary that this type of operating case be handled primarily on the grid side besides the turbine side. As is shown in Figure 3.56, the stator component and the efficiency are controlled on the generator-side converter. The grid-side converter, however, is tasked with keeping the intermediate circuit voltage constant and to feed the reactive current in malfunction operation into the grid. The turbine is pitch-controlled. The power of each generator can only be supplied to the grid when the voltage in the intermediate circuit remains constant [3.38]. A maximum Power Point table or a stored power-rotation curve (Figure 3.36)

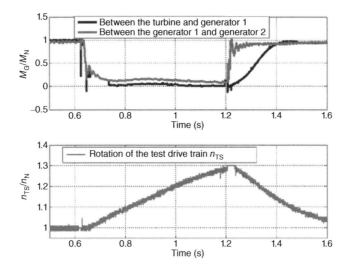

Figure 3.55 Measuring results of generator torque and speed of the test drive train in the dynamic operating condition

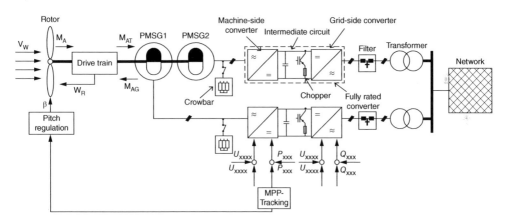

Figure 3.56 Control diagram for a wind turbine with two permanently excited synchronous generators and two fully rated converters

provides the reference values of the efficiency and thus ensures the operation of the turbine near the power optimum. The pitch mechanism changes the set angle when the rated speed of revolution is exceeded in order to reduce the aerodynamic power.

A grid fault results in a decrease of the voltage at the connection point of the generator and thus lead to a quick break-down of the power. The result is that the power fed into the grid by the converter is strongly reduced. At the same time, the generator-side converter transmits the power produced by the generator into the intermediate circuit. This leads to charging of the intermediate circuit capacitors, which again can lead to excess voltage in the intermediate circuit.

The chopper shown in Figure 3.56 serves to convert this excess power into heat. On the other side, the short circuit leads to an increase of the revolutions and thereby to a rise of the voltage required to drive the electrical machine, as long as no measures for flow reduction are available. Because the grid-side converter feeds a reduced amount into the grid during a grid fault, the power fed into the intermediate circuit by the generator-side converter must be reduced in order to protect the electronic components. This means that the stator flow must be weakened or the stator current reduced when the maximum stator voltage is reached and the rotation is increased still further. Because of this weakening of the stator flow, the resulting power equalization resulting from the grid malfunction no longer occurs in the intermediate circuit but in the generator. The weakening of the generator power with constant aerodynamic power of the turbine leads to an acceleration or vibration of the mechanical drive train. The excess power caused by this is converted to rotating energy. The aerodynamic power can be reduced by changes in the pitch angle and thus the acceleration of the generator and the turbine can be prevented [3.38]. However, at this point, it must be stated that the grid malfunction has a short time duration so that the reaction of the pitch mechanism has no great effect. Yet it has been found that torsion vibrations that can be caused by grid malfunctions (150 ms) represent a secondary mechanical load for the shafts of a wind turbine drive train with two permanent magnet synchronous generators. This loading can be accounted for today by normal safety factors and design parameters (e.g. using a crowbar or a damping system in the case of an electrically excited synchronous generator) [3.39].

3.6.4   Ring generator with magnetic bearings (Dissertation K. Messol)

An upscaling of today's normal synchronous generator concept to a power region of 10 MW and more leads over-proportionally to power-increasing gondola masses. Referenced to rotor diameters of plants, the mass rises by the cube and the power quadratically with the diameter. The largest presently available wind generator is the directly driven E-126 plant of the ENERCON Company with its tower head mass of more than 650 tonnes and with a capacity of 7.5 MW [3.40]. Fast running wind turbines with comparable capacity and with a drive are much lighter. For offshore plants, a drive should be avoided for reasons of maintenance and prevention of repairs and possibly required replacement on the high seas.

A new concept of a direct-driven synchronous generator, which is specially suited to application in the offshore region uses various measures for an increase of the power into the 10 MW region with simultaneous reduction of gondola masses in comparison to conventional concepts [3.41]. The following features distinguish the new generator concept:

- Permanent magnet excitation with the use of highly energetic permanent magnetic material in the rotor
- Enlargement of the generator diameter for increasing the circumferential velocity
- Magnetic bearing for stabilizing the air gap
- Polyphase individual windings in the stator

The following sections describe these features.

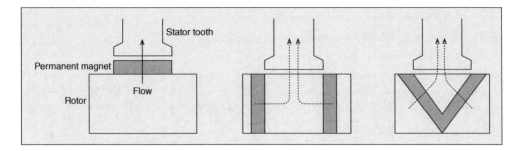

Figure 3.57   Various magnet configurations: flat (a) and collector arrangement (b,c)

3.6.4.1   High-energy magnets in the rotor

Increases in availability and reduction of mass can be shown by the use of permanent magnets in place of electrical excitation. Besides an increase of efficiency of the generator, especially in the part-load region, the high magnetic flow densities in the air gap and thus high power densities, greater pole numbers and more compact design can be achieved. Due to the technical development in the past decades, it is possible today, on the basis of alloys or rare earths (e.g. neodym–iron–boron) to achieve remanence flow densities of up to 1.4 T, greater energy densities and an advantageous de-magnetizing behavior. It is possible to achieve tangential force densities – as characteristic sizes for material utilization of a machine – in a region of $100\,kN/m^2$, whereas with conventional machines, they are only approximately $30\,kN/m^2$ and can be increased to a maximum of $70\,kN/m^2$ only by means of special cooling systems and converters [3.42].

Besides the application of highly magnetic material, an increase in the strength density is also possible by means of the concentration of the magnetic flow. In this, the increase in the field density in the air gap (field focussing) is carried out in contrast to field distribution, in that the cross-sectional area of the magnets, compared to the air gap area, is increased (flow concentration or collector arrangement) as is shown in Figure 3.57.

3.6.4.2   Enlarging the generator diameter

An enlargement of the diameter is a very effective measure for reducing the active mass (permanent magnets, iron and copper coils). With larger diameter D, there is a proportional rise in the circumferential velocity v, when a constant speed of rotation n is conditioned. In order to achieve the same power P, the result with the same force density F of, for instance, $100\,kN/m^2$, sinks in inverse proportion to the circumferential force F

$$P = F \cdot v \tag{3.28}$$

$$= F \cdot \pi n \cdot D \tag{3.29}$$

Proportionally to F, the effective area $A = \pi D l$ and the generator length, $l \sim 1/D^2$ can be reduced.

$$P = F \cdot v \tag{3.30}$$

$$= F'^1 \cdot A \cdot v \tag{3.31}$$

$$= F'^1 \cdot \pi^2 \cdot D^2 \cdot l \cdot n \tag{3.32}$$

A diameter of the generator of approximately 24 m is foreseen for the new generator concept so that for a rotation of $16\,\text{min}^{-1}$ (in the offshore region), circumferential speeds of 20 m/s are reached. The larger diameter leads to the fact that the generator can no longer be accommodated within the gondola of the wind turbine but is fixed as spoke-supported ring-construction outside the gondola (Figure 3.58). The enlarged diameter and the reduced length lead to a reduction of the active mass compared to conventional concepts. However, this effect can only be achieved when, on the one hand, the generator system (coils, magnets, etc.) is protected from environmental influences as with magnetic rail systems (Transrapid) and, on the other hand, the air gap for enlarged generator diameter is kept to a relatively small region. This is possible with the active-gap-limiting measures described in the following.

3.6.4.3 Magnetic bearing of the generator

From an electro-magnetic point of view, the smallest possible air gap of the generator is to be desired. With an increase in the diameter, the size of the air gap is increasingly influenced by the mechanical design of the generator; it increases proportionally with the diameter. Influences are exerted by the centrifugal force-conditioned expansion due to heating, material inaccuracies and operationally caused eccentricities of rotor and stator. A criterion of the dependency of the air gap on the diameter can be found in the following simple formula:

$$\delta \approx 10^{-3} \cdot D \tag{3.33}$$

The air gap mechanically necessary for a generator diameter of more than 20 m would thus be more than 20 mm, which is very large for an efficient magnetic circuit design. For this reason, measures must be taken that permit the air gap to be kept to a smaller dimension. Such a measure is the combination of the generator with a magnetic bearing that permits a resetting into the nominal position in the case of a change in the air gap. The air gap is stabilized by means of a non-controlled stable design with the effect of a resetting force that, with a deflection of the magnetic bearing, pulls it back into the stable middle position (Figure 3.59).

In addition, similarly with the transrapid, a controlled electrically exciter model of a magnetic bearing is possible. Besides the generator and bearing elements, a control electronics is necessary, whereby the effort for the power electronics and control system is low due to the generator velocities.

3.6.4.4 Polyphase individual coils in the stator

The Polyphase system works on the principle that differentiates itself fundamentally from that of the conventional three-phase machines used for the structure of the coils in the ring generator. While in such machines, the coils in the windings overlap and the width of the coil corresponds to the pole division, in this concept each coil surrounds one stator tooth (individual windings). This results in a small winding-head volume, which again, leads to small masses and losses and high current densities. In addition, this helps the segment-type stator that is necessary for this proposed machine size. Individual teeth of the same phase are installed in groups behind each other and are arranged with a spatial displacement that corresponds to the electrical phase displacement. In this way, the number of poles of the rotor and the groove count of the stator are differentiated. The trapezoidal-shaped induced voltage resulting from this form of winding structure leads to a waviness of the torque that can be reduced by an increase in the number of phases of the generator. A change-over to a four- or six-phase current system is recommended in which the torque waviness is clearly reduced in comparison to the usual three-phase systems, and also an increase in the average power can be achieved.

A design concept for this new type of ring generator concept is shown in Figure 3.58. Preliminary investigations within the framework of the *MagnetRing* research project supported by the BMU (Bundesministerium für Umwelt, German Ministry for Environment) have shown that the assumed achievable force density of $100\,kN/m^2$ can be reached within such a machine concept. For this purpose, static test models were developed and erected, in which in a group of similar-phased stator teeth discussed above, the resulting tangential and normal forces were measured. This generator section in a scale of 1:1, which is included in the model, is shown in Figure 3.60. The permanent magnets in the rotor (bottom) were arranged in the form of V with the use of the described collector concept in order to achieve a concentration of the air gap flow. The upper part contains four individual wound stator teeth. Forces can only be taken up in dependence of the air gap length, the tangential displacement of the rotor and the flow in the windings. On the basis of the measurement results as well as finite element simulations, it was possible, after some optimization, to verify the attempted force density of $100\,kN/m^2$. First projections of the active masses as well as the inactive masses necessary for the mechanical design lead one to expect a reduction of the tower head mass of the wind turbine in comparison to the usual concepts.

3.6.5 Compact superconductive and other new generator concepts

In order to further increase the power of the turbine and reduce the wind current costs, many different concepts are being tried out for developing newer generations of generators [3.2]. Besides the reworking of classical generator system, also the multiplication of the generator diameters of permanent magnet machines or increasing the energy density is being addressed with the aid of superconducting coils in order to open new perspectives for wind energy.

Besides direct-drives, one- or two-step drives with the generator form a compact hybrid unit with generator rotations in the region of 150 to 500 revolutions per minute. In new generations of hybrid method of construction, for instance, permanent magnet generators are combined with the drives in one housing. This saves couplings and masses but they make the transfer of flows and heat more difficult. As the cost development of permanent magnet materials is very difficult to estimate, the hybrid drives can offer substantial advantages. They only require approximately 20% of the magnetic materials of direct drives (800 to 1000 kg magnetic material per megawatt) [3.20]. This difference offers new perspectives to electrically excited synchronous generators. The additionally required exciter unit and larger reaction times in the control of the torque are compensated by the advantages of the high temperature sensitivity of permanent magnets.

Figure 3.58   Section of the ring generator with magnetic bearings (IWES)

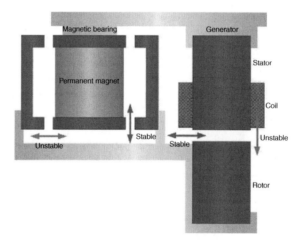

Figure 3.59   Combination of generator and magnetic bearing

Figure 3.60   Section of the ring generator in the static test rigs

Also, the development of slip ring-less double-feed asynchronous generator was taken up again. The stator, as in conventional designs, is directly connected to the grid. The rotor site converter, however, feeds the stator side flows via a coil – instead of via slip rings. Stator and rotor flows induce flows via the air gap to a closed rotor winding that call up a magnetic field and form a torque in connection with the stator field.

This can present advantages – especially in offshore applications.

The use of superconducting exciter coils, e.g. in the rotor of synchronous generators makes directly driven generators similarly compact to those used in drive systems possible [3.2]. The use of superconducting coils permits the exciter current and thus the energy density to be increased by some factors and to achieve very large magnetic fields. In this, with air gaps between rotor and stator of 20 to 50 mm, similar generator dimensions as for conventional drive designs can be achieved. The superconductors consist of a ceramic fixed to a film and covered by it. The low-loss features of the ceramic are achieved at −235 °C, which is ensured by means of a helium-cooling circuit.

3.7   Machine Data

During the planning, design and construction of plants supplying electrical drive and energy, knowledge of experimental values for the electrical machines on the market is of fundamental significance with regard to design, proportioning and economics. For new developments, in particular, manufacturers' and suppliers' ratings are necessary for the design of subsystems. Hence, data overviews are very helpful for the preselection of components and systems, as well as for the creation of preliminary sketches. Important ratings are thus given below as a function of nominal output.

3.7.1   Mass and cost relationships

The mass and cost values in relation to nominal output are shown for asynchronous (Figures 3.61 and 3.62) and synchronous (Figures 3.63 and 3.64) machines for different numbers of poles. The diagrams show that both for synchronous and asynchronous machines the specific mass and cost per kW the machine output falls with increasing size. This trend is explained by the laws of the model. These state that more favourable mass and cost relationships can always be expected for higher-output machines, given the same or similar conditions. If, however, large units give rise to very high costs due to the costs of manufacture, transport, etc., the above relationships could lead to completely different trends. Further differences in mass occur, for example, in asynchronous machines, depending upon whether these are designed with a steel or an aluminium housing.

A comparison of the diagrams in Figures 3.61 and 3.63 shows that synchronous and asynchronous machines exhibit similar mass relationships. On the other hand, the cost relationships shown in Figure 3.62 and 3.64 show considerable differences in the lower output range (e.g. 10 kW). The more favourable values for asynchronous machines can mainly be attributed to the simpler design and mass production. In the higher-output range, both designs show roughly the same cost relationships. It is worth noting at this point that the actual purchase price can deviate significantly from the values given, depending upon the quantity purchased and discount level.

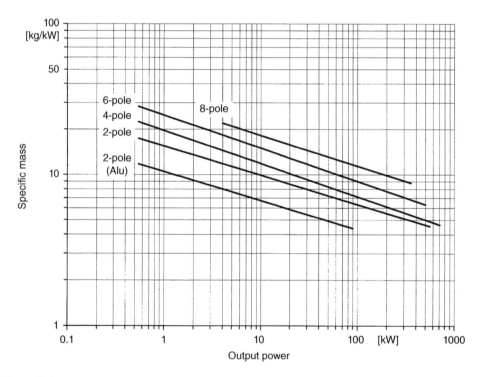

Figure 3.61   Specific mass of asynchronous machines (per kW) as a function of nominal output, with the number of poles as a parameter

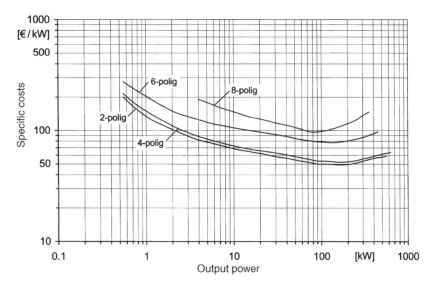

Figure 3.62   Specific costs of asynchronous machines (per kW) as a function of nominal output, with the number of poles as a parameter

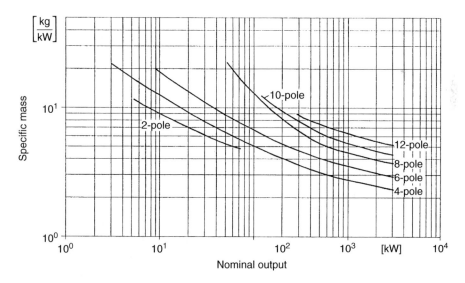

Figure 3.63   Specific mass of synchronous machines (per kW) as a function of nominal output, with the number of poles as a parameter

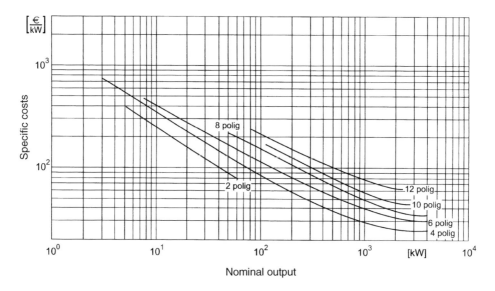

Figure 3.64 Specific costs of synchronous machines (per kW) as a function of nominal output, with the number of poles as a parameter

Furthermore, machines with a higher number of poles must obviously have a correspondingly higher mass due to their high torque, which thus leads to a comparable increase in cost.

3.7.2 Characteristic values of asynchronous machines

When asynchronous machines are used, starting currents (Figure 3.65) must be taken into account in the dimensioning of switchgear and protective gear, as well as nominal currents (see Figure 2.55). If no current limiting measures are taken, these must also be taken into consideration in the design of the drive train.

No-load current (see Section 3.6.1) is an important parameter of asynchronous machines. It is determined primarily by the design of the generator, and its uses include the dimensioning of compensation units, if these are designed for no load running of the machine, as shown in Figure 3.12(a). Figure 3.66 gives approximate reference values for no-load current. These can differ significantly from the values shown, depending upon the design and application.

Further reference values are given in Figures 3.67 and 3.68, which show efficiency and the power factor. Both diagrams show that asynchronous machines with a high number of poles generally lie below designs with a low number of poles, both for efficiency and the power factor, because of their larger air gap. Owing to the relatively large end windings of bipolar machines, their efficiency is less favourable than, for example, that of four-pole machines, but generators with a higher number of poles do tend to produce somewhat lower efficiencies.

As mentioned in Section 3.6, the size of a machine is primarily determined by the inside diameter of the stator. Figure 3.69 gives reference values for the diameter. According to reference [3.43], this is proportional to the fourth power of the machine's output.

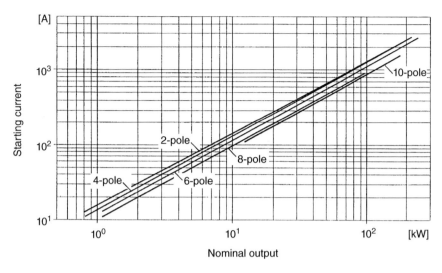

Figure 3.65   Starting current of asynchronous machines (at 400 V) as a function of the power output, with the number of poles as a parameter

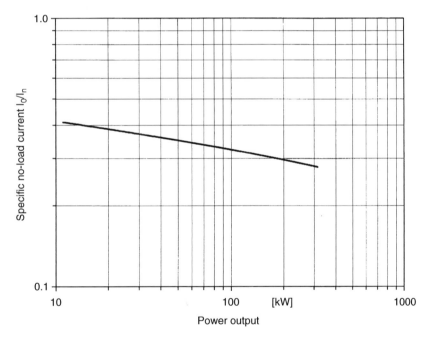

Figure 3.66   Specific no-load current of asynchronous machines as a function of nominal output at 400 V operating voltage

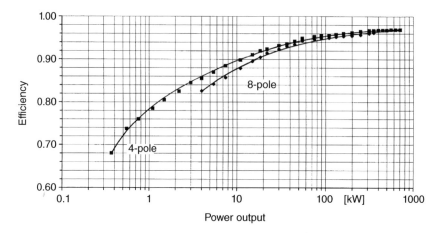

Figure 3.67 Efficiency of asynchronous machines as a function of nominal output, with the number of poles as a parameter

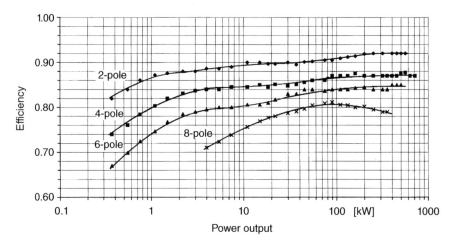

Figure 3.68 Power factor of asynchronous machines as a function of nominal output, with the number of poles as a parameter

3.7.3 Characteristic values of synchronous machines

Reference values for the efficiency of four-pole electrically excited synchronous machines are shown in Figure 3.70. The values are plotted as a function of the utilization P/P_N for nominal outputs between around 100 kW and 3 MW.

Direct-axis synchronous reactance and its transient and subtransient values are determining variables for the electrical and mechanical components of electrical and mechanical processes in synchronous machines. They are therefore shown in Figures 3.71 to 3.73, based on the nominal impedance $Z_N = U_N/I_N$ in relation to nominal output, with the number of poles as a parameter. Reference values can thus be obtained from the diagram and used for initial designs.

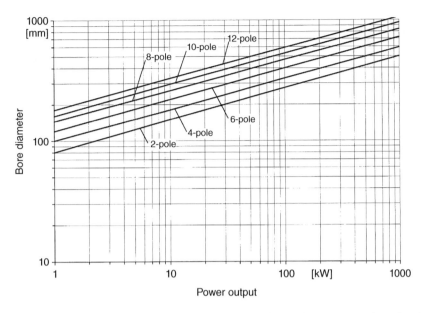

Figure 3.69 Stator bore diameter as a function of machine output, with the number of poles as a parameter

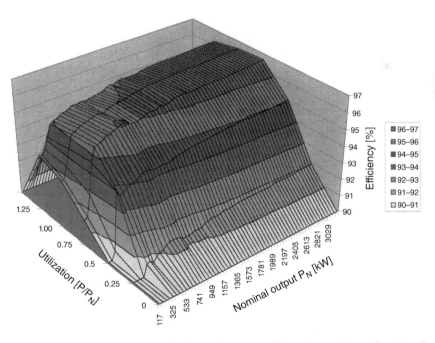

Figure 3.70 Efficiency of electrically excited synchronous machines (four-pole) as a function of output power and utilization (P/P_N)

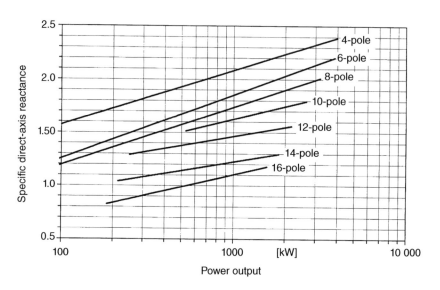

Figure 3.71   Specific direct-axis synchronous reactance of synchronous machines as a function of nominal output, with the number of poles as a parameter

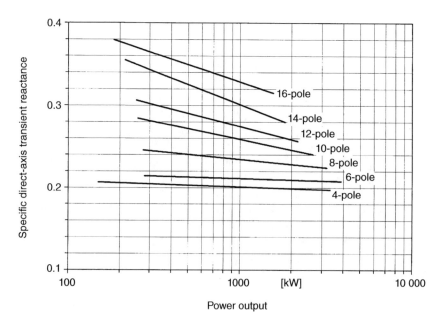

Figure 3.72   Specific direct-axis transient reactance of synchronous machines as a function of nominal output, with the number of poles as a parameter

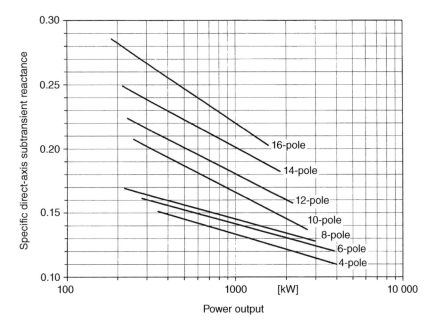

Figure 3.73 Specific direct-axis subtransient reactance of synchronous machines as a function of nominal output, with the number of poles as a parameter

However, the above-mentioned characteristic values can deviate from the guide values given here, e.g. due to construction size graduations and design variants. The machine values can, however, be obtained from the equations

$$X_d = x_d Z_N, \tag{3.34}$$

$$X_d' = x_d' Z_N \tag{3.35}$$

and

$$X_d'' = x_d'' Z_N. \tag{3.36}$$

In steady-state operation, the behavior of currents, voltages and moments of synchronous machines, in particular, is mainly determined by the so-called direct-axis synchronous reactance. As described in Section 3.3, this is made up of the magnetizing reactance in the direction of the rotor field and the leakage reactance of the stator winding. Figure 3.71 shows the increase in specific direct-axis reactance as the nominal output of the machine increases. It also illustrates the fact that generators with a high number of poles show lower direct-axis reactance.

Unlike specific direct-axis reactance, transient and subtransient values decrease in magnitude with increasing nominal output. Higher numbers of poles, on the other hand, give rise to greater values. As shown in the equivalent circuit diagrams in Figures 3.7(b) and 3.36, the synchronous direct-axis reactance is clearly greater than the transient and particularly the subtransient values.

The power conditioning of electrical energy and the connection of generators to the grid will be examined in more detail in the following chapter.

References

[3.1] Vereinigung Deutscher Elektrizitätswerke (VDEW), *Grundsätze für die Beurteilung von Netzrückwirkungen*, 3rd Revised Edn, VDEW, Frankfurt a.M., 1992; corrected reprint 1997.

[3.2] Heier, S. and Kleinkauf, W., Windpark Kythnos, in International Energy Agency (IEA) Conference, Palo Alto, California, October 1983.

[3.3] König, V., *Simulation der dynamischen Betriebseigenschaften einer Asynchronmaschine mit Schleifringläufer beim Einsatz in Windkraftanlagen*, Thesis, Kassel University, 1992.

[3.4] Heier, S., *Windenergiekonverter und mechanische Energiewandler. Anpassung und Regelung, Tagung: Energie vom Wind*, Deutsche Gesellschaft für Sonnenenergie, Bremen, 1977.

[3.5] Voith Turbo Wind: Development WinDrive. Elektronisches Dokument, 2008.

[3.6] Kovacs, K.P. and Racz, J., *Transiente Vorgänge in Wechselstrommaschinen*, Vols 1 and 2, Hungarian Academy of Science, Budapest, 1959.

[3.7] Wüterich, W., Übersicht über die Einschaltmomente bei Asynchronmaschinen im Still-stand, *Elektrotechnische Zeitschrift-A*, 1967, 88, 555–9.

[3.8] Schuisky, W., *Induktionsmaschinen*, Springer, Vienna, 1957.

[3.9] Jordan, H., Lorenzen, H.W. and Taegen, F., Über den asynchronen Anlauf von Synchronmaschinen, *Elektrotechnische Zeitschrift-A*, 1964, 85, 296–305.

[3.10] Keil, F., Zur Berechnung des asynchronen Anlaufs von Schenkelpol-Synchronmaschinen mit massiven Polen, *Elektrotechnische Zeitschrift-A*, 1969, 90, 396–9.

[3.11] Siemens AG, *Formel-und Tabellenbuch für Starkstrom-Ingenieure*, 3rd Edn, Girardet, Essen, 1965.

[3.12] Leonhard, W., Vollstedt, W., et al., *Windenergieanlagen im Verbundbetrieb. Verbundbetrieb von großen Windenergieanlagen mit Gleichstrom- und Drehstromsammelschiene*, Bundesministerium für Forschung und Technologie, Statusreport Windenergie, Lübeck, 1988, pp. 1–20.

[3.13] Caselitz, P., Heier, S. et al., *PV Simplorer*, Zwischenbericht, 2001.

[3.14] Leonhard, W., *Regelung in der elektrischen Energieversorgung*, Teubner, Stuttgart, 1980.

[3.15] Kleinrath, H., *Stromrichtergespeiste Drehfeldmaschinen*, Springer, Vienna, 1980.

[3.16] Richter, R., *Elektrische Maschinen*, Vols I–IV, Birkhäuser, Basel/Stuttgart, 1967.

[3.17] Bödefeld, T. and Sequenz, H., *Elektrische Maschinen*, 8th Edn, Springer, Vienna, 1971.

[3.18] Weiher Electric, *Generatorsysteme für Windenergieanlagen Prospekt*.

[3.19] Weiher, Spezialentwicklung für den drehzahlvariablen WKA-Betrieb. Bürstenloser doppeltgespeister Asynchrongenerator für Pitchanlagen, *Wind Kraft Journal*, 1998, 2/98, 42.

[3.20] Engel, U. and Wickboldt, H., Explosionsgeschützte Drehstrommotoren und die neuen Normspannungen, *Elektrotechnische Zeitschrift-A*, 1991, 112, 1082–6.

[3.21] Weiher Electric GmbH, Weiher Vari Slip-Generator erfolgreich auf einer Megawatt-Stallanlage getestet, *Wind Kraft Journal*, 2000, 6/2000, 44.

[3.22] Seinsch, H.-O, *Oberfeldschwingungen in Drehfeldmaschinen*, Teubner, Stuttgart, 1992.

[3.23] Brach, K., *Wellenspannungen bei Drehstrominduktionsmaschinen mit Käfigläufer*, Dissertation, TH Hannover, 1990.

[3.24] DIN VDE 0530, Beiblatt 2, *Umrichtergespeiste Induktionsmotoren mit Käfigläufer*, Application Guidelines (IEC 60034-17), January 1999.

[3.25] Chen, S. and Lipo, T., Circulating Type Motor Bearing Current in Inverter Drives, in IEEE IAS Annual Meeting, 1996.

[3.26] Binder, A. and Schrepfer, A., Lagerströme bei umrichtergespeisten Drehstrommotoren, *Antriebstechnik*, 1999, 38.

[3.27] Herrmann, T. and Wimmer, J., Wellenspannungen und Lagerströme bei Drehstrommaschinen in Verbindung mit Umrichterbetrieb, *Wind Kraft Journal*, 2000, 6/2000, 30–3.

[3.28] Wimmer, J. and Stadler, H., Lagerströme bei umrichtergespeisten Drehstrommaschinen, *Antriebstechnik*, 2000, 39.

[3.29] Hansberg, V. and Seinsch, H.-O., Kapazitive Lagerspannungen und -ströme bei umrichter-gespeisten Induktionsmaschinen, *Electrical Engineering*, 2000, 82.

[3.30] Vensys brochure.

[3.31] Weh, H., Permanenterregte Synchronmaschinen hoher Kraftdichte nach dem Transversalflußkonzept, *Elektrotechnische Zeitschrift Archiv*, 1988, 10(5), 143–9.

[3.32] Hill, W., *Konzeption und Bau eines getriebelosen Windgenerators*, Thesis, Kassel University, 1995.

[3.33] Heier, S., Hill, W. and Kleinkauf, W., Neuartige Konzeption eines permanenterregten Vielpolgenerators für getriebelose Windkraftanlagen, in 3 Deutsche Windenergie-Konferenz DEWEK 96, Deutsches Windenergie Institut, Wilhelmshaven, 23–24 October 1996, pp. 121–4.

[3.34] Weh, H., Hoffmann, H, Landrath, J., Mosebach, H. and Poschadel, J., Directly-Driven Permanent-Magnet Excited Synchronous Generator for Variable Speed Operation, in European Community Wind Energy Conference, Herning, Denmark, June 1988, pp. 566–72.

[3.35] Enßlin, C., Durstewitz, M., Heier, S. and Hoppe-Kilpper, M., Wind Farms in the German '250 MW Wind'-Programme, in European Wind Energy Association Special Topic Conference '92 on *The Potential of Wind Farms*, Herning, Denmark, 1992, pp. B4.1–B4.7.

[3.36] Jöckel, S., Gearless Wind Energy Converters with Permanent Magnet Generators – An Option for the Future?, in European Wind Energy Confererence, Göteborg, Sweden, pp. 414–7.

[3.37] Heldele GmbH: Konzept einer getriebelosen Windenergieanlage mit Multigeneratorausführung Elektronisches Dokument. Online verfügbar unter http://heldele.de, besucht am 12.5.2010.

[3.38] Michalke, Gabriele: Variable Speed Wind Turbines- Modelling, Control, and Impact in Power System (Dissertation), 2008.

[3.39] Thalemann, Fabian: Untersuchung von Torsionsschwingungen am Triebstrang von Windenergieanlagen bei Netzfehlern, Master Thesis Universität Kassel, 2011.

[3.40] Mascioni, A.: Reduktion der Turmkopfmasse und Erhöhung der Wirtschaftlichkeit bei getriebelosen Windenergieanlagen. In: VDI-Wissensforum, 2010.

[3.41] WEH, H.: Windkraftanlagen mit Ringgenerator. Offenlegungsschrift zur Patentanmeldung, 2007.

[3.42] Jöckel, S.: Calculation of different generator systems for wind turbines with particular reference to low-speed permanent-magnet machines. Dissertation, 2001.

[3.43] Nürnberg, W., *Die Asynchronmaschine. Ihre Theorie unter besonderer Berücksichtigung der Keilstab- und Doppelkäfigläufer*, Springer, Berlin, Göttingen, Heidelberg, 1963.

[3.44] Klamt, J., *Berechnung und Bemessung elektrischer Maschinen*, Springer, Berlin/Göttingen/Heidelberg, 1962.

4

The Transfer of Electrical Energy to the Supply Grid

With regard to the transfer of energy to electrical supply installations, we must differentiate between

- systems with limited supply options that either operate in isolation or supply weak grids and
- unlimited capacity connection with the rigid grid.

Wind energy converters should give reliable operation in both areas of application.

Due to its very high capacity (in comparison with the nominal values of the consumers connected to it), the so-called rigid combined grid can be regarded both as an infinitely rich source of active and reactive current and, for the low-level energy supply devices that wind turbines usually represent, as a sink of unlimited capacity with constant voltage and frequency.

Unlike thermal power plants, wind turbines are usually installed at remote sites with limited supply options. Therefore a weak grid connection is often made using dead-end feeders, which are sometimes long. In large wind energy converters and wind farms, supply power can reach the same order of magnitude as grid transfer power, or can even approach its level, which means that mutual influences must be taken into account. Table 4.1 lists, in simplified form, the requirements and equipment needed for the connection of wind energy converters (WECs) to the grid [4.1].

On the one hand the effects of wind energy converters on the grid are determined, on the one hand, by power conditioning and the resulting grid connection. On the other hand, safety aspects and grid protection can be affected by the influence on protective equipment and short-circuit power, and the function of circuitry can be impaired. Furthermore, grid feedback is possible, which can cause changes to harmonics and voltages and can affect grid regulation.

4.1 Power Conditioning and Grid Connection

Power conditioning, as well as energy conversion, represented a decisive milestone in the development of wind energy technology [4.3]. In the 1980s, generator and grid connection

Grid Integration of Wind Energy: Onshore and Offshore Conversion Systems, Third Edition. Siegfried Heier.
© 2014 John Wiley & Sons, Ltd. Published 2014 by John Wiley & Sons, Ltd.

Table 4.1 Requirements and equipments for the grid connection of wind turbines

Grid connection	Disconnection point in accordance with DIN/VDE 0105 available to the electricity supply company at all times
Switchgear	*Section switch* with at least power circuit-breaking capacity *Design* for maximum short circuit current (WEC, grid) *Inverter*: Connection point on grid side
Protective gear	*Fundamental parameterization of excitation units* [4.2]
	– Undervoltage protection, Range: $0.1 \ldots 1.0 \cdot U_N$
	– Overvoltage protection, Range: $1.0 \ldots 1.3 \cdot U_N$
	– Underfrequency protection, Range: $47 \ldots 50$ Hz
	– Overfrequency protection, Range: $50 \ldots 52$ Hz
Reactive power compensation	Power factor in the range 0.95 capacitive to 0.95 inductive plants power. Switching and type of controls of the reactive power are to be coordinated with the electricity supply company
Connection conditions	Basic limiting values to be adhered to
	– Grid voltage, Range: $U_N > 0.95 U_C$
	– Grid frequency, Range: $47.5 \ldots 50.05$ Hz
	Synchronous generator and double-feed asynchronous generator
	– Synchronizing device necessary
	• Voltage difference: Ä$U \pm 10\%\ U_N$
	• Frequency difference: Ä$f \pm 0.5$ Hz
	• Phase difference: Ä$\varphi \pm 10°$
Grid feedback	*Asynchronous generators*
	– No-voltage connection in the range $(0.95 \ldots 1.05) \cdot n_{syn}$
	– For motorized starting: limitation of starting current
	Voltage swings and flicker
	– Undisturbed operation. Ä$_{ua} \leq 2\ \%$ d. voltage without excitation installation
	– Quick changes, Ä$u_{max} \leq 2\%\ U_C$
	– Long-term flicker strength, $P_{lt} \leq 0.49$
Commissioning	*Harmonics must be limited*
	The operation of ripple control systems must not be impaired Testing:
	– Disconnection devices
	– Meters
	– Short circuit and uncoupling protection
	– Interfaces to grid operator
	– Reducing the feed capacity
	– Monitoring the feed capacitay

Extracts from the technical guidelines *Generating plant of medium-voltage grids (June 2008)*

designs for wind energy converters were based upon conventional electrical energy supply installations and had rigid grid connection systems. Only high-output pilot plants were constructed and operated as variable-speed units. This configuration did not make a breakthrough until electronic power converters were fitted to the 50 kW class of wind power plants at the end of the 1980s. Development progressed from the cheap six-pulse converters with thyristors through quasi-twelve-pulse circuits to the so-called pulse-controlled converters with semiconductor switches operating in the kilohertz range.

For a few years now there has been a trend away from robust single systems, mainly characterized by stall-controlled turbines with asynchronous generators and direct connection to the grid, towards more expensive units. Synchronous machines, often based on gearless, ring-type designs with controlled or machine-commutated rectifiers, direct current links and self-commutated inverters, are favoured in these machines. Double-fed asynchronous generators, on the other hand, permit similar speed-variation ranges with considerably smaller converter systems in reactive current adjustment ranges equivalent to the converter output. The gears necessary in these machines make it possible for manufacturers of what were conventional wind energy converter systems with fixed-speed, usually stall-regulated turbines and asynchronous generators to make an extremely simple transition to innovative variable-speed mechanical–electrical conversion units.

The decisive advantage of significantly lower converter outputs previously had to be traded off against the considerably greater cost for measurement, calculation and control technology, in particular for field-oriented machine supply. The rapid development and enormous price reductions in these fields are, however, opening up, improving prospects for this technology. The use of such systems, even at a high cost, is justified if, by adjusting the turbine speed to the prevailing wind speed, the compatibility of the plant to the environment and to the grid can be improved, leading to a higher energy output and reduced drive-train loading.

This type of system also requires a converter system that is capable of conditioning the variable-frequency electrical energy from the turbine generator for supply to a grid of (almost) constant frequency and voltage.

4.1.1   Converter systems

Power electronic converters, so-called power converters, are the most common solution for the conversion and control of electrical energy. They are also used to an increasing degree in wind energy converters to adjust the generator frequency and voltage to those of the grid, particularly in variable-speed systems. Wind turbines connected through converters, either fully via the stator or partially via the rotor circuit, are increasingly common, particularly in systems of the megawatt class.

Power converters have significant advantages over the rotating transformers based on groups of mechanical components or mechanical commutators that were common in the past, namely

- low-loss energy conversion,
- rapid operator intervention and high dynamic response,
- wear-free operation,
- low maintenance requirement and
- low volume and weight.

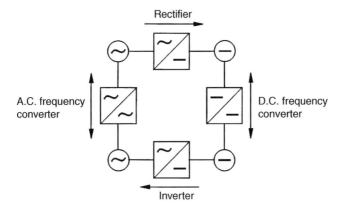

Figure 4.1   Electrical energy conversion by power converters

Figure 4.1 shows the options for low-wear energy conversion using power converters. These are defined as follows. Rectifiers convert single-phase or three-phase alternating current into direct current, with the electrical energy flowing from single-phase or three-phase alternating current systems into direct current systems. Inverters convert direct current into single-phase or three-phase alternating current. The energy flows into the alternating-current side. Direct current conversion is the conversion of direct current with a given voltage and polarity for use in a direct current system with a different voltage and possibly reversed polarity. In alternating current conversion, alternating current of a given voltage, frequency and number of phases is converted for use in an alternating current system with a different voltage, frequency and possibly a different number of phases.

The main components of power conversion systems are the power section, with so-called power converter valves, which carries the electrical power, and an electronic signal processing unit, which performs numerous control, protective and regulating tasks. Over the past few decades, great progress has been made in improving the efficiency of both parts of the converter, owing to the rapid development of semiconductor and digital technology. Basic designs are described in more detail in references [4.4–4.7].

As wind turbines are almost always fitted with three-phase generators, only three-phase converters are relevant for power conditioning. We shall therefore limit the following discussions to these. We differentiate here between

- direct converters and
- indirect converters.

Direct converters are used particularly for the reduction of frequency. In the case of supply from or to a 50 Hz grid, the operating range of 0 to 25 Hz [4.4, 4.8] is preferred. Developments according to reference [4.9], however, also permit the conversion of frequency of the same order of magnitude. Direct converters require two complete antiparallel power conversion bridges per phase to operate the consumer and supply systems (Figure 4.2). This results in high costs for power gates and control elements.

The conversion of grid frequency f_1 into machine frequency f_2, or vice versa, in a direct converter takes place by the selection of voltage sections from the three phases (Figure 4.3) and

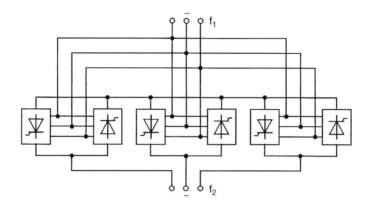

Figure 4.2   Block diagram for direct converters

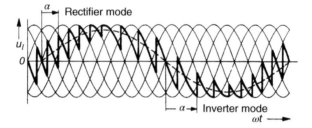

Figure 4.3   Voltage in one phase with a direct converter

by triggering the power converter such that the voltage path after smoothing has the amplitude, phase position and frequency required by the machine. Transfer via converter thus allows active and reactive power to be both supplied and drawn. Figure 4.3 illustrates the schematic voltage path and frequency conversion achieved over one phase (U_1). Rectifier and inverter operation can alternate depending upon the load current after a zero crossing of the current with a certain delay for the duration of a half-period.

Indirect converters consist of a rectifier, constant-voltage or constant-current d.c. link and an inverter. A converter with a constant-current d.c. link will be referred to as an I converter and one with a constant-voltage d.c. link will be referred to as a U converter. Particular characteristics of the link (Figure 4.4) are

- the inductor for current smoothing in the I converter and
- the capacitor for voltage smoothing in the U converter.

Indirect converters have achieved a clear dominance in energy conversion and the connection of variable-speed wind turbines to the grid. Direct converters have only been used in individual cases to supply the rotor circuit of double-fed asynchronous generators. The following discussions therefore concentrate on the indirect converter. We also briefly describe the fundamental characteristics of power semiconductors and important power converter components.

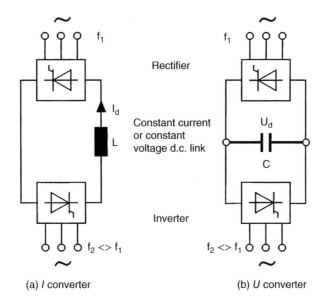

(a) *I* converter   (b) *U* converter

Figure 4.4   Indirect converters

4.1.2   *Power semiconductors for converters*

So-called power converter valves are the main components of the power section of converters. They consist of one or more power semiconductors and conduct electrical current in one direction only. These valves generally alternate periodically between the electrically conductive and nonconductive states, and therefore function primarily as switches. As no mechanical contacts have to be operated they can initiate and/or terminate current conduction very rapidly (i.e. in the microsecond range).

Power converter valves can be either controllable or noncontrollable. Noncontrollable valves (diodes for example) conduct in the forward direction and block in the reverse direction. Controllable valves permit the selection of the moment at which conductivity in the forward direction begins. Thyristors can be switched on by their gate and block if the direction of the current is reversed. Switchable thyristors and transistors, on the other hand, can be switched on by one gate electrode and off by a second, or the same, gate.

4.1.2.1   Semiconductor diodes

Diodes consist of positively (p) and negatively (n) doped semiconductor material with a barrier layer between them that ensures current can flow in one direction only. This is possible if the diode voltage is positive. If the current direction and voltage are reversed, the diode becomes nonconducting and blocks the flow of current. Diodes can therefore only be used in uncontrolled rectifiers and for protective and backup functions, e.g. in the form of freewheeling diodes in direct current circuits or similar circuit elements.

In addition to limit values for current and voltage in the forward and reverse directions, and its thermal behavior, another decisive variable (particularly for protective functions) is conducting-state dynamic behavior. For the effective protection of semiconductor components, so-called fast-recovery diodes with low stored charges are necessary to protect power converter valves from destruction by overvoltage.

4.1.2.2 Thyristors

Thyristors are semiconductor components with four differently (p and n) doped layers. Conventional thyristors, the gate turn-off (GTO) thyristor, the metal oxide semiconductor-controlled thyristor (MCT) and the Integrated Gate Commutated Transistor (IGCT), are the main types used in converters, and are described briefly in what follows.

Thyristors, unlike diodes, do not automatically go into a conducting state when a positive anode–cathode voltage is applied. The transition from blocking to conducting state is initiated by the supply of a current impulse to the gate, and is known as the 'firing' of the thyristor. Once fired, thyristors behave like diodes. They remain in the conducting state as long as a current flows in the positive direction and does not fall below the component's minimum value, the so-called holding current. If a thyristor is in the off-state, it can be fired by a new current pulse or periodic pulse sequences at the gate.

In conventional thyristors it is not possible to stop the flow of current by intervention at the gate. Switchable thyristors, on the other hand, do permit this. The best-known type is the gate turn-off, or GTO, thyristor. This type of thyristor requires a freewheeling arm for uninterrupted current. Despite the option of switching off and the resulting advantages for control and regulation, GTO thyristors have not yet achieved the market share that was expected for them.

The metal oxide semiconductor-controlled thyristor, abbreviated to MCT, behaves in a similar manner to the GTO thyristor. The MCT can be switched on almost without power by a negative voltage (in relation to the anode) at the gate. A positive gate voltage switches it off, and at the current's zero crossover it automatically switches to block operation.

Integrated Gate Commutated Transistors (IGCTs) represent a further development of the GTO thyristor. They combine the very good conducting-state behavior of thyristors with the switching capability of bipolar transistors. Their hard gate control means that when they are switched off they go directly from the thyristor mode into the transistor mode, so that no additional protective circuit is necessary. In the reverse conducting, disc-type variant the freewheeling diode that is generally necessary for converter circuits is integrated into the silicon chip. In the asymmetrically blocking variant an external diode is also required. The rate of change of the current is largely determined by the turn-off behavior of the freewheeling diode and is set by an inductor between the inverter bridge and the d.c. link. Since single switches have no electrical insulation from earth they are mounted on separate coolers. These are electrically isolated. The firing circuits are incorporated into the IGCT. Further tuning is thus unnecessary. The fact that the materials in the discs of the IGCT are not rigidly connected means that they can move slightly in relation to one another. This leads to a high service life with some 100 000 temperature cycles.

The MCT has not yet achieved any significant degree of popularity. However, its further development to higher off-state voltages could lead to it becoming a good alternative to the insulated-gate bipolar transistor(IGBT) [4.10, 4.11] in the middle-output range, which will be briefly described in the next section.

4.1.2.3 Transistors

Transistors are semiconductor components with three differently (p and n) doped layers. Bipolar, metal oxide semiconductor field-effect and integrated gate bipolar transistors are the main types used in converters. As valve components they function exclusively as switches.

Bipolar transistors (BPTs), in their function as power semiconductors, are usually used in the emitter mode. This allows a high level of power amplification to be achieved. Almost like switches, they become conductive when a control current is passed through the base electrode. When switched off, the on-state of the transistor is terminated and the flow of current blocked. In order to achieve a low on-state voltage, and thus low losses, transistors are operated with a relatively high base current. The transistors therefore operate in the so-called saturation range.

Much smaller control currents are needed for Metal Oxide Semiconductor Field-Effect Transistors (MOSFETs) than those for bipolar transistors. MOSFETs can be switched almost without power by voltage control at the gate. This, however, requires that the internal capacitances of the transistor be recharged. Increasing the switching frequency causes more frequent charge reversals and thus higher losses in the driver. MOSFETs are used in the lower-output range at high switching frequencies (see Table 4.2) for switched-mode power supplies and converters, and have advantages over bipolar transistors and IGBTs, particularly at high switching frequencies.

Insulated gate bipolar transistors (IGBTs) combine the advantageous characteristics of MOSFETs and bipolar power transistors. The field-effect transistor at the control input facilitates rapid switching at very low driving power. IGBTs automatically limit increases in the current at the output. This results in good overcurrent and short-circuit characteristics. Integrated freewheeling diodes protect the transistor in the off-state mode. Different types of IGBT are used as individual transistors or are connected together in modules of two to six transistors to form bridge connections. Normally, transistors are built into modules with driver

Table 4.2 Characteristics and maximum ratings of switchable power semiconductors (maximum rating is currently not achieved simultaneously)

	Component					
	BPT	IGBT	MOSFET	MCT	GTO	IGCT
Symbol						
	Maximum rating					
Voltage (V)	1200	6500	1200	6000	6000	6000
Current (A)	800	3600	700	600	6000	6000
Output (kVA)[a]	480	4000	70	2400	24 000	24 000
Turn-off time (μs)[b]	15–25	1–4	0.3–0.5	5–10	10–25	10–15
Frequency range (kHz)	0.5–5	2–20	5–100	1–3	0.2–1	–2
Drive requirement	Medium	Low	Low	Low	High	High

[a]Ideal switching capacity per switch ($U_{d\,max}\,I_N$).
[b]Including delay times and partial current phase.

switches, protective switches and electrical insulation. IGBTs can be connected in parallel. However, this requires that all transistors exhibit the same thermal behavior.

The development and availability of new power electronic semiconductor components has given new impetus to converter technology and its application in the field of drive and power engineering. Particularly in the small- and medium-output range, new components (IGBTs) have largely pushed transistors and GTOs out of the market. IGCTs will gain more significance in this output spectrum in the future. Table 4.2 shows the symbols, maximum ratings and characteristics of the switchable power semiconductors described briefly above [4.12] that are currently available on the market.

4.1.3 Functional characteristics of power converters

The main components of power converters are the power converter valves and their electrical connections and trigger equipment. Also necessary are protective elements, energy stores, auxiliary devices and devices for commutation, filtering, cooling and protection, and usually also transformers. It is not possible to describe the many connection options and the complex relationships for the construction and operation of power converters within the framework of the discussion below. This will be limited to the basic three-phase current variants that can be used in wind energy converters. In addition, the most important fundamental circuits will be briefly illustrated, based on the control and timing of converters (Figure 4.5).

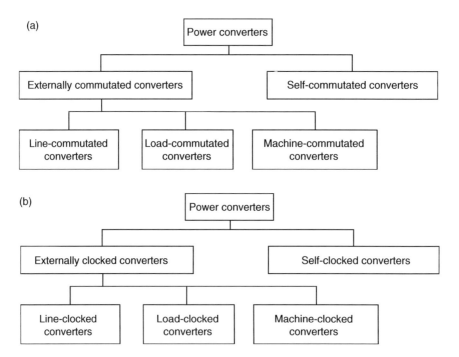

Figure 4.5   Differentiation between power converters according to the origin of (a) the commutation voltage and (b) the elementary frequency

Power converters must be commutated in their voltage and clocked with the corresponding frequency. The origin of the commutation voltage and commutation reactive power at the current transition to another valve is decisive for voltage commutation. Externally commutated power converters operate using natural commutation. They require a grid, load or machine that specifies the voltage and can supply reactive power (Figure 4.5). Self-commutated converters, on the other hand, operate with forced commutation. The required reactive power is provided by capacitors.

The internal function of power converters must also be differentiated with regard to the origin of the clock frequency. Externally clocked power converters take their control pulse from the system with which they work in parallel. Line clocking is the adjustment of the zero-crossings or phase intersections to match those of the grid voltage. Thus the load- or machine-clocked power converter orients itself to the load or machine voltage. Self-clocked power converters have an internal clock generator and are thus not dependent upon external frequency information.

As well as the commutation voltage and clock frequency, the so-called pulse number, the number of nonsimultaneous current transitions (commutations) from one valve to another within one cycle, is an important parameter of power converter circuits. Three and six (Figure 4.6), as well as twelve, pulse connections are normal for three-phase systems. The pulse number is characterized by the number of sine peaks (pulses) of the unsmoothed direct current.

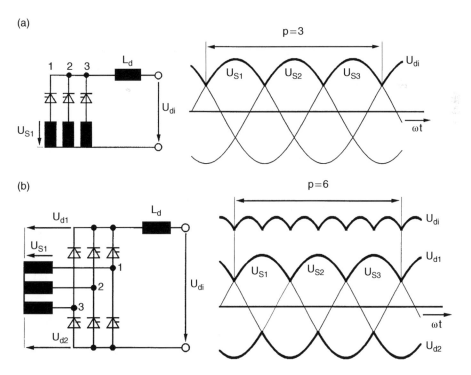

Figure 4.6 Circuit and pulse number of power converter circuits: (a) a three-pulse star connection and (b) a six-pulse bridge connection

One of the most important and commonest power converter fundamental circuits is the six-pulse three-phase bridge connection. Figure 4.6(b) shows the block diagram of this circuit with thyristors and the corresponding uncontrolled voltage path. The fully controlled six-pulse three-phase bridge circuit represents the most common variant of line-commutated power converters and of rectifiers for frequency conversion.

Commutation, the transfer of current between the individual valves, can occur in different ways. If the conducting valve is turned off before the next valve is fired then for a brief period no current flows in the connection. As gaps occur in the direct current, this process is known as the intermittent flow. In contrast, it is possible to fire a second valve while the valve to be turned off is still conducting. This creates a temporary short-circuit between two alternating current lines. The current in the valve to be turned off is quickly forced beneath its holding point. This interrupts the short-circuit before the operating current is exceeded. This changeover is known as the commutating operation.

4.1.3.1   Line commutation

Thyristors, unlike diodes, can block positive voltages if they are not fired. By firing the thyristors, it is thus possible to delay the connection or current flow by a specified time $t_v \geq 0$, or by the angle $\alpha = \omega t_v$, compared with the natural point of ignition determined by the grid. Thus three-phase bridge connections can function as rectifiers or inverters. Their operating state is dependent upon the trigger delay angle α. This is the same in all three phases for steady-state operation (Figure 4.7).

Thyristors can only allow current to flow in one direction. The voltage direction depends upon the trigger delay angle. In the range $0 \leq \alpha \leq 90°$ the power converter works in the rectifier mode (Figure 4.8). The (arithmetic) mean values of direct voltage and direct current have the same polarity sign. Thus a connected consumer is supplied. For firing angles $\alpha > 90°$, the direct voltage mean value is reversed. The power converter thus operates in the inverter mode.

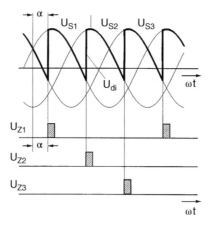

Figure 4.7   Path of unsmoothed direct voltage U_{di} and firing pulses U_{Z1}, U_{Z2}, U_{Z3} and the valve trigger delay angle α

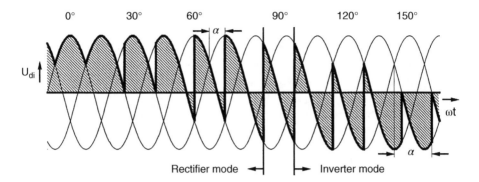

Figure 4.8   Voltages and trigger delay angles of power converters in the rectifer and inverter modes

In principle, the trigger delay angle α can be varied between 0 and 180°. However, before valves can accept a positive blocking voltage, they need a so-called circuit-commutated recovery time. If this is not adhered to, there are still enough charge carriers in the thyristor to reignite a positive voltage immediately after return. The current then commutates back to the valve, which is not completely turned off, and the inverter shoots through. The maximum value of the trigger delay angle is called the inverter stability limit, and in normal operation this lies at around $\alpha_{max} = 150°$. If voltage changes are expected in grids supplied via converters then this value must be reduced accordingly. In the event of 15% voltage dips, the inverter stability limit lies at approximately $\alpha_{max\,15} = 138°$.

If the converter has to facilitate current as well as voltage reversal, antiparallel valves, as shown in Figure 4.9, are necessary. The black part of the valve is for one current direction and the white part for the other direction. They must also be triggered accordingly.

In the steady-state mode all valves of a bridge connection operate with the same delay angle, as shown in Figure 4.7. Thus, in the course of one period, each thyristor in the positive and negative bridge half carries the current for one-third of the duration of the period in the triggered state. For the ideal, smoothed direct current (with the aid of inductance L_d as shown in Figure 4.6), the grid current in one phase is found to be rectangular blocks of 120° length, which are positive during the first half-period and, after a 60° interruption, are switched over to the

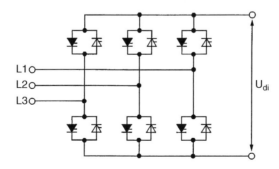

Figure 4.9   Two-way inverter with antiparallel valves

negative side of the alternating current line. This current form deviates significantly from the sinusoidal path. The grid current thus contains strong harmonics. This results in corresponding grid effects (Section 4.3.5).

Line-commutated power converters require commutation reactive power and phase-control reactive power for the control of energy systems. In driving the power converters, the current flow is delayed in relation to the grid voltage by the trigger delay angle α. Due to this phase shift, phase-control reactive power is required, which, if we disregard the commutation reactive power, is roughly proportional to the cosine of the delay angle α. As the phase shift φ of the fundamental frequency current, compared with the adjoining voltage, is approximately equal to the delay angle α, the rectifier is operated close to $\alpha = 0$ and the inverter is operated as close as possible to the inverter stability limit in order to keep the requirement for the fundamental component reactive power low.

One significant disadvantage of the line-commutated power converter is that intervention for the purpose of regulation is only possible at a few discrete points in time within a cycle. Control commands between these times will only be reacted upon after a delay – a so-called dead time. Their statistical mean is half the period divided by the pulse number. For six-pulse three-phase bridge connections at 50 Hz, the mean dead time is 1.67 ms.

4.1.3.2 Self-commutated systems

In addition to control by the control unit, self-commutated power converters require that the necessary commutations can take place independently of the available back-electromotive force (e.m.f.). The power converter valves must therefore be switchable at any chosen point in time.

When using conventional thyristors, which cannot switch themselves off automatically, a special reset switch is necessary. This disconnects the current-carrying valve at any selected point in time. Reset circuits can have different layouts. They are made up of a commutating capacitor, a GTO thyristor and a commutating inductance, arranged in parallel to the main thyristor.

Self-commutated power converter systems can benefit from the inclusion of transistors. At the appropriate operating frequency, the grid effects and effects on the supplied machines can be made much more favourable. The present market situation, with a clear trend towards transistor converter systems, reflects this. As the self-commutated power converters that come into consideration here are designed with a d.c. link, this layout will be discussed further in the next-but-one section.

4.1.4 Converter designs

The grid connection of wind turbines is basically determined by generator characteristics and permissible grid effects. When using converters, the conversion losses and the extra costs of the technology must be covered by correspondingly higher energy yields. Therefore, low losses and low costs must be the aim for feeding power. Grid characteristics at the point of connection take on a particular significance in this context.

The power supplied to the grid should approximate a sinusoidal path, exhibiting low harmonics. The control and stabilization of the tendency to oscillate, particularly in permanent-magnet synchronous machines, by the appropriate influencing of the generator current is necessary to guarantee reliable plant operation. Moreover, it is desirable to implement grid frequency converters in the form of active reactive power filters (filter consisting of semiconductor devices) or as a unit to support or limit the grid voltage.

Frequency converter systems are used to supply the current generated over a wide range of frequencies by wind generators operated at variable speeds to a constant-frequency grid. Very different demands can be placed on the frequency converter depending upon the generator design. As described in Chapter 3, synchronous machines are mainly used in this application.

Generators with excitation windings allow voltage to be controlled at the terminals or in frequency converter branches by excitation devices. Permanent-magnet machines do not provide this option.

Depending upon the design of the frequency converter, the operation of the generator and the d.c. link can be influenced. This is not immediately possible when uncontrolled rectifier bridges are used (Figure 4.10(a)). Controlled thyristor rectifiers facilitate control interventions, which are able to adapt a power output and energy situation in the d.c. link within a certain range (Figure 4.10(a) and (b)). Using pulse frequency converters, on the other hand, the generator states can be used on the basis of the magnitude and phase of the generator current, to change and regulate the d.c. link voltage (Figure 4.10(c)).

For cost reasons, generators in the 100 kW and megawatt range are designed for output voltage in the low-voltage range (below 1000 V). Dimensioning for between 500 and 800 V is common. Generator designs in the medium-voltage range for offshore use [4.12] facilitate new, economical grid supply concepts. This is only possible if semiconductor components with the required current and voltage data for the selected frequency converter design are available. If this is not the case, several valves or power converter branches, for example, must be connected in parallel (see Figure 4.16). The parallel connection of identical frequency converter units can achieve further operating advantages, as described in Section 4.1.5.5, and grid effects can be minimized owing to the phase differences between the individual power converters.

Further system criteria, determined by the requirements of the generator and grid, must be observed in the design of the frequency converter. Electromagnetic compatibility, system losses, protective measures in normal operation and in the case of failure, and the control of energy flow into the grid are of particular importance here. The concepts and designs that follow relate to the most commonly used type – the indirect converter.

4.1.5   Indirect converter

As shown in Figure 4.11, the power branch of systems for the grid connection of wind energy converters that operate using a constant-current or constant-voltage d.c. link can be subdivided into three subsystems. These will be described briefly in what follows. A good converter design is also based upon the recording of state variables, as well as turbine management and monitoring.

A generator-side rectifier (RE) converts the generator voltage or current to a d.c. link value and controls the generator operation and thus also the wind turbine. Decisive variables here are the turbine and generator driving torque, rotation speed, rotation angle, and generator voltage and current. The d.c. link (DCL) with energy storage enables the grid frequency to be

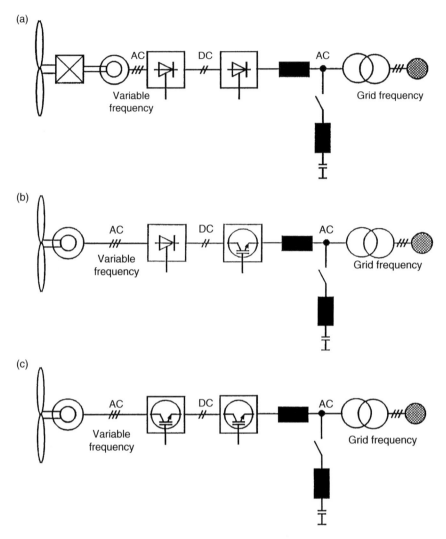

Figure 4.10   Converter systems (with a d.c. link) for the grid connection of wind turbines: (a) an uncontrolled or controlled thyristor rectifier with a line-commutated thyristor inverter; (b) an uncontrolled or controlled thyristor rectifier with a self-commutated switching transistor inverter; (c) a controlled or self-commutated switching transistor rectifier and inverter

disconnected from the generator frequency. The performance of the d.c. link is influenced by the voltage and current levels within it. The grid is supplied via the inverter. Its operating state is characterized by the voltage, current and frequency, and the resulting active and reactive power values.

Power converters can therefore, depending upon design, be made up of different subsystems. Table 4.3 shows machine-side and grid-side power converters for three-phase systems

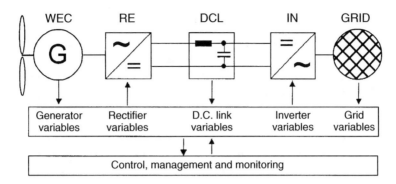

Figure 4.11   Grid supply of the wind energy converter (WEC: turbine, generator) via an indirect frequency conversion system

Table 4.3   Basic circuits of power converters [4.14]

Type	Name	Abbreviation	Energy flow
Rectifier with load-independent direct current	Current-controlled rectifier	CCR	
Rectifier with load-independent direct voltage	Voltage-controlled rectifier	VCR	
Inverter with load-independent direct current	Current-source inverter	CSI	
Inverter with load-independent direct voltage	Voltage-source inverter	VSI	

with constant-voltage and constant-current d.c. links, and gives the load-independent values for each. The normal abbreviations are also given. Rectifier and inverter subsystems can be constructed using thyristor or transistor components, and the construction is, in principle, identical for both types. The drive and the pulse pattern [4.13] of the frequency converter are decisive for their function.

A particular characteristic of the frequency converter with a constant-current d.c. link is the high series inductance in the d.c. link, as shown in Figure 4.4(a). As a result, the d.c. link current can be considered to be constant and load-independent during the short commutation period. In contrast to this, the frequency converter based on a constant-voltage d.c. link (Figure 4.4(b)) has a high case capacitance in the d.c. link. This means that the d.c. link voltage can vary only relatively slowly.

The most important rectifier systems that are relevant to the conversion of wind energy will be briefly described below. The supply line between the generator and rectifier is, however, of particular importance here.

4.1.5.1 Supply line between the generator and rectifier

As well as ohmic losses, wiring also has inductive and capacitive reactance components. These affect the transmission of three-phase current. The resistance per unit length R', the inductance per unit length L' in the longitudinal direction, the conductance per unit length G' and the capacitance per unit length C' in the shunt arm in relation to the line length can be represented for a line element, as in Figure 4.12. A long line can be viewed as a corresponding number of line elements connected in series.

In wind energy converters the generator is normally at the top of the tower, but the grid connection and the management, control and frequency conversion systems are located at or in the foot of the tower. Therefore the supply line in plants, which currently have a tower height of up to 120 m, has an influence on the frequency converter.

For the low-loss connection of the generator current to a controlled rectifier with an operating frequency of up to 20 kHz, very steep connection gradients are necessary. This causes high-frequency interference signals in the supply line, which give rise to transient overvoltages and electromagnetic emissions. The sudden increase in voltage brings about polarization effects in the insulation of the winding, which give rise to a capacitive displacement current in the insulation. If no measures are taken to limit this current such as a voltage rise filter, also known as a dU/dt filter, it will damage the insulation over the long term.

Overvoltages caused by switching operations can be reduced by using long build-up times. However, this increases switching losses and reduces the clock frequencies that can be achieved. Similar effects can be achieved by reducing the line resonance to low frequencies using inductors at the rectifier input, thus limiting the rate of increase of the current.

The use of LC low-pass filters, which allow the maximum generator frequencies to pass undamped and block the rectifier pulse frequency, facilitates better overvoltage limitation. However, this does influence the shape of the generator current.

Moreover, the rectifier pulse pattern can be used to limit the overvoltage caused by long generator supply lines. Predetermined prepulses and afterpulses work against the overshoot in the line, thus creating defined voltage states [4.15, 4.16]. The additional impulses, however, give rise to increased switching losses.

4.1.5.2 Rectifiers

Electrically excited and permanent-magnet synchronous generators in variable-speed operation supply different frequencies from those of the grid. The rectification of the generator output voltage or its current, and the filtering effect of storage elements in the d.c. link, have

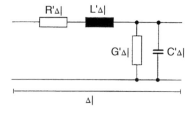

Figure 4.12 Equivalent circuit of a line element

the effect of largely disconnecting the generator frequency from the grid frequency. The quality of the d.c. link is basically dependent upon the rectifier type and storage elements used.

The main functions of the rectifier in a frequency converter system, apart from the conversion of variable-frequency electrical energy, is to exert influence over and control the generator and turbine train by controlling the output and the protection and electrical disconnection of the generator in the case of a fault.

Rectifiers can be basically divided into

- uncontrolled diode rectifier bridges,
- diode bridges with series-connected direct current (d.c./d.c.) regulators and
- controlled rectifiers.

Uncontrolled diode rectifier bridges or continuously triggered thyristor rectifier bridges (see Figure 4.6), which exhibit the same normal operating behavior, can be used if the generators permit voltage regulation by the excitation unit. Thyristors can protect the frequency converter rapidly and effectively from short-circuits in the event of faults by blocking the firing pulses.

When using permanent-magnet generators, there is no excitation effect on uncontrolled rectifiers. Operating behavior can only be influenced via the power output of the grid-side frequency converter and by mechanically controlling the output or speed of the turbine. Very rapid control interventions are not possible due to the time constants of the control system.

We must also take into account the fact that the combination of synchronous generators with rectifiers under dynamic load can activate unstable behavior [4.17]. This is caused by the steep current–voltage curve of the d.c. link. This can make the generator tend to vibrate at frequencies below 10 Hz. Stable operation can be achieved using constructional measures on the machine to damp the vibrations or by using controlled rectifiers [4.18].

The combination of an uncontrolled rectifier with a d.c./d.c. converter (Figure 4.13) represents an alternative to the controlled rectifier. The d.c. chopper controller determines the value of the d.c. link voltage by altering the timing (clock signal) relationships. It must be capable of holding the speed-dependent generator voltage or the rectified value almost constant. Using a so-called step-up converter, a d.c. link voltage that is too low can be increased to the level required for the grid-side frequency converter without reducing the utilization of the generator. The disadvantage is that the phase position between the generator current and voltage cannot be influenced. For synchronous generators with excitation windings, this process represents a low-loss option for controlling the d.c. link voltage.

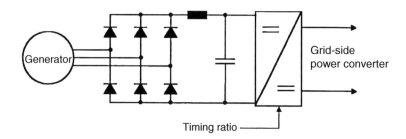

Figure 4.13   Uncontrolled rectifier with a d.c./d.c. regulator

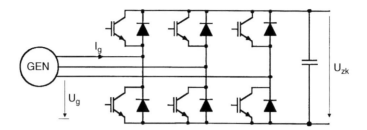

Figure 4.14   Pulse-controlled a.c. converter in the form of a step-up converter

A controlled rectifier with non-switchable power converter valves (see Figure 4.6) can be operated as a system with a load-independent current. The lower dead time of the frequency converter, which is in the millisecond range, compared with the mechanical time constants of generators means that parasitic oscillations and oscillating torques can be prevented by intervention. To maintain a control range in both directions, the firing angle must have an initial value of, for example, $\alpha_0 = 15°$, so that only a correspondingly lower d.c. link voltage can be expected. Thus, both a positive and negative control range is possible. This, however, reduces machine utilization. This type of rectifier can be used for electrically excited and permanent-magnet generators.

Controlled rectifiers with switchable valves can fulfil further demands and desires regarding wind energy converter control and operation. Pulse-width-modulated control of the constant-current or constant-voltage D.C. link facilitates near-optimal generator control by creating a pulse pattern. Along with rapid influence of the current path, harmonics in the generator current can be largely suppressed. Moreover, because of the elementary frequency of the rectifier in the kHz range, ripples in the d.c. link are of a relatively high frequency (in comparison with diode or thyristor bridges). The need for storage elements in the d.c. link is reduced accordingly.

In conjunction with the generator inductive reactances, which serve as short-term energy stores, it is also possible to operate the d.c. link as a three-phase step-up converter (Figure 4.14). For this, the d.c. link voltage U_{zk} must always lie above the generator peak value to prevent uncontrolled rectifier operation through the freewheeling diodes. At low speeds, with correspondingly low generator voltages, the step-up converter increases the voltage, thus ensuring the flow of current in the d.c. link.

The shape of the control pulse pattern must be adapted to the generator design. Generators with trapezoidal-induced voltage thus require square-wave currents. For variants with sinusoidal voltage, on the other hand, the pulse pattern is based upon a sinusoidal current reference.

A comparison of losses shows that diode bridges or continuously triggered thyristors exhibit on-state losses only. Both on-state and switching losses occur in the current path of a step-up converter, because of the pulsing at a semiconductor switch, e.g. the IGBT. In addition, the current always flows through several semiconductors at the same time. The switching losses are dependent upon the type of operation and the pulse frequency.

To sum up, we find that controlled rectifiers with switchable semiconductor components have a higher technical cost. However, by using them significant advantages can be achieved, particularly when operating with permanent-magnet generators. They facilitate the adjustment and control of the amplitude and phase positions of the generator current, as well as the resulting torque. Better regulation and protective functions (e.g. in the drive train) can be achieved than is the case with conventional systems, which means that the use of controlled rectifiers in d.c. link converters is very advantageous, despite the higher losses.

4.1.5.3   Thyristor inverters

With the development and introduction of thyristors, the power converter made its breakthrough in the form of controlled rectifiers and line-commutated inverters, which will be briefly considered here. They have found a broad field of application in drive and energy technology, and for a long time dominated the market for d.c. link converters. In six-pulse inverter bridges (see Figure 4.15(a) and (b)), however, the high harmonics that typically occur are very disruptive to the grid owing to the significant deviation of line current blocks from the sinusoidal form. Costly filter devices, mainly for the very strong fifth and sixth harmonics, must also be installed. It is also impossible to adjust the frequency and phase position of the line current. These are predetermined by the grid and supply system, and are dependent upon the firing angle.

Significantly better grid compatibility can be achieved with the twelve-pulse design. This is normally achieved by connecting two six-pulse systems in parallel, which are then brought together, via a magnetic circuit, by one common transformer with two secondary windings,

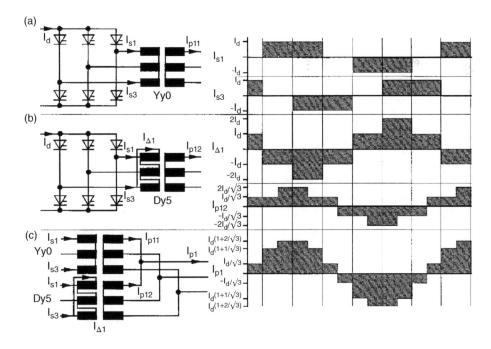

Figure 4.15 Connection and current path of inverters: a six-pulse inverter with (a) star and (b) delta transformer coupling and (c) a twelve-pulse design with the transformer circuit as in (a) and (b)

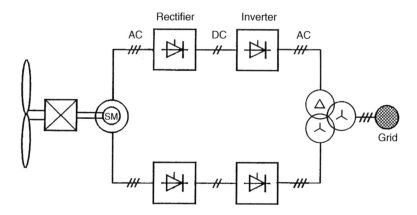

Figure 4.16   Twelve-pulse grid supply via two six-pulse frequency converters connected in parallel

phase-shifted by 30° (Figure 4.15(c)). This brings the line current much closer to the sinusoidal form. The fifth and seventh harmonics almost disappear compared with the six-pulse design. Only the significantly reduced amplitudes of the eleventh and thirteenth harmonics and the subsequent weakened components must be taken into account.

These quasi-twelve-pulse frequency converters, however, generally necessitate specially designed generators and grid transformers (for connection at the medium–high voltage levels) with two separate windings in the three-phase system (Figure 4.16).

4.1.5.4   Pulse-controlled inverters

Favourable conditions, in particular with regard to grid effects, are created by the use of self-commutated frequency converters, particularly in designs that operate with a pulse modulation process: so-called pulse-controlled inverters. The cheap availability of efficient power converter relay valves and the progress in signal processing meant that, just a decade after a short introductory period, self-commutated frequency converters based upon IGBT technology represented the state of the art due to the impetus provided by wind power technology. Higher outputs can also be achieved by parallel connections, as shown in Figures 4.16 and 4.21. At the moment, 5 MVA frequency converter systems are already being planned and constructed for wind energy converters. Conventional and GTO thyristor inverters currently still dominate the market for high outputs in the high megawatt range. New market opportunities are opening up for IGCT semiconductor switches. They offer advantages in terms of switching and service life. It is anticipated that IGBT modules offering a further increase in frequency converter output will soon be available. The discussion below is limited to these.

As a pulse-width-modulated (PWM) inverter with a load-independent voltage, the circuit shown in Figure 4.17 achieves rapid regulation due to its high pulse frequency. Six electronic power circuit elements (IGBT) with an integrated freewheeling diode can be used for this. This circuit permits highly dynamic regulation and the supply of near-sinusoidal currents. Moreover, both the supply of pure active power and the compensation of reactive power from consumers and other generators in the grid or in grid branches can be selected.

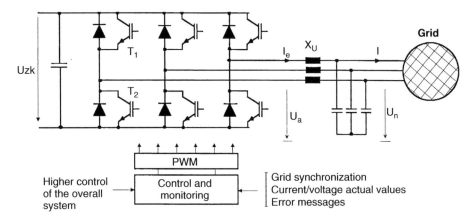

Figure 4.17   Block diagram of a pulse-controlled a.c. converter connected to the grid via a constant-voltage d.c. link

The transistors are switched by the control device, which is synchronized with grid voltage U_n, such that the grid current I or I_e has a near-sinusoidal path. The inverter inductors, with a series reactance X_u, limit the rate of change of the current. The difference between the momentary voltage values at the power converter output and the grid determines the flow of current. The combination of inductors with series-connected capacitors represents a low-pass filter, which serves to reduce the harmonics of the current.

The characteristics of the pulse-controlled a.c. converter are, as mentioned above, principally determined by the controlling pulse pattern. Figure 4.18 shows the path of line current I, which is regulated within the tolerance band ΔI, and illustrates the influence of the switching cycles of the transistors T_1 and T_2 of a bridge branch (Figure 4.17). The desired value can be preset as a sinusoidal variable in the middle of the tolerance band, synchronized with the grid voltage.

The transistor T_1 is switched on during a positive increase in current. If the upper limit of the tolerance band is reached, T_1 switches off and T_2 switches on. The current drops to the lower limit, which initiates another changeover.

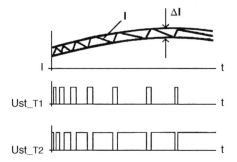

Figure 4.18   Tolerance band regulation of current and control signals for a bridge branch of a pulse-controlled inverter

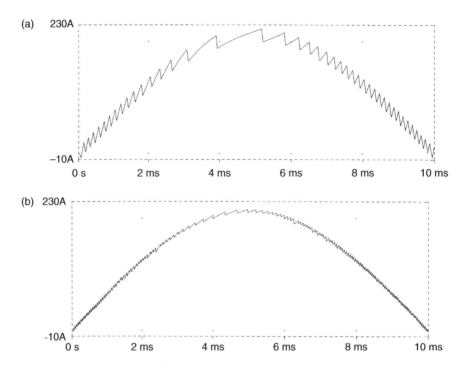

Figure 4.19   Graph of output current of a pulse-controlled a.c. converter: (a) with a large tolerance band $\Delta i/I_{\text{eff}} = 15\%$; (b) with a small tolerance band $\Delta i/I_{\text{eff}} = 5\%$

Current continuously oscillates between the maximum and minimum value in relation to the desired value. The tolerance band therefore determines the required timing. As shown in Figure 4.18, the switching rate can be freely adjusted as a function of current and the selected tolerance band. In other procedures the switching rate can also be preset. The desired values, the trigger procedure and the resulting switching rate, however, influence the total harmonic distortion (Figure 4.19). High-quality requirements with regard to output current thus result in higher losses in the frequency converter.

The value and phase angle of the output current of a self-commutated pulse-controlled inverter can be freely selected by regulation within certain design-dependent limits. Its use in wind energy converters thus permits a contribution to grid support in grid operation.

Along with the pure active power operation, the current can also be adjusted to lead and lag in relation to voltage, as shown in Figure 4.20. The supply of capacitive and inductive reactive power is thus possible. Due to its short access time, this type of plant can, at high operating frequencies, be used for dynamic reactive power compensation, as well as energy supply. If necessary it can take on an active filtering role.

If the clock frequency is drawn from an internal clock generator, systems with pulse-controlled a.c. converters can also help to stabilize the grid frequency, particularly in weak isolated grids. The connection of wind turbines via pulse-controlled a.c. converters thus has clear advantages with regard to high grid utilization.

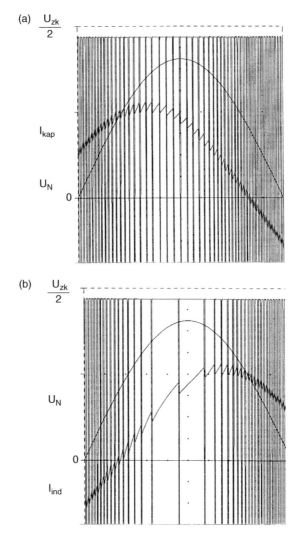

Figure 4.20   Current and voltage path of a pulse-controlled inverter for (a) capacitive and (b) inductive supply

4.1.5.5   Parallel operation of frequency converters

The maximum output of conversion and transmission systems can be limited by the available components. Thus, for example, when they were introduced the use of IGBT pulse-controlled inverters with individual components was limited to systems up to around 300 kW nominal output. For wind energy converters in the 500 kW class, output therefore had to be split. However, further advantages can be achieved by such junctions (branching points) of the energy circuit. For example, one frequency converter junction can be disconnected in the partial-load range. On the other hand, if one part of the frequency converter group fails, the wind energy converter can continue to operate at around half the nominal output. The superposition of frequency converter harmonics operating in parallel reduces the grid influence by partial or extensive cancelling out. The use of identical modules for several output classes makes development, manufacture and storage, as well as service, much easier and reduces costs.

There are various options for the parallel operation of frequency converters. Figure 4.16 shows the normal connection in conventional thyristor frequency converter technology, which is also used for switching transistor systems. This requires a six-phase design of the generator winding. The parallel frequency converter units are thus electrically separated, giving favourable conditions for high reliability of operation and low grid influence. Further variants are possible with three-phase generators, with the energy flow divided on the three-phase side or in the constant-current d.c. link. In addition to magnetic coupling by means of partial transformers on common iron cores, the electrical energy can also be brought together on the three-phase side by bus section reactors in all six phases or connection to the grid in pairs via centre tapping of three reactors. In the case of electrically coupled units, increased cost is incurred in the splitting of the current flow in order to prevent the overloading of one frequency converter branch.

4.1.5.6 Frequency converter concepts

Frequency converters have the task of conditioning the electrical energy from wind generators and supplying it to the grid. The influence on the grid should be kept as low as possible. The characteristics of the supplying generator and the grid have a decisive influence on the way in which frequency converters are constructed. Advantageous functional characteristics and favourable types of behavior in components and subsystems can be taken into account in the design of the frequency converter.

Active and reactive power control with rapid intervention, low grid influence, etc., can be achieved using pulse-frequency converters in the grid with IGBT valves and a constant-voltage d.c. link. High operating frequency, a low triggering requirement and the capacity to disconnect overvoltages are decisive advantages of IGBT frequency converters. By contrast, the low pulse frequencies of GTO thyristors, which are also available in significantly higher power ranges, bring about a poorer-quality feed current.

Figure 4.21 shows the frequency converter design for wind energy converters that have generators with sinusoidal synchronous internal voltage. The constant voltage d.c. link is supplied by the generator via a controlled rectifier. This is current-controlled, so the value and phase of the generator current is determined by the triggering of the rectifier. By phase-shifting the generator current $I_{Generator}$ or I_1 in over- and underexcited operation, as shown in Figure 3.10, the voltage level at the generator terminals can be altered and matched to the d.c. link. The rectifiers and inverters are of largely similar construction.

For generators with trapezoidal-induced voltage, the rectifier must be designed differently, because the generator only carries current in two winding phases at the same time. Moreover, the output voltage must be stepped up by a step-up converter for current to flow in the d.c. link. The six-pulse bridge shown in Figure 4.22 functions using the generator winding inductances as storage reactors in step-up operation. In this system the rectifier must be placed right next to the generator, since high-frequency oscillations down the long line through the tower could cause disruptive overvoltages.

Figure 4.23 shows the layout of a frequency converter with an uncontrolled rectifier and series-connected step-up converter in the d.c. link. This type of system was used in

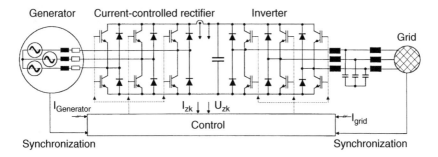

Figure 4.21 Frequency converter concept for an electrically excited or permanent-magnet generator with sinusoidal voltage

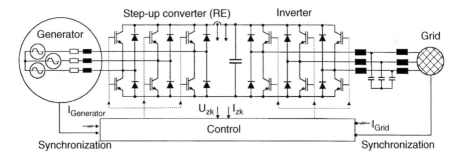

Figure 4.22 Frequency converter design for generators with trapezoidal voltage and a step-up converter in the rectifier circuit

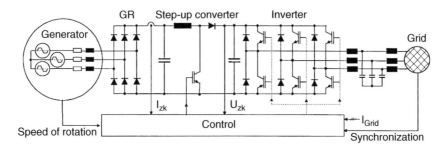

Figure 4.23 Frequency converter design for generators with trapezoidal voltage, a square-wave current path and a step-up converter in the d.c. link

gearless wind energy converters with generators that require electrical excitation and exhibit a trapezoidal voltage path. The two variants described previously are, however, preferred in the new systems.

4.1.6   Electromagnetic compatibility (EMC)

Semiconductor valve switching on and off processes in frequency converter systems cause interference emissions over a wide signal spectrum. These must be limited to such a level that other electrical devices can be operated in the vicinity without problems [4.19]. Furthermore, the function of wind power plant frequency converters must not be disrupted by low levels of external interference that could give rise to dangerous operating states.

Interference signals are transmitted by different routes. This transmission can occur by an electrical connection via the impedances between the (emitting) interference source and the (receiving) potentially susceptible equipment. Moreover, inductive and capacitive connections via magnetic and electrical fields are also possible. In addition, electromagnetic waves can be emitted and received by radiation.

The 'law of electromagnetic compatibility (EMC) of devices' has been in force (implemented) in Germany since November 1992. Since January 1996, the emission of interference and immunity to interference have been regulated by law. The term 'device' includes all electrical equipment, systems and components, and therefore includes wind turbines and their frequency converters. Different EMC limits apply, depending upon the power range and application.

Frequency converter units for wind power plants and their regulation and control components are electromagnetic sources. Their connection and supply lines carry current pulses of high clock frequency. These pass out of the range of the protected junction boxes, etc., via generator and grid lines. The frequency converter supplies the grid with current afflicted by harmonics. This is distributed over all voltage levels and causes a voltage drop at grid impedances. The shape of the grid voltage is distorted by the influence of harmonics. The EMC standards DIN VDE 0838 for general application, IEC 61000-3-6 or VDE 0838-6 for medium-voltage and high-voltage grids and IEC 77A/169/CDV or VDE 0838-4 for low-voltage grids relate to harmonics (VDE 0838-7 and IEC 77A/136/CDV or VDE 0838-5 relate to voltage fluctuations and flicker in the medium-voltage and high-voltage grids or low-voltage grids) and set down the permissible limit values. These can be adhered to by the creation of a pulse pattern giving as near to sinusoidal current as possible. The frequency converter timing gives rise to high-frequency current components. Compact filters in the grid line can reduce these to permissible values.

Rectifiers generate current pulses in the supply line on the generator side of the frequency conversion system. In uncontrolled operation, only current peaks with a relatively low pulse frequency occur during the charging of the smoothing capacitors. Controlled rectifiers create interference signals due to the pulsing of current into the generator supply line. These can be significantly reduced by a spatial arrangement in which the rectifier is located directly adjacent to the generator. This allows the electrical energy to be transmitted to the grid-side power converter via the direct-voltage level. Filters at the rectifier input also facilitate a reduction in interference signals. Depending upon the ratio of interference frequency to generator frequency, however, filters lead to reduced rates of current rise.

Broadband emissions of interference signals, caused by computing and control units, can be limited by fitting these devices in switchgear cabinets. Mutual interference of these components must also be avoided to ensure the function of the system as a whole.

High-frequency radiation can induce disruptive voltages in lines that are not screened (e.g. control lines). This can be compensated by input filters and blocking capacitors in the circuits, as can interference originating from grid supply lines.

Direct or indirect lightning strikes are potentially very hazardous for components, and can occur in the wind turbine, the wires or the surrounding area. The incorporation of lightening arrestors and a risk-minimizing wiring layout, e.g. not parallel to the lightning conductor and earth wiring, very effectively limit the damage that occurs.

The wide range of electronic devices in industry, business and households necessitates that all devices and equipment keep within predetermined interference limit values. To avoid complicated and expensive measures for the limitation of interference, possible effects during operation must be taken into account during the system development phase. The supply, transmission and use of electrical energy can thus be achieved safely.

4.1.7   Protective measures during power conditioning

Faults in the grid and generator or the failure of components may give rise to voltages or currents in the frequency converters of the wind energy converter that are so high that functional groups can be destroyed. Suitable measures are therefore necessary to protect the turbine and grid from damage. This requires the constant monitoring of all relevant variables. Moreover, control devices such as rectifiers and inverters must recognize deviating plant conditions and initiate suitable protective measures.

4.1.7.1   Generator side

A short-circuit between the input terminals of the rectifier creates an overcurrent in the generator. The sharp increase in current or the collapse of voltage must be recognized as a fault by the frequency converter controller. The tripping of fuses or safety switches terminates the generator-side short-circuit. The control system must turn down the grid-side power converter and the plant management must then brake the plant mechanically. Measures for the prevention of bearing and shaft currents have already been described in Section 3.6.1.5.

4.1.7.2   Frequency converter

Rapid switching processes in the power converter give rise to significant overvoltages in inductive devices. These must be limited to protect electronic components. Protective circuitry [4.20] made up of resistors, capacitors and diodes acts against harmful voltages and converts the energy stored in inductive reactances into heat.

4.1.7.3   D.C. link

In the event of a short-circuit in the d.c. link the voltage in the d.c. link collapses. The rectifiers and grid-side frequency converters supply the short-circuit via the freewheeling diode, which means that intervention is no longer possible. Therefore the grid and generator must be disconnected from the frequency converter by the system management.

4.1.7.4   Grid side

The regulations and guidelines of the power supply companies and manufacturers' associations must be adhered to for the supply of wind energy to the public grid [4.21–4.23]. According to these regulations, frequency converters on the grid side must recognize voltage or frequency changes to prevent unintentional isolated operation. In the case of overvoltage and undervoltage outside the stipulated limits or rapid auto-reclosing in the grid, wind turbines and their frequency converters must be disconnected from the grid (within 100 ms in the event of overvoltage and in 3 to 5 s in the event of undervoltage). Systems for the recognition of these faults can be integrated into the plant control system or designed as external units.

Grid-side shorts to earth and short-circuits lead firstly to an increase in current and secondly to a grid voltage dip in one or more phases. This gives a clear indication of a fault. Immediate control interventions, or an immediate blocking of the power converter, cut off the supply to the short-circuit. Disconnectable switch components have significant advantages for protective functions, due to their short intervention time. In particular, when using IGBT frequency converters, the advantages already mentioned come to bear due to their limiting characteristics. As well as being influenced by control and power conversion devices, the grid connection must also be protected by fuses.

In permanent-magnet generators a grid failure leads to an increase in generator and d.c. link voltage. This must be limited by intervention at the rectifier or in the d.c. link until mechanical countermeasures can be taken and the system brought into a safe operating condition. The inertia of the rotating masses counteracts a rapid increase in rotation speed, but also hinders a rapid shut-down. Overvoltages caused by the effects of lightning on the generator supply lines or the grid connection can be effectively limited by lightning arrestors on the rectifier input and inverter output.

Up until now we have largely based our discussions concerning the electrical energy supply, the planning of energy distribution and the layout of grid protection devices on the assumption of central electricity suppliers and the distribution of energy to decentralized end users. However, if decentralized wind turbines are integrated into the supply structure possible changes must be taken into account, particularly with regard to protective devices.

4.2   Grid Protection

Grid protection encompasses measures to protect against excessively high currents and voltages in the individual supply levels that could damage components and devices. Measures to protect against overvoltage are familiar and therefore will not be discussed further.

Wind turbines supplying the distribution system can impair the function of (overvoltage) fuses, hamper the coordination of reconnection devices, cause undesired isolated operation in the case of auto-reclosing or grid failure and thus put previously operational safety devices out of action.

4.2.1   Fuses and grid disconnection

If grid faults occur between a grid fuse and the associated reconnection device, e.g. due to a short-circuit, the generator can continue to supply current to the load side. Under realistic conditions, the currents generated by synchronous and asynchronous generators are not normally capable of tripping the existing fuses during the first two cycles [4.24].

Asynchronous generators only supply a small continuous short-circuit current ($I_k \approx 0$) in a three-phase short-circuit, since the field excitation is weakened. The worst case is a two-phase short-circuit [4.25–4.27].

Due to their high run-up time constants (see Figure 2.70), wind turbines with asynchronous generators that are run up by the grid in motor mode require electrical connections to be designed in accordance with the starting currents that will occur. Fuses and wiring must be dimensioned for five to eight times the nominal current. In small (by today's standards) wind energy converters the power and speed of the generating system is often graduated in that the running up of the wind turbine is governed by a small generator, designed for weak winds at low speeds (see Section 3.6.1.2), which reaches the nominal current of the generator dimensioned for full-load operation during the running-up period. This arrangement offers very favourable grid connection possibilities. The demands placed upon the drive train are also much lower with this running-up procedure.

The use of synchronous generators usually necessitates modifications to the reconnection device [4.25–4.27]. For the medium-voltage level, isolating switches should be used that can disconnect the decentralized generators within two cycles, i.e. significantly quicker than isolating switches on the transformer stations (>10 periods) [4.28].

The additional energy supplied as a result of the connection of wind turbines increases the short-circuiting power in the supplied grid branches, which will be described in more detail below. This can lead to the breaking power of connected devices (fuses, isolating switches, etc.) being exceeded, which may cause significant damage and reduce the safety of a system. However, it also increases the capacity and productivity of the grid.

4.2.2   Short-circuiting power

The short-circuiting power of a machine, a grid branch or a grid can be defined as the apparent power that will be converted at the machine or grid impedance in the case of a short-circuit.

When determining and assessing grid feedback, the special grid short-circuiting power is largely the defining factor for obtaining permissible or limiting values. S_k can be obtained for a node from the simple relationship

$$S_k = \frac{U_N^2}{Z_k},$$
(4.1)

where U_N is the nominal voltage and Z_k the short-circuit impedance between the source and the grid node under consideration, which can be defined using phase values or characteristic three-phase values. Using this simplified assumption, the specified limit values can be safely adhered to in estimates, even in unfavourable cases.

Unlike the assessment of grid feedback, the initial symmetrical short-circuit power of rotating machines (subtransient value)

$$S_k'' = \frac{cU_N^2}{Z_k}$$ (4.2)

forms the basis for the design and rating of protective devices. The supplying nonmeasurable initial voltage during the first period must be taken into account here. For this purpose, an appropriate magnification factor c is introduced (DIN VDE 0102). We assume that

- $c = 1.1$ for high-voltage and medium-voltage grids and
- $c = 1.05$ for low-voltage grids.

Using this magnification factor and by multiplication by the so-called withstand ratio $k = 0.1$ to 2.0, the greatest possible short-circuit current is taken into account in the design of protective devices. This is based upon the fact that, for the generator supply, the short-circuit current is not usually driven by the terminal voltage at the level of the nominal value, but the somewhat higher induced voltage.

Figure 4.24 shows the grid connection of a currently standard wind turbine in the 0.5 to 3 MW range. Its connection point (CP) is generally on the low-voltage side (400 to 960 V). The supply connection to the point of common coupling (PCC) – to which, in addition to the grid, consumers are also connected or could be connected – is via a transformer (0.63 to 4 MVA) and a medium-voltage cable primarily in the 20 kV level, which has ohmic, inductive and capacitive components [4.23].

The capacity and productivity of the grid is characterized by the short-circuiting power at the connection point S_{kCP}''. This depends upon the resistance and reactance components of the connecting transformer and connecting cables (R_{L+T} and X_{L+T}) and the short-circuiting power at the point of common coupling S_{kPCC}'' of the higher-level grid. The equivalent circuit in Figure 4.25 illustrates these relationships and forms the basis for the following calculations and representations.

For further discussion we have selected a transformer at the connection point ($S_N = 2.5$ MVA, $u_k = 6.8\%$, $U_N = 20/0.4$ kV, $P_k = 20$ kW) and overhead power transmission

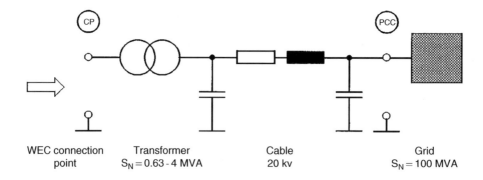

Figure 4.24   Low-voltage-side grid connection of a wind turbine

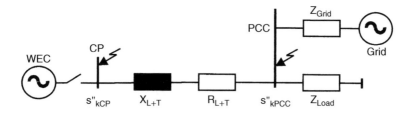

Figure 4.25 Equivalent circuit to determine the short-circuit power S''_{kCP} at the connection point

lines (Figure 4.26(a)) or cables (Figure 4.26(b)), each with a cross-section of 120 mm^2. The calculation results for the short-circuit power at the connection point are thus shown in relation to the short-circuit power at the point of common coupling. Furthermore, we have assumed distances of 1, 2.5, 5 and 10 km between the connection point and the point of common coupling, these being connection distances that are relevant to practical situations. We thus find resistance to reactance ratios of $R/X = 0.14$ (1 km) to 0.25 (10 km) for the overhead power transmission lines and $R/X = 0.14$ (1 km) to 0.33 (10 km) in the case of cable connection. Furthermore, it should be noted that the impedance ratios at the connection point, and thus the value of its short-circuit power, is dominated by the characteristic values of the grid transformer.

The diagrams in Figure 4.26 show the clear influence that the short-circuit power at the point of common coupling and the length and type (overhead power transmission line, cable) of connection to the connection point have on its short-circuit power. An economical increase in the short-circuit power at the point of common coupling from, for example, 100 to 200 MVA is associated with a significant increase at the connection point for all cable and overhead

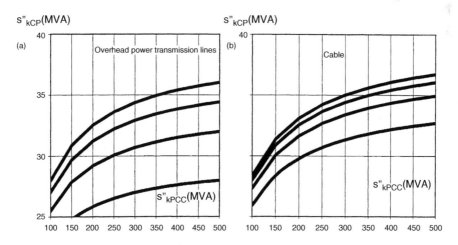

Figure 4.26 Short-circuit power at the connection point in relation to the short-circuit power at the point of common coupling for a 2.5 MVA transformer over 1 to 10 km connections using (a) an overhead power transmission line and (b) a cable

power transmission line connections. Doubling this value from 250 to 500 MVA, on the other hand, brings about significantly lower changes, particularly if long lines are used. This comparison illustrates, among other things, the possibilities and limitations of grid amplification measures at the point of common coupling for increasing the short-circuit power at the connection point.

Furthermore, the influence upon the relationships at the connection point by the selection of supply lines in the medium-voltage range must also be taken into account. A comparison of Figure 4.26(a) and (b) shows that for short connections the type of line and – as also shown by further calculations – its cross-section (within permissible limits) have only a slight influence on the short-circuit power S''_{kCP}. However, at greater distances the short-circuit power falls most when overhead lines are used. Different conduction materials (aluminium, copper) and the cross-section of lines and cables have, as shown by further calculations, less effect. The short-circuit power in the grid falls from the point of common coupling with increasing line length. However, if energy suppliers are connected, it increases again in the direction of the grid connection point.

The impedance of generators and transformers is usually very high compared with the impedance of the line. The distribution of generators thus has no dominant influence on short-circuit power along the line [4.24]. To simplify the determination of short-circuit power, as an approximation the generators can be combined as one characteristic point that is selected to assist the calculation as much as possible. In a short-circuit of rotating machines, significantly altered impedances – compared with normal operation – are, in effect, what influence the subsequent short-circuit power.

4.2.3 Increase of short-circuit power

Like motors, synchronous and asynchronous generators in operation increase the short-circuit power of the grid [4.29]. The calculations that were normally used for short circuit power up until until 1990 disregarded machines that were not constantly connected to the grid. This could lead to false estimates with regard to the dimensioning of disconnection devices and to the grid effects that could be expected.

When calculating the initial impedance Z_k that is effective in a short circuit, the fact that the rotating electrical machines will exhibit different impedances in their subtransient time response compared with normal operation must be taken into account. The simplified short-circuit equivalent circuit diagram in Figure 4.27(a) without and Figure 4.27(b) with a connected wind energy converter can serve as a generally valid starting point. The influences due to winding and earth capacitances in the case of a short-circuit, which are very low, are disregarded. The short-circuit power contribution to the grid by rotating machines is independent of whether the machine is working in the motor or generator mode at the moment of the short-circuit.

In synchronous machines the subtransient direct-axis reactance can form the basis for the determination of the short-circuit reactance X_k in the equivalent circuit diagram:

$$X_k = X''_d = x''_d \frac{U_N^2}{S''_N}. \tag{4.3}$$

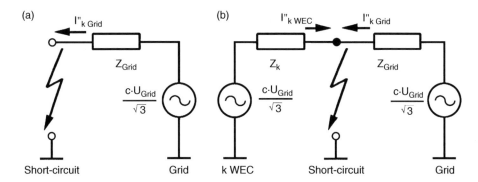

Figure 4.27   Equivalent circuit diagram for a generator in (three-phase) short-circuit (a) without a wind energy converter and (b) with a wind energy converter

In this equation x''_d represents the ratio X''_d/Z_N (see Section 3.7.3). To estimate the maximum values of short-circuit currents, however, the lower saturated value of subtransient direct-axis reactance $X''_{d\ total}$ must be taken into account. Depending upon the machine size, the ohmic component R_k can be assumed to lie between 5% of short-circuit reactance for high-output, high-voltage machines and around 15% for low-voltage machines.

In asynchronous machines the effective subtransient reactance in the subtransient time period is determined by the subtransient impedance

$$Z_{an} = \frac{U_N}{I_{an}\sqrt{3}} = \frac{I_N}{I_{an}}\frac{U_N^2}{S_N} = \frac{I_N}{I_{an}}Z_N. \tag{4.4}$$

This also characterizes the short-circuit of the machine, for which the leakage reactance of the stator and rotor winding $X_{1\sigma}$ and $X'_{2\sigma}$ respectively and the corresponding winding resistances R_1 and R'_2 are decisive machine parameters (see Section 3.3), where

$$X_k = X_{1\sigma} + X'_{2\sigma} \tag{4.5}$$

and

$$R_k = R_1 + R'_2. \tag{4.6}$$

The ratio of starting current to nominal current lies, in accordance with reference [4.27] and Figures 2.55 and 3.64, approximately in the range of

$$5 \le \frac{I_{an}}{I_N} \le 7,$$

where the active component makes up around 10 to 20% of the short-circuit reactance, depending upon the machine variant.

These discussions show that rotating electrical machines can contribute a multiple of their nominal output to the increase in the short-circuit power of the grid owing to their significantly reduced impedance in a short-circuit. Disregarding the relatively low active component, we can roughly estimate the magnification factor and the increase in short-circuit power for short-circuits near the terminals in synchronous machines from

$$u_{k\ SM} = \frac{1}{x''_d} \quad \text{and} \quad S_{k\ SM} = u_{k\ SM} S_{N\ SM} \tag{4.7}$$

and for asynchronous machines from

$$u_{k\ ASM} = \frac{I_{an}}{I_N} \quad \text{and} \quad S_{k\ ASM} = u_{k\ ASM} S_{N\ ASM} \tag{4.8}$$

For a short-circuit at some distance from the generator, the additional line and transformer impedances between the generator terminals and the short-circuit point must be taken into account. Depending upon the distance and impedances, correspondingly lower magnifications will result.

In accordance with reference [4.23], if the short-circuit currents of an in-plant generation unit are not known, we can use the following as an estimate for the effective values:

- eight times the sum of generator nominal currents for synchronous machines,
- six times the sum of generator nominal currents for asynchronous machines and
- the sum of generator nominal currents for generators with power converters.

For long-term (three-phase) short-circuits, i.e. those that exist over several cycles, significant differences exist, depending upon whether supply is by a synchronous or an asynchronous generator [4.30]. Synchronous machines can supply a short-circuit continuously if excitation is supplied independently of the grid, but if the generator excitation current is drawn from the grid the short-circuit leads to a collapse of the terminal voltage. Asynchronous generators, on the other hand, cannot supply a short-circuit over a long period and so the sustained short-circuit current in this case is equal to zero. The increased short-circuit power

$$S''_{k\ inc} = S''_{KN} + S''_{KM} \tag{4.9}$$

is thus found from the sum of contributions from the grid and the connected machines or wind energy converters.

A more precise method for determining short-circuit power, or changes to it, is possible by calculating the short-circuit current according to the superposition principle [4.31]. This procedure is particularly suited to the conditions in the case of branched structures. The calculation for a multiple-supplied grid with a corresponding number of sources takes place with the same number of calculations as for single-feed grids. Thus the usual processes can be used for grid calculation. The partial results for current and voltage can be superposed and summed up at each grid or supply point of interest to determine the state variable.

Line-commutated power converters, unlike rotating electrical machines, do not contribute significantly to the increase in short-circuit power. Designs with nondisconnectable valves cannot be immediately switched off in the event of a short-circuit. They can only be relieved of current after a time delay via intercepting circuits. Thus this type of inverter contributes to the initial short-circuit current of the grid.

In normal operation a line-commutated inverter (in the ideal case) supplies square-wave current to the three-phase grid. These currents have a sinusoidal fundamental component. They are predetermined by the grid, which provides the commutation voltage and commutation reactive power for the power converter valves. Since the grid voltage collapses in the case of a short-circuit, the power converter can no longer commutate. The short-circuit current is therefore no longer periodically separated, but flows on in the phase that led when the short-circuit was initiated. Line-commutated inverters thus cannot contribute to the grid symmetrical short-circuit current. The short-circuit current flowing before regulation only contributes to the sudden short-circuit value. This, however, is disregarded in the calculation of short-circuit power.

Self-commutated pulse-controlled inverters can maintain their voltage even in the case of a short-circuit and continue to provide sinusoidal current to the grid for as long as the feed capacity of the system permits. However, since they work with a modulation frequency in the kHz range, intervention is possible within a few microseconds in the case of a fault. Therefore pulse-controlled inverters can be regulated so quickly upon recognition of a short-circuit that they make no contribution even to the initial symmetrical short-circuit current of a grid.

Self-commutated pulse-controlled inverters are thus an ideal grid link, increasing the output capacity of the grid by the frequency converter feed component in normal operation and protecting the grid from any faults.

4.2.4   Isolated operation and rapid auto-reclosure

If the grid is disconnected, the function of protective devices can be impaired by the continued operation of wind power plants, leading to uncontrolled isolated operation, e.g. in disconnected dead-end feeders. Voltage-regulated synchronous generators and self-commutated pulse-controlled inverters are capable of creating grids in the order of magnitude of their capacity. They can therefore, within certain limits, support the grid in the case of voltage failure or maintain a supply in the case of a grid interruption. Synchronous machines coupled direct to the grid are, however, almost never used in wind power plants. Self-commutated pulse-controlled inverters are currently used in variable-speed wind energy converters both in synchronous and asynchronous generators.

Asynchronous generators, which predominate, and synchronous machines with a grid-commutated frequency converter supply, which were commonly used until about 1990, do not have this capability. However, in connection with turbine regulation they can supply the grid with active power and support the grid frequency within the range of their capacity. The combination of this type of wind power plant, particularly with plant-determined and consumer-specific reactive-power compensation equipment, maintains the grid in the case of interruption (similarly to synchronous generator supply), with frequency and voltage drift usually being dependent upon the load state.

As well as intentionally induced auto-reclosing described, lasting for about half a second, isolated operation over a longer period is possible. This is the case if wind turbines are disconnected from the grid but continue to supply their power into the fault-free line section and if the momentary output of the installed wind turbines can come close to covering the grid load on the isolated line. Possible reasons include

- opening the lines for maintenance reasons,
- correcting a fault before reconnection of the grid voltage and
- three-phase cut-off due to a single-phase short-circuit in the grid when wind turbines continue to supply in the fault-free phases.

In wind farm configurations and in the case of separate installations, wind power plants with different equipment can supply limited grid sections in combination. When turbines that have asynchronous generators directly coupled to the grid work together with plants that are connected to the grid via frequency converters, grid interruption can lead to various equilibrium states, depending upon the wind turbine control system. If insufficiently high loads are present in the disconnected part of the grid, continued energy supply by variable-speed turbines connected through frequency converters can lead to motor-mode operation of the asynchronous generators. The wind turbines coupled to them can thus be accelerated up to the aerodynamic braking mode.

Grid interruption in the case of the combined operation of wind turbines that are all fitted with asynchronous generators but have different nominal and operating slip values, which is the case for wind turbines of differing size or generator designs, will cause the turbines with smaller slip values to be driven in the motor mode at the rotary frequency of the generators with high slip values. Reconnection to the grid after isolated operation can, in the worst case, occur in phase opposition, which then leads to a high inrush current in the grid and generator, as well as extreme loading of the drive train. Asynchronous generators that have been disconnected from the grid can usually be reconnected without any danger of damage if the voltage has dropped to below a quarter of the nominal voltage in the grid [4.32].

In three-phase short-circuit, almost the same short-circuit time constants can be expected in the vicinity of the generator terminals as at other points on the line, because the total impedance of the fault and the line is usually small in comparison with the generator impedance. For generators in the 100 kW range, these time constants lie in the 100 ms range.

In the case of dead short-circuits, extra capacitors for reactive power compensation and grid filter reactances only have a small influence on time constants, which means that the voltage in the disconnected subsystem will have decayed significantly after only a short interruption. Therefore, no damage can be expected during the reconnection of asynchronous generators after a short-circuit. Asynchronous generators carrying approximately a half-load can, however, in combination with compensation devices, maintain their voltage at above 25% of the nominal voltage during a short interruption [4.24], thus causing high currents and torques upon reconnection, which may damage circuitry and the drive train.

Wind turbines with synchronous generators and direct grid connection, which up until now have only occasionally been used, usually have a torsionally elastic shaft. Turbines in the MW range can achieve shaft torsions of a few degrees at nominal output. If the load is released (e.g. due to auto-reclosing) then the shaft relaxes. The generator rotor thus accelerates according to its flywheel effect component (k_T) at the rotor time constant T_R (see Section 2.4) and the gear transmission (e.g. between 1:50 and 1:100 in plants in the MW range). This leads to torsion at the generator, which can lead to a full rotation of the generator shaft due to mechanical effects, and can even lead to two rotations in four-pole machines, due to electrical effects. An overshoot can lead to almost double these values [4.33]. Owing to the large component of the rotor time constant of $(1 - k_T)T_R$, which depends upon the wind wheel (see Sections 2.4 and 2.5), the turbine speed alters very little. The variation in the generator rotor speed, on the other hand, is large. It leads to oscillations around the synchronous speed.

To avoid unfavourable connection conditions (e.g. in phase opposition) the reconnection of wind turbines that have synchronous generators after load shedding or short-circuit should only take place after successful synchronization.

4.2.5   Overvoltages in the event of grid faults

In three-phase lines with distributed supply generators, resonance effects can appear in single-phase short-circuits. Due to the combined effect of capacitive compensation devices and the inductances of coupled generators, and in connection with user and grid filter reactances, overvoltages can occur in the fault-free phases.

Calculations and measurements [4.25], which we shall not describe in more detail here, for this type of grid fault in the medium-voltage range, in which the capacitance of the zero system is connected in series with the inductances of the positive and negative sequence system, yielded overvoltages that reached around twenty times the nominal voltage at low load. Experiments yielded values approximately ten times the nominal voltage. At no load or low load, the single-phase short-circuit current is thus sufficient to extend the electric arc of a fault and to convert a brief fault into a long-term fault if the generator is not disconnected from the grid before reconnection after auto-reclosing. Surge-Voltage Protectors (SVPs) can limit this type of overvoltage and must be capable of absorbing the resulting energy, which in the case of resonance can lie significantly above the values expected in normal operation.

4.3   Grid Effects

The connection of wind turbines to the electricity supply grid affects the grid [Chapter 3, Reference 1]. These interactions between turbines and grid include

- changes in short-circuit power,
- output variations,
- voltage fluctuations and possible flicker effects resulting from these output variations,
- voltage asymmetry,
- harmonics,
- subharmonics and
- other interference emissions.

The investigations described here are mainly limited to output variations and harmonics. The other subdomains will only be briefly outlined.

4.3.1   General compatibility and interference

Public grids must be protected from the disruptive effects caused by wind turbines. Overvoltage, short-circuit and generator protection serve this purpose. In the case of voltage and frequency deviations from normal operation, rapid disconnection from the grid must be ensured. Motor-mode operation should only be permitted for brief periods. The reactive power requirement of wind turbines must be kept within the limits of the grid-specific power factor limits.

The starting current should be kept as low as possible whether the generator is connected with the turbine stationary or at operating speed (see Sections 3.4.2.1 and 4.2.1), in the former case to protect the components of the wind energy converter (from the electrical circuitry through the mechanical drive train and mountings for the generator and gearbox to the turbine) from high shock loads and in the latter case to prevent emitted interference and grid voltage dips.

4.3.2   Output behavior of wind power plants

Apart from connection procedures, emergency disconnections and other changes of condition, wind turbines in normal operation at partial or full load normally give rise to electrical output variations in the range of 10 Hz to 1 Hz, depending upon the system design, as well as fluctuations over longer periods. Output fluctuations are determined by their amplitudes and rates of change and are caused mainly by periodic and nonperiodic wind-speed gradients. These are influenced by the distribution of air flows over the entire turbine area and the transmission behavior of the drive train and generator, which can be varied by changes to rotor angle displacement in synchronous machine systems and by slip variation in asynchronous machine systems.

4.3.2.1   Short-time behavior of a wind farm

In order to quantify the output behavior of wind turbines when connected together, investigations were carried out in the short-time range at the Westküste wind farm [4.34] with a measuring duration of 1 minute and a sampling frequency of 5 Hz. This showed that in normal operation, e.g. at medium wind speeds of approximately 12 m/s, variations occurred between approximately 10 and 15 m/s, with the wind direction remaining almost constant (see Figure 4.28(a) and (b)). In the 50 kW range (30 and 55 kW respectively), for largely fixed-speed turbines with asynchronous generators coupled directly to the grid, only slightly higher output fluctuations and output gradients can be expected than for variable-speed units with synchronous machines, rectifiers and inverters (cf. Figure 4.28(c) and (d)). In the case of variable-speed turbines, more favourable values can be achieved by control engineering measures.

However, in larger wind turbines (particularly those in the megawatt range) there are fundamental differences. A smaller slip range in larger asynchronous machines, and thus a more rigid grid connection, can bring about higher output fluctuations. On the other hand, higher rotor time constants, and the resulting improved smoothing effect, in variable-speed systems bring about increased output smoothing and permit constant power output, e.g. at wind conditions below the nominal range.

The total output of all fixed-speed or all variable-speed turbines as shown in Figure 4.28(c) and (d), measured at the appropriate feed transformers, illustrate the smoothing effect and its dependency upon the number of turbines. Therefore, based on the output behavior of the individual turbines as shown in Figure 4.29, the different total outputs shown in Figures 4.30 and 4.31 will be considered below, including a consideration of the influence of the turbine configuration. Figure 4.29 clearly shows the very different output behavior of individual wind turbines in relation to

- wind speed and
- turbine location (for geometry, see Figures 4.30 and 4.31).

The fluctuation range for each turbine over a measuring period of 1 minute is represented in the form of a rectangular plate, with output on the vertical axis and wind speed represented in the different sections. Maximum and minimum values are represented by the top and bottom of the side of the cuboid plates, the standard deviation is characterized by the dotted area and the mean value is represented by a thick line in the inner region. The electrical output values

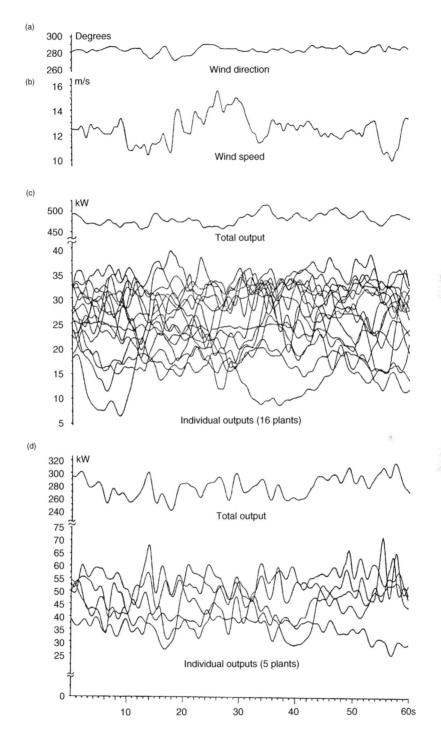

Figure 4.28   (a) Wind direction and (b) wind speed, as well as individual and total output of plants connected at (c) fixed speed and (d) variable speed at the wind farm

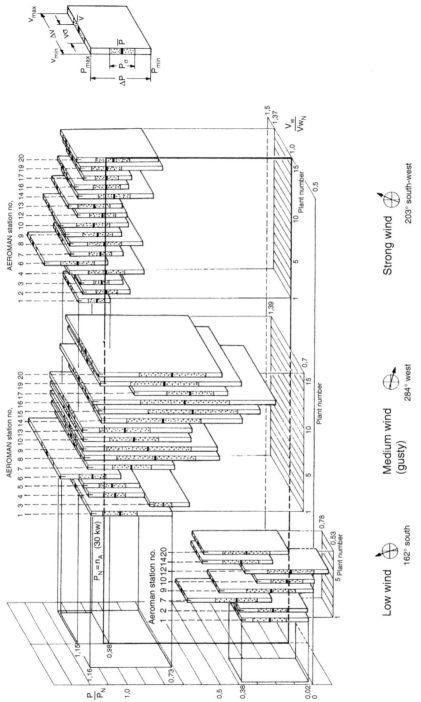

Figure 4.29 Output fluctuations with maximum value (P_{max}), minimum value (P_{min}), standard deviation (P_σ) and mean ($\bar{P}$) of the individual wind turbines in the wind farm at different wind speed ranges (for layout geometry, see Figure 4.30)

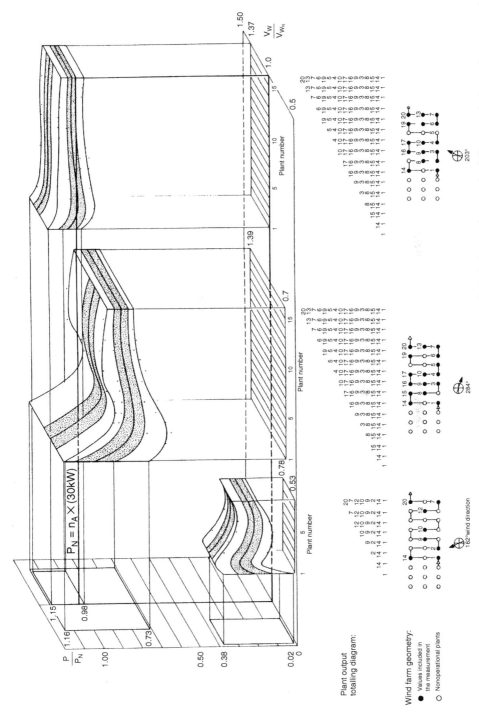

Figure 4.30 Output smoothing with maximum value, minimum value, standard deviation and mean in different wind-speed ranges with wind farm output added in columns

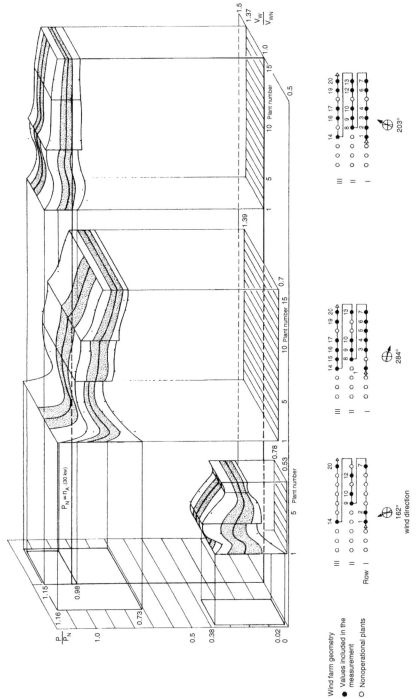

Figure 4.31 Output smoothing with maximum value, minimum value, standard deviation and mean in different wind-speed ranges with wind farm output added in rows

were measured at the same time for each individual turbine; the wind conditions were detected by an anemometer and direction gauge in an undisturbed airflow area at the outer corner of the wind farm close to turbine 7. This wind measurement was used for all turbines, divided into three categories of 'low wind', 'average wind' and 'strong wind', related to a nominal speed of $v_N = 11.9$ m/s, at which the wind turbines reach their nominal output of 30 kW. The values for individual plants can be determined by projecting the values of individual plates on to the side and lower datum levels. In the 'low-wind' range, for example, an output fluctuation of (0.02 to 0.38) $P_N(P_N = 30$ kW) is detected for the first turbine. Standard deviations and mean values can be determined in the same manner.

Consideration of the wind-speed ranges represented shows that output fluctuations are high, particularly at speeds just below the nominal operating speed. Fluctuations are also more pronounced in the back rows (clear from the turbine number). In the lower partial-load range and in the nominal-load range, when operating at above nominal wind speeds, on the other hand, only small output variations occur.

Output smoothing is strongly dependent upon installation geometry and the associated local wind conditions. Wind farm geometry is represented in Figures 4.30 and 4.31 with the prevailing wind direction at the lower edge of the figure. In order to show the differences, the smoothing effect for wind turbines with a fixed-speed grid connection is shown in columns (see Figure 4.30) or rows (see Figure 4.31) for comparison. In addition, the simultaneously measured individual outputs are summed in the step-by-step calculations according to the addition diagram in Figure 4.30. Each calculation is represented by the fluctuation width, standard deviation and mean. For example, in the lower wind-speed range, when only eight plants were operating, based on the 1st plant, the total output is shown for the 1st and 14th plants, the 1st, 14th and 2nd plants, etc., up to the 1st, 14th, 2nd, 9th, 10th, 12th, 7th and 20th plants. Only the plants marked in black were in operation when the measurement was taken. The values in Figure 4.31 were obtained in a similar manner. The dependence of output changes on wind speed or its fluctuation range is particularly clear here.

A comparison of Figures 4.30 and 4.31 shows, owing to the reduced fluctuation range, a clear smoothing of output, particularly in the high wind-speed ranges, i.e. during operation close to nominal output. Moreover, if the output is summed in rows, the row step changes are found again in the fluctuation range and standard deviation. If columns are added, such transitions disappear. As shown in Figure 4.28, the outputs of individual turbines yield a total output supplied into the grid with a much lower fluctuation range. The fluctuation ranges of individual turbines are influenced by the position of the turbine in the wind farm, as shown in Figure 4.29. Both methods of addition demonstrate that by connecting around six plants together, a final value for the fluctuation range of approximately ±6 % of the mean value can be achieved. Due to the spatial offsetting of the individual plants, the airstream meets each turbine at a different time. This yields the largest smoothing effect at the highest individual turbine fluctuations, which occur at low and medium wind speeds.

Due to the wind-speed conditions and wind-speed fluctuations prevailing here, these considerations are valid for short-time investigations. The layout geometry and timing of a wind event must be taken into account here; e.g. in this case a 600 m wind farm depth and 10 m/s wind speed were found in an observation period of 60 seconds. Transferring the results to larger areas and correspondingly longer time periods is only possible to a limited degree. Further measurement results for a layout of plants on a correspondingly larger area with longer measuring periods are discussed below.

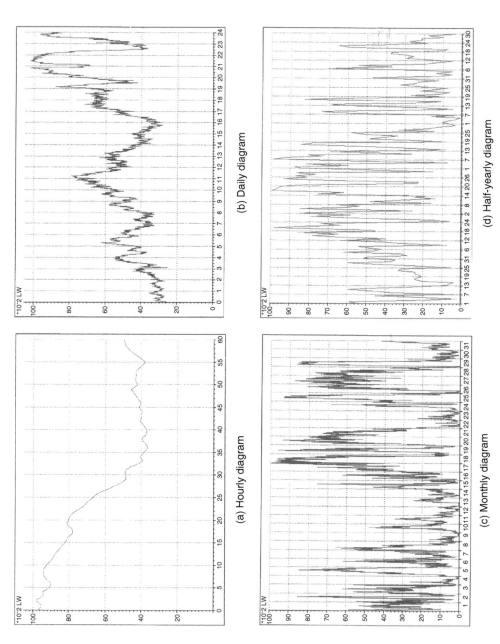

(a) Hourly diagram

(b) Daily diagram

(c) Monthly diagram

(d) Half-yearly diagram

Figure 4.32   Graph of output of a 10 MW wind farm (VEW Waldeck)

4.3.2.2   Long-time behavior of supply areas

In the VEW Waldeck-Frankenberg supply area in Germany, 33 wind turbines of nominal output between 150 and 500 kW are installed in the Diemelsee area at distances of up to 6 km apart, i.e. around ten times the distance considered above. The total output of the wind farm is 10 MW, of which 70% of the output is provided by turbines with asynchronous generators with a rigid connection to the grid and 30% by variable-speed systems with gearless drive train, synchronous generator and pulse-controlled frequency converter. Figure 4.32 below shows plots of the total wind farm energy over time. Figure 4.32(a) shows a plot of the total output of all plants over one hour, Figure 4.32(b) shows this for one day and Figure 4.32(c) for one month in March 1995. Figure 4.32(d) shows the output diagram of the wind farm during the first half of 1995. These figures illustrate the limited output contributions of wind power in the supply area considered here. Furthermore, by breaking the figures down into the time periods that are of interest, relatively precise output gradients can be determined and measures for output smoothing derived. These results can provide characteristic values for the selected time period.

Precise statistical or time-dependent predictions can, however, only be made using the appropriate methods. This requires large amounts of data and long time periods (preferably several years), to obtain reliable results. These can be obtained for the whole of Germany, or for wind-rich regions, within the framework of the 250 MW wind program of scientific measurement and evaluation, which has been set up for a ten-year measurement period. The results of these investigations [4.3, 4.35–4.37] are shown in Figure 4.33 for the North German region. The power duration curves for individual turbines illustrate the duration of the output in question, which is dependent upon local wind conditions; e.g., a turbine at Fehmarn achieves its nominal output for around 1000 hours and operates for approximately 8000 hours in a year (8760 hours). Turbines in the North German area, however, never all reach their nominal output simultaneously. Their energy contributions range over the entire year and 10% of the nominal output is supplied for more than 6000 hours. In addition to the evaluation of measured data, predications can also be made based upon the above-mentioned program, and these will be briefly described in the following.

4.3.2.3   Wind power predictions

The German interconnected network grid has the greatest installed wind power capacities anywhere in the world. The power contributed by wind generators exceeds the consumed values at off-peak periods in some areas of the grid. Wind power thus plays quite an important role in terms of the operation of grids, load control and generation schedules [4.38–4.42].

In addition to power station failures and stochastic load variations, unpredictable changes in the supply of wind generated power is one of the most common reasons for the use of expensive compensation and control power in power system management. Current consumption over time – the so-called load schedule – can be predicted relatively precisely for the near future by means of the prediction procedures currently in use. Great importance must therefore be attributed to achieving as precise a prediction as possible of wind power, which is subject to weather-dependent stochastic processes and is supplied to the grid in fluctuating form [4.43–4.47]. In particular, precise knowledge of the wind power available on the following day, for example, is of great importance in the planning of generation schedules.

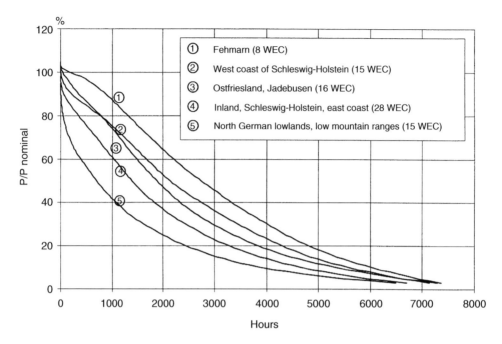

Figure 4.33 Power duration curves of individual wind turbines and a combined group of turbines in the North German area. Reproduced by kind permission of ISET

Backed by the German Ministry for Finance and Technology, the Institut für Solare Energieversorgungstechnik (ISET), in cooperation with the power supply utilities, weather service, etc., has developed a numerical model for the prediction of wind power. This model is based upon

- a prediction of wind speed and direction for representative selected locations,
- the transformation of the predicted data to take account of local conditions using a three-dimensional atmospheric model (Figure 4.34),
- determination of the associated wind power with the aid of an 'artificial neural network' and
- extrapolation of the wind power (Figure 4.35) to the total supply in the power supply area in question using an on-line model.

The prediction model supplies the graph of expected wind power over time in the supply area for up to 48 hours in advance. This was achieved by installing measuring equipment at 16 representative wind farms or farm groups. Measured wind and power data obtained in the past was used to improve the predictions produced by neural networks and to help us learn about the relationships between wind speed and wind farm output [4.48]. Figure 4.36 shows the good correlation between measured and forecast wind power by the association of power stations. A further refinement of the procedure will improve the accuracy of the predictions still further. The forecasting model, when implemented at the power distribution board of the power supply grid, will permit enormous progress to be made in the drawing up of a precise

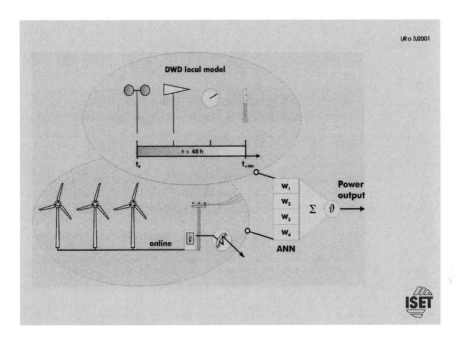

Figure 4.34   Wind power measurement and weather forecasting

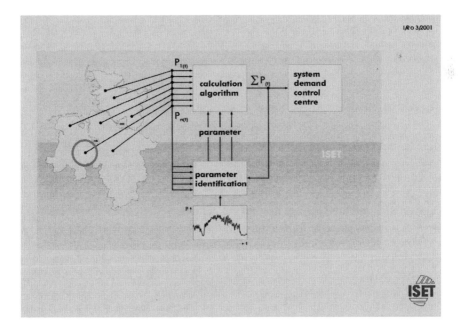

Figure 4.35   Representative wind farm power output measurement and forecasting for supply areas

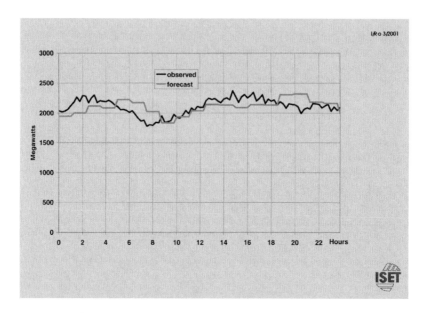

Figure 4.36   Comparison between wind power measurements and 24 hour forecasts over one day

load schedule and the cost-oriented planning of power station use, and will lead to overall cost savings in the operation of power stations and of the supply grid.

In view of a large-scale use of wind power the expected power compensation possibilities (in particular for fossil fuel electricity generators) represent a very important aspect for the future and are crucial to the value of wind power.

4.3.2.4   Short-time behavior in a small isolated system

Small energy generation units, e.g. hybrid systems, are normally made up of only two to three wind turbines, with a lead acid battery as a short-term storage and a diesel unit as a back-up system. Such plant configurations are used in isolated operation, e.g. on islands, which have no connection to the interconnected grid [4.49].

Due to its coastal location, the meteorological conditions at the Westküste (west coast) wind farm considered above are comparable with the conditions on islands, which means that the readings for the output behavior of individual turbines and small groups of plants can be transferred to isolated network systems. In this context, the output behavior of individual turbines, of two or three turbines in combination, and of different layouts will be considered at three characteristic wind speeds with mean values $\bar{v} = 7.7$, 11.6 and 14.6 m/s. Here $\bar{P}$ represents the mean value, ΔP the fluctuation range and P_σ the standard deviation of output and Δv and v_σ the corresponding wind-speed values. Table 4.4 summarizes the statistical evaluation of 5 Hz readings over a period of 60 seconds for different configurations and flow directions. A comparison of characteristic values illustrates increased output smoothing with larger numbers of turbines and increasing distance between turbines. The greatest smoothing effects are found, as expected, for flow normal to the wind turbine rows. Similar effects were achieved for

Table 4.4 Characteristic values for the output behavior of wind turbines for different turbine numbers (1, 2, 3), distance between turbines, wind speed and wind direction

Arrangement	Wind conditions in m/s		
	Low	Medium	High
$\bar{v}$	7.7	11.6	14.7
Δv	2.95	8.13	4.35
v_a	0.7142	1.53	0.81
$\bar{P}/P_N$	0.35	0.74	0.97
$\Delta P/P_N$	0.44	0.64	0.44
P_a/P_N	0.118	0.17	0.09
$\bar{P}/P_N$	0.22	0.96	1.02
$\Delta P/P_N$	0.24	0.15	0.26
P_a/P_N	0.068	0.035	0.052
$\bar{P}/P_N$	0.38	0.89	1.04
$\Delta P/P_N$	0.36	0.42	0.216
P_a/P_N	0.098	0.102	0.046
$\bar{P}/P_N$	0.16	0.87	1.03
$\Delta P/P_N$	0.205	0.458	0.148
P_a/P_N	0.046	0.109	0.029
$\bar{P}/P_N$	0.29	0.767	1.031
$\Delta P/P_N$	0.196	0.45	0.175
P_a/P_N	0.06	0.132	0.040
$\bar{P}/P_N$	0.16	0.95	1.022
$\Delta P/P_N$	0.2	0.15	0.23
P_a/P_N	0.046	0.031	0.05
$\bar{P}/P_N$	0.28	0.99	1.025
$\Delta P/P_N$	0.22	0.24	0.184
P_a/P_N	0.05	0.052	0.042
$\bar{P}/P_N$	0.22	0.91	1.037
$\Delta P/P_N$	0.28	0.25	0.22
P_a/P_N	0.075	0.05	0.043

a triangular shape. In the medium wind-speed range particularly high output fluctuations are found for individual turbines, and good smoothing effects are shown for two or three turbines [4.50].

4.3.2.5 Frequency behavior of wind power plants

The results represented above in the minute, hour, day, month and annual time scale can be described as long-term considerations in relation to electrical smoothing processes in the hertz range. As well as this type of behavior, short-time investigations are also of great interest.

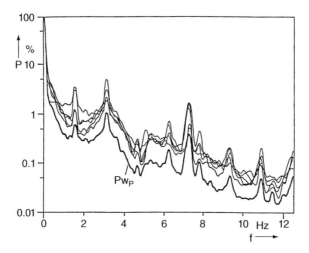

Figure 4.37 Scaled amplitude spectra of the output of five individual wind turbines and wind farm output P_{WF} [4.51]

Figure 4.37 shows the scaled amplitude spectrum of individual turbine outputs together with the spectrum of wind farm output [4.51]. The value at 0 Hz corresponds with the output (active) power. As five turbines of the same type are considered, the alternating components of the individual outputs are also found in the total output. However, due to the smoothing effect between the turbines, their amplitudes lie clearly below the individual values. It should be noted that the maxima are clearly visible due to the logarithmic representation in the diagram. They have only a small effect on the corresponding periodic components in the path of output over time.

Dominant maximum values are brought about by asymmetries and tower effects. Moreover, tower and blade frequencies can also be determined. Accordingly, each plant has a characteristic amplitude spectrum, which can be used to identify it.

Changes in the spectrum can thus also be attributed to the transition to different system behaviors. This permits faults and defects in plant components to be recognized early, which means that this type of frequency spectrum can be used for fault prediction (see Section 5.7.5) [4.52].

Long-term and short-term changes in wind turbine output can affect the voltage of the grid that they feed, as considered below.

4.3.3   Voltage response in grid supply

The effects of wind turbines on the grid and the power that can be drawn or supplied are determined primarily by the type of grid connection and voltage levels. Changes and periodic fluctuations in voltage and flicker effects must be taken into account, as well as any asymmetries.

4.3.3.1   Voltage levels

The electrical components of wind turbines are designed for low voltages up to 1000 V for cost reasons. System configurations in the medium-voltage range represent the exception as yet. As turbine sizes increase, however, these could become more important. For turbines, connection to the grid takes place at the low-voltage level, represented in Figure 4.38 as the lowest block. Systems of the 100 kW to the MW class, on the other hand, supply to the medium-voltage range via a transformer, as shown in Figure 4.24. For large wind farms (e.g. 50 MW) a high-voltage grid connection (middle block in Figure 4.38) can be created in an economically viable manner, and is thus a much cheaper grid connection option. The 220 and 380 kV levels are not currently an option for the grid supply of wind turbines. These will, however, become increasingly important for the connection of large-scale offshore wind farms, which in the future will match the size of conventional power stations at several hundreds of megawatts.

4.3.3.2   Voltage asymmetries

Only three-phase generators with symmetrical windings are used in grid-operated wind turbines. In machines in full working order, asymmetric effects on the grid during supply can thus be disregarded in normal operation.

Asymmetric loads in the grid can, however, impair the function of lines and transformers as well as that of synchronous and asynchronous generators. Asymmetries can, for example, be caused by the failure of individual thyristors in bridges or asymmetric frequency converter triggering. Asymmetries cause currents to flow in the countersystem. These give rise to a field rotating against the excitation field at twice its speed. This causes additional losses, particularly in the rotor, which reduce the machine's load limit. Local asymmetries can be taken into account, if they are known, in the dimensioning of the generator.

4.3.3.3   Voltage changes, voltage fluctuations and flicker

If a large proportion of the grid load is supplied by wind turbines, output variations due to wind-speed changes can cause voltage fluctuations and flicker effects in normal operation, as described in the preceding section. These are mainly determined by the path of the apparent power of the supply over time, its relation to grid short-circuit power and the corresponding phase angle. Harmonic power can be disregarded here. Voltage changes can occur in specific situations, e.g. as a result of load changes, load or generator connection or release, switching between generator levels, wind-speed fluctuations, tower effects, etc. [4.53].

In the case of severe output fluctuations, the standalone operation of small wind turbines in low-output isolated grids and the connection of large turbines to low-output points in the combined grid can lead to voltage changes. These can be expected in particular in the case of generators connected to the grid at a fixed speed. Large asynchronous generators, in particular, pass output changes caused by wind-speed fluctuations directly on to the grid because of their low slip values [4.54].

In a flexible-speed connection between the wind wheel and grid, on the other hand, a significant smoothing of output can be achieved by the short-time intermediate storage of energy

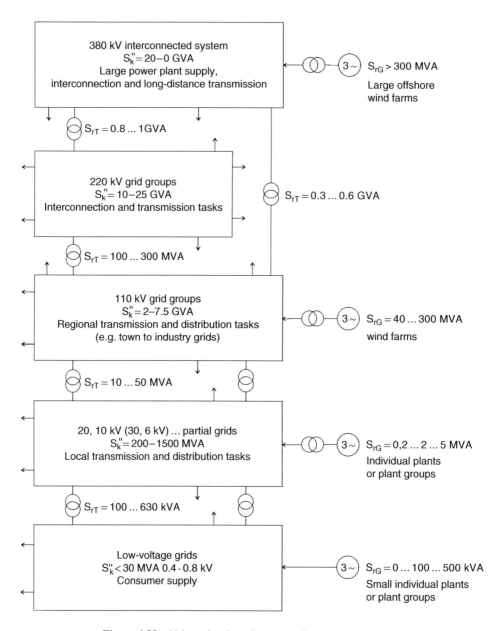

Figure 4.38   Voltage levels and outputs of energy supply grids

in the rotating masses of the drive train (see Section 2.4). As described in Section 2.5, wind turbines exhibit higher rotor run-up time constants with increasing size and nominal output (Figure 2.70). Larger turbines can thus achieve significantly better output smoothing using variable-speed operation, particularly in the short-time range (Hz range), than smaller units.

The speed regulation range is also a contributory factor to the degree of smoothing, with a large speed variation range being more capable of suppressing output fluctuations. Start-up currents do not play a role for turbines coupled via frequency converters, as they contribute less than 1% of nominal current and can be run up. Short-time voltage changes and flicker are therefore of secondary importance in this type of turbine [4.55].

Small- and medium-sized wind turbines can normally only significantly influence the grid when connected in large groups. The connection of many units, which can exhibit a high level of output fluctuations individually, nevertheless brings about better smoothing of the total output as the number of turbines increases (see Section 4.3.2). Therefore, in normal operation, no harmful voltage fluctuations can be expected [4.56].

The standalone operation of large wind turbines with high output fluctuations can lead to voltage changes in weak grids. The output fluctuations of large units should therefore be as low as possible. This can be achieved by variable-speed turbine and generator systems owing to the large rotor run-up time constant associated with the high centrifugal mass.

Similar grid effects can also be expected for small wind turbines that supply very low-output isolated grids. Due to the low rotor time constant, smoothing effects comparable to those of large turbines can be achieved for small units by increasing the speed adjustment range [4.57].

In contrast to the guidelines aimed mainly at consumers [Chapter 3, Reference 1], in determining the voltage changes caused by the connection of wind turbines the specific requirements of generating systems with regard to the direction of so-called 'voltage drop' in transformers and lines must be taken into account (see Figure 4.39) [4.58, 4.59].

This representation shows that in the motor mode, motor voltage $\underline{U}_{Mot}$ can generally be determined with adequate precision as the difference between the contribution of grid voltage $\underline{U}_{Grid}$ and that of voltage drop $dU = (R_{L+T} + jX_{L+T})\underline{I}_{1Mot}$. The vector diagram for generator operation, on the other hand, shows that grid voltage $\underline{U}_{Gen}$ and generator voltage $\underline{U}_{Grid}$ have roughly the same values. However, the voltage drop here causes a clear phase difference between the two voltages, the so-called line angle λ. It is thus clear that the generator voltage can exceed or fall below the grid voltage, depending upon the position of the generator current I_{gen} and the ratio of ohmic to inductive components R_{L+T}/X_{L+T} in the lines (overhead power transmission lines, cables) and transformers.

Voltage change, voltage rise
Normally, when power is supplied load components are low in relation to the supply power. Figure 4.40 illustrates in simplified form the grid configuration and associated voltage relationships for the following calculations. Based upon the grid voltage U_1 or the nominal value U_N at the higher Grid Point (GP), the corresponding value U_{PCC} is found at the point of common coupling (PCC) given the voltage drop dU with or without supply by a wind turbine at the Connection Point (CP). Here ΔU_r characterizes the voltage rise and $\Delta U_{r\,perm}$ characterizes its permissible values (e.g. 2% in the medium-voltage network).

Figures 4.41 and 4.42 show the relative change in voltage at the point of supply to the grid in relation to the power factor (cos φ capacitive or cos φ inductive) [4.60, 4.61] if the supply power $S_{supply} = 1, 2, 4$ and 10% of the grid short-circuit power S_k or, in accordance with Equation (4.13), the reciprocal $K = k_{Kl} = S_k/S_{supply} = 100, 50, 25$ or 10. This illustrates the influence of supply ratios on voltage changes particularly well. This calculation is based upon a ratio of active to reactive components for transformers and lines of $R_{L+T}/X_{L+T} = R/X = 1(a), 0.5(b), 0.25(c)$ and $0.1(d)$. A value of R/X of 1(a) characterizes a weak grid connection

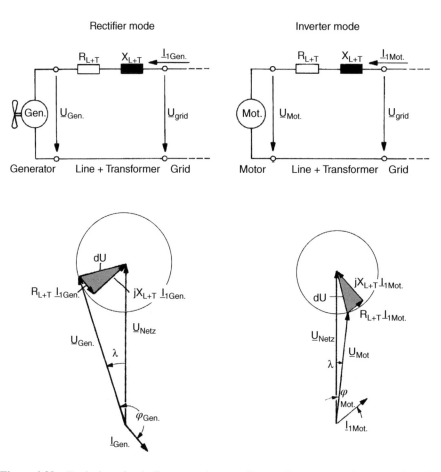

Figure 4.39   Equivalent circuit diagram and vector diagram for generators/motors at the grid

and a value of 0.1(d) characterizes a strong grid connection with correspondingly large (a) or small (d) voltage changes at different short-circuit power ratios ($K = S_K/S_{supply}$).

Figure 4.42 shows, firstly, specifically the ratios according to Figure 4.41 for a pure effective power supply at $\cos \varphi = 1$ and, secondly, the ratios for a voltage-neutral supply, at $dU = 0$ for ratios of grid short-circuit to supply power $K = S_k = S_{supply} = 10, 25, 50$ and 100 at ratios of effective to reactive components $R_{L+T}/X_{L+T} = R/X = 1(a), 0.5(b), 0.25(c)$ and $0.1(d)$. The range of voltage change is limited here to the permissible value of 2%. It is clear from this that a near-voltage-neutral effective power supply is only possible in transmission links with low ohmic components. Voltage changes can thus be reduced or avoided by a supply that is matched to system and grid data, even in the event of power fluctuations.

In the integration of supply systems that permit the phase angle φ_{Gen} to be freely adjusted (e.g. synchronous generators, pulse-controlled inverters) it is thus possible to influence the voltage of grid points. Inductive supply leads, according to the above figures, to a reduction in voltage. Capacitive supply, on the other hand, leads to a voltage rise.

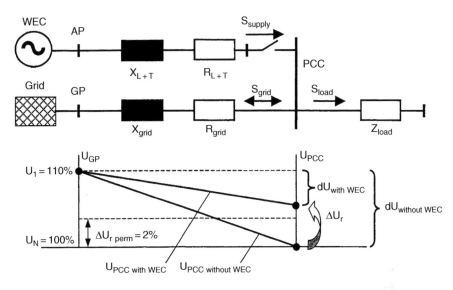

Figure 4.40 Equivalent circuit diagram and voltage change at the point of common coupling of a grid with and without wind energy converters

Furthermore, the ratios between supply power and load power at the supply point or point of common coupling should also be taken into consideration. According to Figure 4.40 we find

$$U_{PCC} = U_1 - \frac{(P_{load} - P_{supply})R_{grid} + (Q_{load} - Q_{supply})X_{grid}}{U_1},$$
(4.10)

where

$$S_{grid} = S_{load} - S_{supply} = (P_{load} - P_{supply}) + j(Q_{load} - Q_{supply}).$$
(4.11)

In the same way, the voltage for the point of common coupling U_{PCC} can also be found in accordance with Figure 4.40, taking into consideration the resistance and reactance components of the line (Z_L). Figure 4.43 illustrates the supply-dependent and load-dependent voltage changes at the point of common coupling for the (in practice often striven for) case of supply at $Q_{supply} = 0$ or $\cos \varphi = 1$. The representations relate (a) to strong grids ($R_{grid}/X_{grid} = 0.1$) with low voltage changes at large power differences and (b) to weak grids ($R_{grid}/X_{grid} = 1.0$) with significantly greater effects. Accordingly, the right-hand side of the voltage area is changed according to reactive power. This property of the grid at the point of common coupling thus permits voltage changes to be kept within the permissible range ($\Delta U_{PCC} \leq 2\%$), for example, by a suitable choice of the ratio of effective to reactive power supply.

The voltage change can be tuned at the supply point of the wind energy converter (connection point), the grid connection to the consumer (point of common coupling) or at a higher grid connection point by voltage regulation in the grid (see Section 4.6 on 'grid control'). The last-mentioned requires the monitoring of voltage readings and their transmission from the reference point to the supply point. In this manner, voltage fluctuations at the higher-voltage point, which can occur particularly in weakly loaded grids due to a reversal of the energy flow direction caused by the supply of wind power, can be minimized.

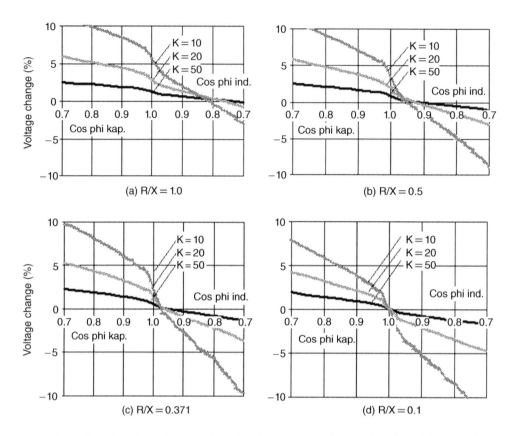

Figure 4.41 Relative voltage change at the point of common coupling as a function of the power factor at different supply ratios $K = S_K/S_{supply}$

Figure 4.44 shows the voltage deviation in relation to the supply current angle for the motor and generator mode of a machine in a weak grid with a short-circuit power of 8 MVA and, for comparison, a stronger grid (32 MVA). This clearly shows that a generator in the grid configuration selected here can supply inductively at around 18° and a motor can be operated capacitively at 23°, in order to maintain grid and machine voltage at the same level (e.g. $dU = 0$) [4.30]. Away from these supply or reference angles, large changes in voltage are found at low short-circuit power and small effects are found for strong grids. This neutral-voltage angle and the corresponding power factor are dependent upon the ratio of active to reactive components of the transformer and line for the transmission links.

Evaluation criteria and limit values for voltage changes at the point of common coupling are cited in reference [Chapter 3, Reference 1] with 3% recommended for low-voltage ranges and 2% for medium-voltage ranges. This includes the provisions according to EN 50 160 as well as the technical connection regulations of the power supply company. An assessment must take place in cooperation with the power supply company.

According to reference [4.23] the voltage rise at the point of common coupling can be approximated as a function of the maximum apparent power of the turbine $S_{A\,max}$, the grid

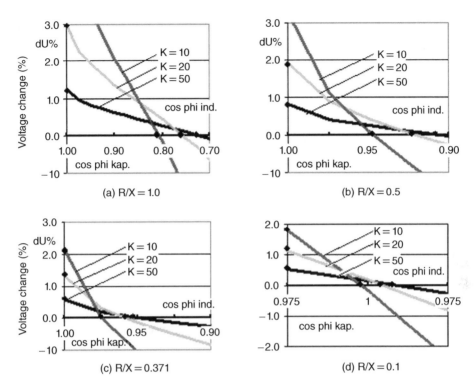

Figure 4.42 Determination of the relative voltage change dU with an effective power supply (cos φ = 1) and of the power factor cos φ at a voltage-neutral supply (dU = 0)

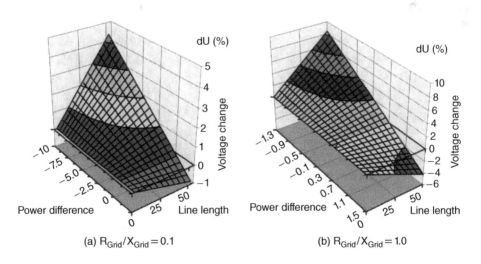

Figure 4.43 Relative voltage change at the point of common coupling as a function of the power difference ($P_{load} - P_{supply}$ in MW, supply and reference pure active power) and the line length (in km) at different line ratios R_{grid}/X_{grid}

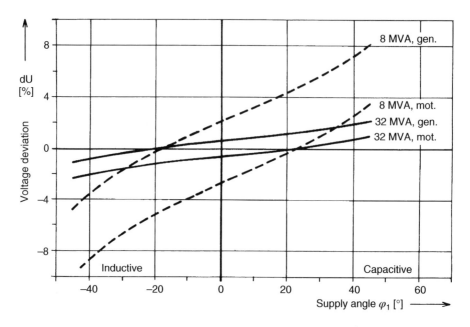

Figure 4.44 Voltage deviation as a function of the supply angle, with grid short-circuit power as a parameter

impedances R_{kV} and X_{kV} at the point of common coupling and the phase angle φ of the in-plant generator according to

$$\Delta u_a = \frac{S_{A\,max}(R_{kV}\cos\varphi - X_{kV}\sin\varphi)}{U^2}, \tag{4.12}$$

where U is the nominal voltage of the medium-voltage grid.

The assessment of voltage changes is usually based on the ratio of grid short-circuit power to maximum turbine output, the so-called short-circuit power ratio

$$k_{kl} = \frac{S_{kV}}{\sum S_{A\,max}} \geq 50, \tag{4.13}$$

where S_{kV} represents the short-circuit power at the point of common coupling. This evaluation only applies for radial networks and leads to an absolute consideration which supplies a good approximation for motors and other consumers. In the generator mode, however, there can be significant deviations from the results according to Equation (4.12), as shown in Figure 4.36.

According to reference [Chapter 3, Reference 1], the magnitude and direction of the active and reactive components in accordance with the vector diagram (Figure 4.36) can be included in the evaluation of the voltage changes. This generally yields lower grid effects.

If the grid impedance angle ψ_{kV} and the grid short-circuit power S_{kV} at the point of common coupling are known, the voltage rise is simple to calculate, taking into account the supply angle φ, by

$$\Delta u_a = \frac{S_{A\,max\,1\,min}\cos(\psi_{kV} + \varphi)}{S_{kV}}. \tag{4.14}$$

Here

$$S_{A\,max\,1\,min} = \frac{P_{NG}p_{1\,min}}{\lambda} \tag{4.15}$$

represents the maximum turbine mean value of apparent power for the duration of one minute, P_{NG} represents the nominal output of the generator, λ represents the power factor, which generally corresponds to the fundamental value ($\lambda = \cos\varphi$), and $p_{1\,min}$ represents the maximum relative active power over one minute (e.g. $p_{1\,min} = 1.3$). We will therefore not go into more detail on this subject at this point.

In addition to voltage changes related to turbine and primary available output, switching processes, tower effects, etc., lead to aperiodic or periodic voltage changes and flicker, particularly in fixed-speed wind turbines.

Switching-dependent voltage changes

Voltage changes due to the connection and disconnection of individual systems and generator systems may not generally exceed

$$\Delta u_{max} = k_{i\,max}\frac{S_N}{S_{kV}} \le 2\% \tag{4.16}$$

if switching processes take place no more frequently than once every 1.5 minutes. At very low switching frequencies the power supply companies may permit greater values. In Equation (4.16), S_N represents the nominal apparent power of a single unit and $k_{i\,max}$ represents the ratio of start-up to nominal current of the generator. If no figures are available, a maximum value of $k_{i\,max} = 8$ can be assumed for asynchronous generators, for example. If they are connected at close to their synchronous speed ($0.95n_1 \ge n \ge 1.05n_1$) a value of $k_{i\,max} = 4$ can be assumed. In the case of ideally synchronized synchronous generators and inverters, by contrast, calculations can be based upon $k_{i\,max} = 1$.

The 'grid-dependent switched current factor' $k_{i\psi}$ should be provided and proven by turbine manufacturers or included in testing reports supplied with the wind turbine. This figure evaluates the level of current and its path over time during the switching process and is quoted as a function of the grid impedance angle ψ in the testing report. This allows the fictitious 'equivalent voltage change'

$$\Delta u_{eq} = k_{i\psi}\frac{S_N}{S_{kV}} = \frac{k_{i\psi}}{\lambda} \times \frac{P_N}{S_{kV}} \quad \text{where } \lambda = \cos\varphi \tag{4.17}$$

to be determined, which again may not exceed a limit value of 2%. The switching processes of several generators should be staggered over time. In the event of maximum voltage changes the interval must be at least 1.5 minutes and at half the permissible value it may be a maximum of 12 seconds.

4.3.3.4 Flicker

Periodic and aperiodic fluctuations in grid voltage cause corresponding changes in brightness in filament bulbs, which causes a 'flickering' of the light. Due to the high sensitivity of the human eye to changes in brightness, deviations of the voltage supplied to consumers from the steady-state value (usually nominal voltage) must be limited. An assessment plan, evaluation criteria and permissible limit values, which characterize the subjective impression of brightness changes, are given in detail in reference [Chapter 3, Reference 1].

In lighting engineering, brightness variations are called 'flicker'. Thus flicker must in the future be related to the voltage deviations that cause it. When discussing flicker disturbance factors, we must differentiate between short-term mean values

$$A_{st} = P_{st}^3 \tag{4.18}$$

in a 10 minute interval (with the short-term flicker level P_{st}, which will not be considered further here) and the 2 hour mean value A_{lt} that prevails in the long term,

$$A_{lt} = P_{lt}^3 = \frac{1}{12} \sum_{i=1}^{12} A_{st\,i} \le 0.1, \tag{4.19}$$

where the long-term flicker value is

$$P_{lt} = \sqrt[3]{\frac{1}{12} \sum_{i=1}^{12} P_{st\,i}^3} \le 0.46. \tag{4.20}$$

EN 50 160 allows $P_{st} = 1.0$ (Europe wide). The voltage deviations due to flicker are thus made up of 12 sequential short-term values P_{st} over a period of 120 minutes. Furthermore, P_{st} is proportional to the value of the voltage change. A_{st} is inversely proportional to the repeating rate of the voltage.

For wind turbines the turbine flicker coefficient c and the flicker phase angle $\varphi_f > 0$ are quoted in test reports or the manufacturer's data. These turbine-specific values, along with the nominal power of the individual turbine S_N and the short-circuit power at the point of common coupling S_{kV} according to

$$P_{lt} = c \frac{S_N}{S_{kV}} |\cos(\psi_{kV} + \varphi_f)| \tag{4.21}$$

determine the flicker generated by the plant at the point of common coupling.

When several individual turbines and wind farms are connected, the long-term flicker level $P_{lt\,i}$ should be determined separately for each turbine and the resulting value for the flicker disturbance factor calculated using

$$P_{lt\,res} = \sqrt{\sum_i P_{lti}^2}. \tag{4.22}$$

If we are considering n identical plants, the resulting flicker disturbance factor can be calculated as follows:

$$P_{lt\,res\,estimated} = \sqrt{n}\ P_{lt} = \sqrt{n}\, c\, \frac{S_N}{S_{kV}}. \tag{4.23}$$

If the limit value is exceeded, the flicker-related phase angle φ_f according to

$$P_{lt\,res} = P_{lt\,res\,estimated} |\cos(\psi_{kV} + \varphi_f)| \tag{4.24}$$

can be taken into account.

The above discussions show that disruptive voltage fluctuations can be minimized, or even avoided altogether, in all configurations by suitable turbine design and by control engineering intervention [4.59]. If necessary, measures for voltage support can be taken (see Section 4.6), either by rotating or power electronic phase shifters or directly by turbine frequency converters.

The desired speed variations in wind turbines can be achieved in the case of fixed-frequency coupling by changing the gear transmission or in fixed transmission by adjusting the generator frequency with the aid of electronic power frequency conversion systems. However, the grid compatibility of these components with regard to harmonics and subharmonics, which will be described briefly below, is of particular importance in this context.

4.3.4   Harmonics and subharmonics

Wind energy can make a useful contribution to the supply of electrical power if high levels of power can be achieved over the usable land area. Existing or new grid connections must be used as cost-effectively as possible. This can be achieved by connecting several or a great many wind turbines together to form wind farms. With the power concentrations found here, however, the compatibility of the supplying turbines with the grid must be ensured, since the grid's capacity is limited. Clear differences are evident here, depending upon the design of the system and the construction of components. Therefore, tendential relationships and problems that can be expected in the connection of different configurations will be examined below.

The combined effect of the units can give rise to

- smoothing or weakening effects or
- amplifying influences

on the harmonics and subharmonics in the grid, and these effects must be assigned particular importance. A basic discussion of circuitry and the design of power converters, generators, etc., including likely grid effects, has already been given in the preceding sections. Therefore, we will limit the following discussion to some comments on fundamental oscillations and harmonics.

Harmonics are sinusoidal oscillations of currents and voltages, the frequencies f_v of which are integer multiples of the fundamental frequency of the grid. These occur if currents or voltages have a nonsinusoidal form. They can be caused in electricity supply grids by

- consumers with nonlinear impedance characteristics,
- saturation influences in electrical machines and
- power electronic equipment.

The following effects in particular should be highlighted:

- functional disturbances in power electronic equipment that is not resistant to jamming,
- premature ageing of dielectrics and insulation materials, which may for example, lead to the destruction of capacitors,
- disturbance of audio-frequency ripple control systems and
- possible faults in protective devices.

Periodic nonsinusoidal paths of deformed currents and voltages over time (see Figure 4.16) can be considered to be the superposition of the fundamental with different harmonics, and may be represented, for example, for voltage in the form of an equation

$$u(t) = U_0 + \sum_{v=1}^{\infty} U_v \sin(\omega_v t + \varphi_v).$$ (4.25)

Similar equations can be derived for current. Here U_0 represents the direct-voltage component, U_v the amplitude, ω_v the angular frequency and φ_v the phase displacement angle of the vth harmonic, where

$$\omega_v = 2\pi f_v = 2\pi v f_1$$ (4.26)

characterizes the multiple of fundamental frequency f_1 and the corresponding ordinal number

$$v = \frac{f_v}{f_1}$$ (4.27)

characterizes their integer multiples. These parameters can be found mathematically by a Fourier analysis and using measuring techniques with the aid of frequency-selective evaluation.

Sinusoidal oscillations with frequencies that do not correspond to integer multiples of the basic frequency f_1 are called subharmonics. This type of oscillation is mainly caused by frequency converters, but can also be caused by rotating electrical machines. For differentiation from the integer ordinal number v of harmonics, the noninteger factors

$$\mu = \frac{f_\mu}{f_1}$$ (4.28)

are used to characterize the subharmonics. The subharmonic frequencies mainly occur as upper and lower sidebands of harmonic frequencies and can be lower than the fundamental frequency (i.e. $\mu < 1$).

Ensuring the quality of supply in the public grid necessitates the restriction of harmonics and subharmonics to a grid-compatible level. EN 61000-2-2 stipulates the so-called compatibility level u_{vVT} as limit values for individual harmonic frequency voltages, which may not be exceeded, according to the equation

$$\frac{U_v}{U_1} = u_v \leq u_{vVT}$$ (4.29)

for the harmonic levels U_v in relation to the fundamental U_1. U_v and U_1 represent root-mean-square (r.m.s.) values. For the sake of simplicity, U_1 can be replaced by the nominal value of the grid voltage for the electrical supply grid. The compatibility levels for harmonics are listed in Table 4.5; for subharmonics, a limit value of

$$u_{\mu VT} = 0.2\%$$

is generally specified.

The compatibility levels are valid in the low-voltage and medium-voltage grid, and must be maintained at the so-called point of common coupling and the point of supply. They facilitate the evaluation of individual harmonic levels.

Table 4.5 Compatibility levels $U_{\nu VT}$ in the medium-voltage grid for harmonics up to the ordinal number $\nu = 25$ (EN 6100-2-2 or VDE 0839 Part 88, IEC 77A)

Odd values of ν that are not divisible by 3

Ordinal number ν	5	7	11	13	17	19	23	25	>25
$U_{\nu VT}(\%)$	6.0	5.0	3.5	3.0	2.0	1.5	1.5	1.5	$0.2 + 1.3 \times 25/\nu$

Odd values of ν that are divisible by 3

Ordinal number ν	3	9	15	>15
$U_{\nu VT}(\%)$	5.0	1.5	0.3	0.2

Even values of ν

Ordinal number ν	2	4	6	8	10	>10
$U_{\nu VT}(\%)$	2.0	1.0	0.5	0.5	0.5	0.2

In wind turbines with frequency converters, the harmonic currents should be documented by the manufacturer (e.g. during type testing). According to reference [4.23], with only one point of common coupling in the medium-voltage grid the permissible harmonic currents found are from the specific values $i_{\nu,\mu,\,\text{perm.}}$ according to Table 4.6 multiplied by the short-circuit power at the point of common coupling, i.e.

$$I_{\nu,\mu\ \text{perm.}} = i_{\nu,\mu\ \text{perm.}} S_{kV}. \tag{4.30}$$

If several turbines are connected at the point of common coupling, the permissible harmonic currents for a turbine are found by multiplying by the ratio of turbine apparent power S_A to the

Table 4.6 Total harmonic currents permissible at a medium-voltage grid, related to short-circuit power, that are generated by directly connected turbines [4.23]

Ordinal number ν, μ	Permissible specific harmonic current $i_{\nu,\mu\ \text{perm.}}$ (A/MVA)	
	10 kV grid	20 kV grid
5	0.115	0.058
7	0.082	0.041
11	0.052	0.026
13	0.038	0.019
17	0.022	0.011
19	0.018	0.009
23	0.012	0.006
25	0.010	0.005
>25 or even number	$0.06/\nu$	$0.03/\nu$
$\mu < 40$	$0.06/\nu$	$0.03/\nu$
$\mu > 40^a$	$0.18/\nu$	$0.09/\nu$

^{a}Integer and noninteger within a bandwidth of 200 Hz.

power that can be drawn or the planned supply power S_{AV} at the point of common coupling, according to

$$I_{v,\mu,A\ perm} = I_{v,\mu\ perm.} \frac{S_A}{S_{AV}} = I_{v,\mu\ perm.} S_{kV} \frac{S_A}{S_{AV}}. \tag{4.31}$$

When several wind turbines of the same type are connected together the turbine output is replaced by the sum of the individual turbine outputs

$$S_A = \sum S_{nE},$$

i.e.

$$I_{v,\mu,A\ perm} = I_{v,\mu\ perm.} \frac{S_{nE}}{S_{AV}} = I_{v,\mu\ perm.} S_{kV} \frac{S_{nE}}{S_{AV}}. \tag{4.32}$$

In turbines that are not of the same type this value represents only an upper estimate according to reference [4.23].

Medium-voltage to low-voltage voltage transformers that do not transmit a zero phase-sequence system are normally used in public supply grids. In this type of grid connection, the values given in Tables 4.5 and 4.6 for the nearest order can be taken as the basis for harmonics with ordinal numbers not divisible by three.

One measure of the distortion of grid voltage is represented by the harmonic distortion factor. In this context we must differentiate between the harmonic distortion factor k_u, as defined in the past by DIN EN 61000-2-2, and the total harmonic distortion k_{THD}. The harmonic distortion factor gives the ratio of the actual value of the oscillation without the fundamental to the actual value with the fundamental for voltage, according to

$$k_u = \sqrt{\frac{\sum_{v=2}^{50} U_v^2}{\sum_{v=1}^{50} U_v^1}} = \sqrt{\frac{U_2^2 + U_3^2 + U_4^2 + \cdots}{U_1^2 + U_2^2 + U_3^2 + U_4^2 + \cdots}}. \tag{4.33}$$

Total harmonic distortion, on the other hand, relates the actual value of the oscillation without the fundamental to the fundamental of voltage, according to the equation

$$k_{THD} = \sqrt{\frac{\sum_{v=2}^{50} U_v^2}{U_1^2}} = \sqrt{\frac{U_2^2 + U_3^2 + U_4^2 + \cdots}{U_1^2}}. \tag{4.34}$$

The two factors are the same at low values of up to around 20%. From 30%, small differences are found, e.g. where $k_{THD} = 0.5$, $k_u = 0.45$, or where $k_{THD} = 1$ the corresponding value of k_u is 0.7 [4.62].

In practice, owing to the limited bandwidth of the grid transmission elements, the spectrum of harmonics is considered up to an ordinal number $v = 40$ or up to a frequency $f_v = 2$ kHz.

Of all the different methods of connecting wind turbines to the grid, line-commutated frequency converters in particular create harmonics. These can be attributed to the approximately square-wave grid currents and the commutation grid voltage dips. However, such systems were only used in wind turbines up until the end of the 1990s. Today IGBT frequency converters are used, which due to their high switching frequencies can supply near-sinusoidal currents.

An inverter of pulse number p generates – if we disregard commutation effects and assume an ideally smoothed direct current in the d.c. link – a harmonic spectrum with the ordinal numbers

$$v = kp \pm 1, k = 1, 2, 3, \ldots. \tag{4.35}$$

The amplitudes or r.m.s. values of the current

$$I_v = \frac{I_1}{v} \tag{4.36}$$

decrease according to the ordinal number v. For a six-pulse inverter, the 5th and 7th harmonics are thus the strongest. These do not occur, however, for a twelve-pulse inverter, which means that the ordinal numbers 11 and 13 are the first harmonics. All the subsequent harmonics (17, 19, 23, 25, 29, 31, etc.) exhibit correspondingly lower amplitudes. Taking commutation effects into account, harmonics of other ordinal numbers also occur in practically measurable spectra. In addition, some levels deviate from the values given in Equation (4.20). The emphasis is placed upon the realization of practical findings in the descriptions below.

For wind turbines with asynchronous geenerators and direct connection to the grid, harmonics have been measured in the normal form up to over 2000 Hz and their amplitude spectra are represented logarithmically up to the 40th harmonic. This showed that a greater number of turbines caused lower amplitudes of harmonics and subharmonics, in particular those of a low order (cf. Figure 4.45).

This phenomenon can be attributed to an increase in the capacity of the grid due to the increase in

- short-circuit power and
- the filtering effect of the wind turbine with the capacitive compensation system,

which increase with the increasing number of turbines. Increases of individual harmonics (here the 11th), which sometimes occur, can be reduced, if this is necessary, by the design of the magnetic circuit and the coil to achieve the desired values.

In contrast to this, turbines coupled to the grid via six-pulse thyristor power converters usually exhibit different behavior with regard to harmonic effects. For a larger number of converters, and thus higher power components in the grid supply, the amplitudes of harmonics are higher over the entire frequency spectrum, as shown in Figure 4.46.

We therefore find that for wind turbines that supply their energy directly to the grid from asynchronous generators the harmonic component is reduced as the number of turbines, and thus the output, increases. On the other hand, for turbines fitted with variable-speed generators and line-commutated power converters, a more marked influence on the grid by harmonics can be observed as the number of turbines, and thus the output, increases.

The contrasting behavior of different types of coupling can, as shown in the following discussion, be used to reduce the grid effects caused by harmonics, when different systems are connected together.

The measurements in Figure 4.47 show that by the combination of wind turbine systems that have variable-speed synchronous generators and are connected to the grid via a power converter with those that have fixed-speed asynchronous generators and a direct connection

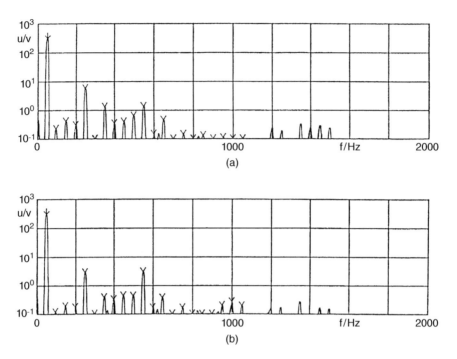

Figure 4.45 Voltage spectra of wind turbines with asynchronous generators and direct connection to the grid: (a) in solo operation (one turbine) and (b) in combined operation (18 turbines)

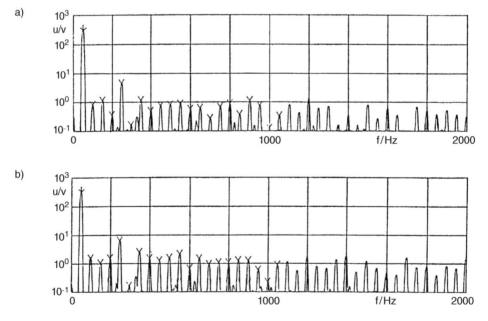

Figure 4.46 Voltage spectra of wind turbines with variable-speed generators and grid connections via line-commutated power converters (a) in solo operation (one turbine) and (b) in combined operation (five turbines)

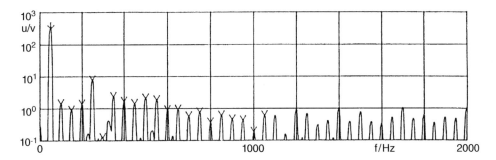

Figure 4.47 Voltage spectra for the combined operation of wind turbines with different grid connections (nine synchronous generators connected via power converters and two asynchronous generators coupled directly to the grid)

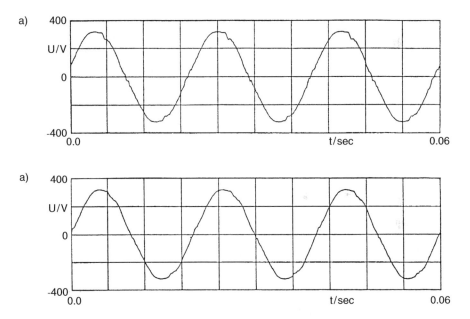

Figure 4.48 Graph of the grid voltage in phase L_1 for supply by (a) five power-converter-coupled turbines (275 kW) and (b) nine power-converter-coupled turbines (320 kW) in combination with two turbines with asynchronous generators (330 kW) coupled directly to the grid of around the same total output

to the grid, a clear reduction in harmonic and subharmonic amplitudes can be achieved, thus improving voltage characteristics (Figure 4.48). Upon closer inspection, lower commutation effects can be seen in Figure 4.48(b). This type of configuration therefore allows grid effects to be significantly reduced.

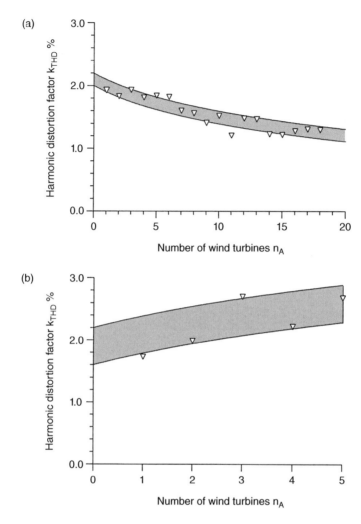

Figure 4.49 Harmonic distortion factor as a factor of the number of wind turbines for (a) asynchronous generators coupled directly to the grid and (b) synchronous generators operating through rectifiers, constant current d.c. links and inverters

Data collated using random samples on states at different wind speeds and fluctuations at corresponding times and grid conditions show the following trends, as illustrated in Figure 4.49, for the harmonic distortion factor, this being the characteristic variable for harmonics. In general, the path of the harmonic distortion factor k_u and the total harmonic distortion k_{THD} for low turbine numbers up to around $n_A = 20$ can be approximated in the form

$$k_{THD} = k_{THD0} + k_{THD1} \ln (1 + k_{THD2} n_A), \qquad (4.37)$$

where k_{KHD0} represents the initial value as the axis intercept, which is highly dependent upon the grid short-circuit power and the grid state. Typical values are

$$k_{THD0} = 1.5 \text{ to } 3\%.$$

The gradient is represented by k_{TDH1}. In wind turbines supplied via power converters, this usually takes on positive values, e.g.

$$k_{THD1} = 0.85$$

and in turbines with generators coupled directly to the grid it usually takes on negative values, e.g.

$$k_{THD1} = -0.55.$$

The factor k_{THD2} influences the elongation of the approximation function (Equation (4.37)) for values in the range

$$k_{THD2} = 0.2 \text{ to } 0.4.$$

Pulse-controlled inverters are used for variable-speed systems in the current generation of wind turbines. Due to the relative high-frequency switching processes in the kHz range in power technology, currents and voltages can approximate the sinusoidal form very well by the selection of pulse pattern or so-called bandwidth control. Harmonics with low ordinal numbers are thus largely avoided. In the evaluation of compatibility levels the fixed or variable pulse frequencies of such systems generally lie at the upper end of the range of ordinal numbers and thus usually exhibit low amplitudes. If necessary, these can be reduced by the use of compact filters.

4.3.5 Voltage faults and the fault-ride-through (FRT)

Falling branches onto lines, ageing insulation failures in cables, insulators, etc., and cable or line damage during construction, lightning strikes, etc. lead to short circuits and earthing and result in a voltage fault at the affected phases. In order to be able to keep the grid in operation, it is necessary to maintain the operating voltage or support it. For this purpose, provision must be made for the plants feeding in the energy to provide reactive power.

The behavior of voltage in the grid, however, is dependent on the respective conditions of the grid such as the voltage level, y-point handling, size of the grid, etc. Thus, the requirements of the grid operator in the form of their given grid codes can vary greatly. The leader in this was the German E.ON grid operator.

Already in the decade before the application of the grid codes, it was shown that wind turbines are also able to support the grid (see Section 4.6.2) when the converters permit this. For this reason, due to the high installation power of wind turbines in Northern Germany, the grid codes were strongly oriented to maintaining the grid and were also extended to other systems besides wind turbines.

4.3.5.1   Behavior of the feed systems in the case of grid faults

Because converter-fed supply systems such as large wind turbine and photovoltaic plants have a different behavior at the grid than, for instance, directly grid-coupled generators of conventional power stations, etc., a basic difference is made between these two energy-generating plants.

Three-phase short circuits in the grid usually result in the largest short circuit flows. They represent the largest loading for the affected components. For this reason, they are often selected as reference cases for malfunctions.

Direct-coupled synchronous generators (Type 1) are the most common types used in conventional power stations. They are also used increasingly in wind turbines with three-phase variable drives (Windrive). In a three-pole fault, these types of feed systems must remain on the grid for the first 150 ms to maintain the normal operating condition and thus contribute to grid stability. Figure 4.50 shows the course of the voltage as a borderline of the generator operation for a three-phase fault. Above this curve, the generator must maintain its operation. Only after the fault clearing time of 150 ms, may the plant separate itself from the grid in order to prevent damage to electrical components.

If, in case of a fault, the plant would additionally load the already loaded grid due to its own requirement, then separation from the grid within 100 ms is permissible.

All other generator types and grid connection concepts (Type 2) are of great importance for wind turbines. Modern designs with electrical or permanently excited synchronous generators or asynchronous generators, all in connection with fully rated converters (Figure 3.4(i) to (k)) as well as double-feed asynchronous generator with part-load converters (Figure 3.4(f)) are also considered here. All other transformer systems to Figure 3.4 have either lost their importance (a) or were unsuccessful in the market (c,d,e,h and l).

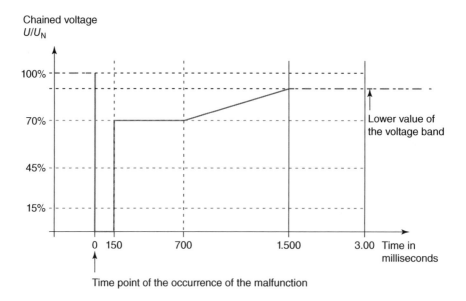

Figure 4.50   Borderline curve for direct-coupled synchronous generators [4.63]

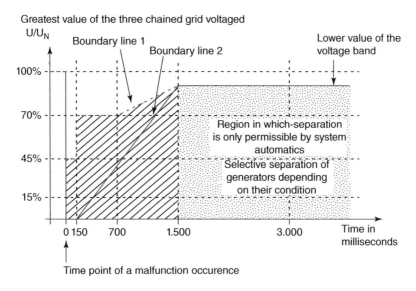

Figure 4.51 Boundary line for generating units type 2 [4.63]

Therefore, there must be no separation of the wind turbine from the grid for wind turbines with fully or partly rated converters for malfunctions that occur outside of the protective region of the plant. The short circuit current I_k'' must be fed in for the duration of the fault. Because of the current path via converter, this should be fed in the dimension $I_k/I_N = 1$ (1 pu) as reactive power and not greatly exceed the rated current amount I_N (to 1.2 pu)

Figure 4.51 shows the borderlines 1 and 2 that determine the operating range of wind turbines. The wind turbine must not be separated from the grid above the borderline 1, and the generator must not reach an unstable operating condition. In the hatched area between the borderlines 1 and 2, the wind turbine must remain at the grid; it must pass through the fault and must contribute to the voltage stabilization by means of reactive power feeding. Only after falling below borderline 2 is a separation of the wind turbine from the grid permissible. The switch-off conditions are dependent on the duration of the malfunction and from the amount of the remaining grid voltage. In addition, the power factor of the generator or converter must be considered.

4.3.5.2 Voltage support

If the voltage in the grid falls below 85% of the rated voltage ($0.85 \cdot U_N$), then this corresponds to an under-excited generator operation (Figure 3.10) and a wind turbine must be separated from the grid after a delay time of 0.5 s. With the delayed grid separation, the borderlines are meant to be adhered to and the voltage support in the case of a malfunction is to be ensured. Therefore, the result must be a reactive power feed besides the provision of a short circuit current.

The **voltage controls** that are to be maintained when passing through a fault must be carried out in accordance with the parameters shown in Figure 4.52. These are meant to ensure that

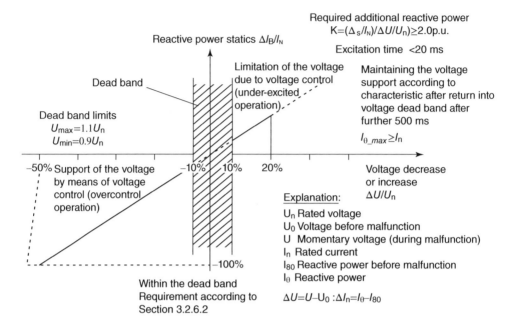

Figure 4.52 Definition of the voltage support by the generator [4.135]

sufficient reactive power is fed into the grid within 20 ms after fault recognition. In this, a displacement of the rated current to a reactive power of 2% per percent of the voltage dip is undertaken, meaning that with 50% grid voltage dip, the generator or converter must deliver 100% of the rated voltage as reactive power.

Further conditions are that with the increase of the voltage after clearing the fault of more than 90%, meaning within the dead band of the controls, the voltage support of the wind turbine must remain active for a further 500 ms. In this way, balancing processes after voltage return are caught up. With the operation of several wind turbines at a common grid connection point (GCP), a quarter of the plants must be separated from the grid after 1.5 s when the rated voltage falls and remains at 80% of the rated voltage. The further three-quarter remaining at the grid is separated from it after 1.8, 2.1 and 2.4 s. With the re-start of the wind turbines, a grid synchronization must first be undertaken and the rise of the fed-in active power must not exceed 10% of the grid connection capacity per minute.

4.3.5.3   Grid codes in comparison

Besides the Germans, many foreign grid operators have also promulgated guidelines for the ride through a grid fault. The individual grid codes differ mainly in the amount and dura-tion of the voltage dip and the resulting borderlines. Figure 4.53 shows the voltage profile of low-voltage ride-through (LVRT) of various national and international grid operators. The bor-derlines of European grid operators are very close together. Thus, they make similar demands on the wind turbines to be connected. The reason for this is the similar grid topologies within

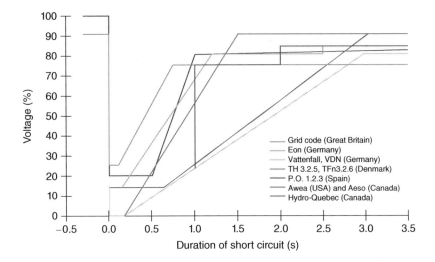

Figure 4.53   Voltage profile for LVRT – requirements of various countries

the Union for the Co-ordination of Transmission of Electricity (UCTE) and the European Network of Transmission System Operators for Electricity (ENTSO-E) grid.

A substantial difference between the national grid codes is in the types of faults. German grid codes with test requirements are based on the simulation of two- and three-phase short circuits without earthing.

The three-phase short circuit provides the greatest short circuit current of all types of faults and is thus the largest load for testing. A two-phase short circuit, in contrast, usually results in the largest short circuit current of asynchronous faults. If the wind turbine to be tested can withstand this fault, then it is generally assumed that other malfunctions will not result in higher loads for the plant and there is therefore no danger for a safe ride-through of grid voltage collapses.

Foreign grid operators partly demand further types of fault such as, for instance, *National Grid Electricity Transmission PLC* (UK) requires the verification of withstanding single-phase earthing which, up to now has not been tested as the generator or converter recognizes this on the low-voltage side as a two-phase voltage collapse [4.64]

4.3.5.4   Plant testing and short circuit simulation

The aim of a plant test is to check, by means of a grid fault, that the energy-generating plant to be tested fulfils the grid code requirements. Here, the plant must provide the required reactive power as well as a targeted grid separation and re-coupling. Although the sequences of the grid fault especially load the electrical components of the wind turbine, they must not suffer any damage because of this. In order to prove this, a voltage is applied to the connection terminals that simulates the real fault.

In order to prevent real short circuits and voltage dips in the grid, as there is a danger of damage to the grid components up to grid failure, use is made of test circuits for these test sequences. These are inserted in the current path between wind turbine and the common GCP.

Figure 4.54 Fraunhofer IWES converter-based FRT plant (30kW-class)

The test circuits permit targeted voltage dips to be simulated without impairing the operation of the grid. Test circuits for this type of grid voltage collapse investigation can be structured on the basis of self commutated converter systems.

- Transformer with stepped outputs
- Inductive or
- Ohmic voltage controllers.

The cost differences between these systems are enormous. According to the order shown above, the converter-based systems provide the most accurate and broadest investigation possibilities but at by far the highest costs, whereas the voltage splitters have the lowest variation possibilities at the lowest cost. Thus, the test systems should have at least three times the plant power. For a 7,5 -MW wind turbine, the test system should have a capacity of 22.5 MW (as short circuit power). This means that especially converter and also transformer systems become very expensive due to their large overdimensioning.

Converter-based test plants (Figure 4.54) are used especially for powers up to the 100-kW region for machines and inverter investigations for photovoltaic plants and small wind turbines of the 1 to 10 kW classes. Transformer-based test units are used mostly for stationary test installations because of the high transformer weight. Test containers (Figure 4.55) based on voltage splitters are used for wind and photovoltaic plants in the megawatt class.

4.4 Resonance Effects in the Grid During Normal Operation

If the grid is considered at the point of connection of one or several wind turbines, the total impedance of the configuration is decisive for behavior during supply, and this can be determined based on the prevailing combination of ohmic resistances, inductances and capacitances in the supplier, distributor and consumer systems. Depending upon the quantity, type of connection and impedance level of these individual elements, self-resonant frequencies can be

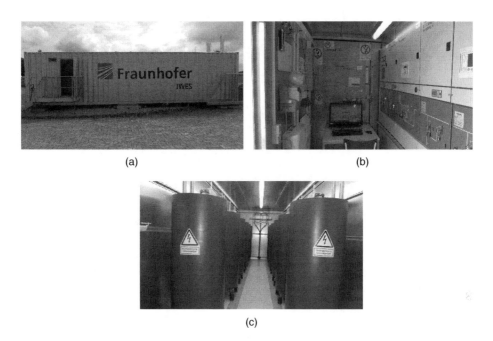

Figure 4.55   Fraunhofer IWES FRT Test Container (MW-Class): (a) Container; (b) Switching cabinett; (c) Inductivities

determined for the part of the grid under consideration, which can cause rises in the current

$$I_v = \frac{U_v}{Z_v} \text{ where } Z_v \to 0 \qquad\qquad (4.38)$$

or the voltage

$$U_v = I_v Z_v \text{ where } Z_v \to \infty \qquad\qquad (4.39)$$

in connected components when excited. Such excitation can be brought about in the electrical supply grid by nonlinear suppliers and consumers, e.g. due to saturation effects in transformers, reactors and rotating electrical machines and in particular by power converter units, because in addition to the fundamental frequency these systems also give rise to higher-frequency components due to nonsinusoidal current and voltage paths.

The harmonic content in the grid can be kept low by the appropriate design of electric machines and can even be reduced if machines are selected specifically to achieve this (see Section 4.3.4). Therefore, in addition to power electronic consumer systems, resonance excitation in the grid can be particularly attributed to supply via power converters, which was common in wind turbines and controllable drives with line-commutated frequency converters until approximately 1990.

Taking the example of the Westküste wind farm at the beginning of the 1990s, the equivalent circuit diagram shown in Figure 4.56 [4.65] is derived from the block diagram in Figure 4.57. Using a program developed for this purpose [4.66], the impedance path can be graphically represented up to the 40th harmonic, as shown in Figure 4.58, for the point of connection to the harmonic source, shown here as a wind turbine with frequency converter supply.

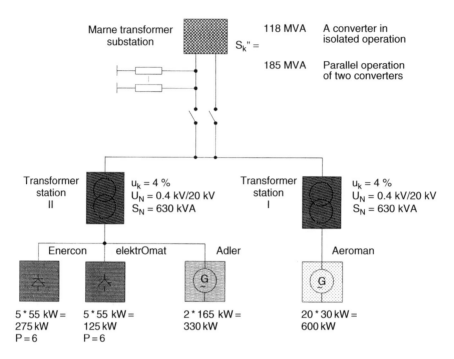

Figure 4.56 Equivalent circuit diagram of the Westküste wind farm [4.65], where 1 to 20 indicate electrical sections of the wind turbine with a compensation device (1990)

Figure 4.58 illustrates noncritical operation in all frequency ranges. Relevant resonance points with local maximum impedance lie above the 13th harmonic (650 Hz), which means that at most one-thirteenth of the maximum value of the fundamental component of the current would be effective. Thus voltages capable of causing damage to components do not occur at any frequency.

The impedance path calculated in Figure 4.58 is shown for a short-circuit power of the Marne grid of $S_K'' = 185$ MVA. If the Westküste wind farm were to supply a grid with around a quarter of the capacity, the impedance path shown in Figure 4.59 would prevail. This clearly illustrates that at lower short-circuit power the resonance point is moved to lower frequencies (above the 11th harmonic) and leads to higher impedances and thus a greater voltage drop at the point under consideration. With decreasing short-circuit power, therefore, operation becomes increasingly critical.

If the grid is connected via overhead power transmission lines, the stronger inductive behavior of the overhead power transmission line slightly increases the resonant frequency compared with cables at the same maximum impedance. If individual wind turbines with asynchronous generators and compensation units are disconnected from the grid, small changes occur in the impedance path, as shown in Figure 4.60. Only in the transition between the connected groups can clear differences be recognized due to different line lengths.

Thus, for example, if the first turbine is connected with the corresponding grid section, the high maximum resonance at 1741 Hz ($\mu = 34.82$) is considerably reduced. Similarly, this trend is evident for the connection of the fifth turbine with its grid branch at the mean ordinal number

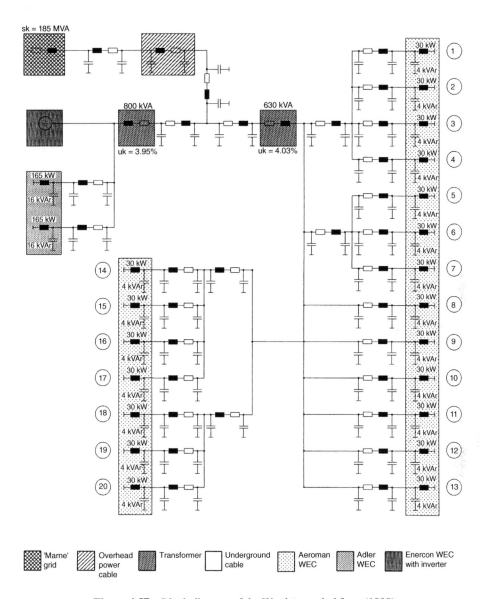

sk = 185 MVA

800 kVA
uk = 3.95%

630 kVA
uk = 4.03%

165 kW
16 kVAr
165 kW
16 kVAr

30 kW
4 kVAr

'Marne' grid	Overhead power cable	Transformer	Underground cable	Aeroman WEC	Adler WEC	Enercon WEC with inverter

Figure 4.57 Block diagram of the Westküste wind farm (1990)

($\mu = 23.20$) and for the connection of the first turbine of the last group of three (18th turbine) at the lower resonance point ($\mu = 14.04$). In contrast to this, if the stub cable is connected with the 14th turbine, a further resonance point ($\mu = 17.20$) is created.

In a summarized comparison between 20 turbines connected to the grid and their complete disconnection, the relevant impedance values are found to be only slightly changed. The slightly marked resonance point of 20 turbines at 675 Hz ($\mu = 13.50$) completely disappears when all turbines are disconnected. In contrast to this, the slightly higher value at 843 Hz

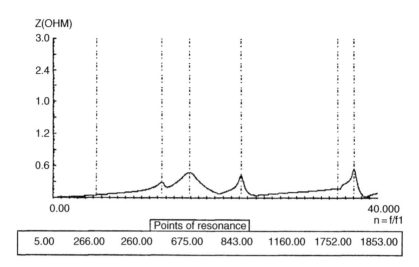

Figure 4.58 Impedance path of the network at the Westküste wind farm with a short-circuit power $S''_K = 185$ MVA

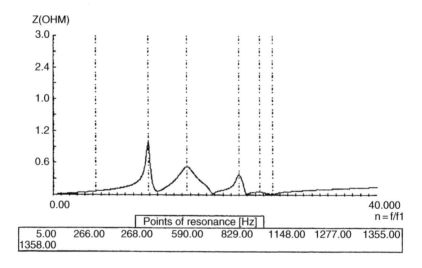

Figure 4.59 Impedance path in a weak grid with a short-circuit power $S''_K = 50$ MVA

($\mu = 16.86$) is moved insignificantly to 830 Hz ($\mu = 16.60$). The next highest resonance point at 1160 Hz ($\mu = 23.20$) is retained at 1158 Hz ($\mu = 23.16$). The highest value, which occurs in the range up to the 40th harmonic, has not, however, moved significantly from 1853 Hz ($\mu = 37.06$), being 1741 Hz ($\mu = 34.82$).

If approximately the same cable impedances are summarized in the equivalent circuit diagram shown in Figure 4.56, the simplified structure in Figure 4.61 is the result. The resulting impedance path as shown in Figure 4.62 corresponds very well to Figure 4.58. Only at the maximum value, in the region of the 37th harmonic, which is of little relevance due to the

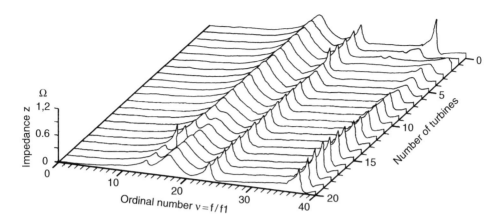

Figure 4.60   Impedance path of the grid at the Westküste wind farm ($S_K'' = 185$ MVA) as a function of the ordinal number and number of plants (0 to 20 plants of 30 kW each)

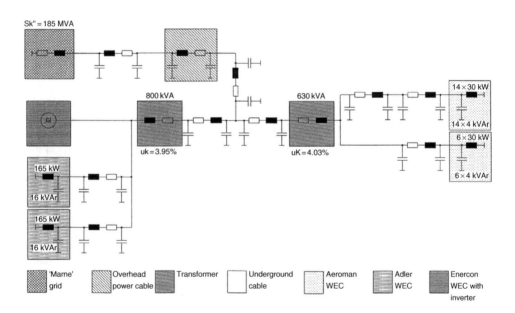

Figure 4.61   Simplified equivalent circuit diagram for the Westküste wind farm

high ordinal number and thus low amplitude, the deviation can be recognized. This type of approximation is largely adequate for determining any resonance points that may occur and could take on critical values.

Further discussions below will cover attempts at remedial measures against resonance effects so that, taking into account the grid control, noncritical operation of wind turbines connected to the grid can always be achieved.

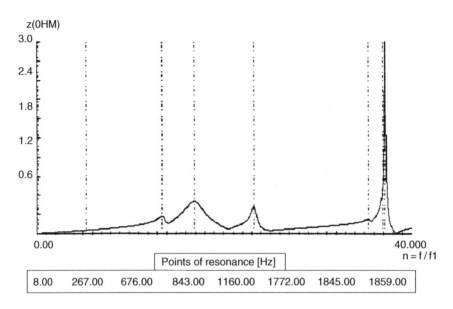

Figure 4.62 Impedance path of the simplified network at the Westküste wind farm

4.5   Remedial Measures against Grid Effects and Grid Resonances

The use of electronic power conversion systems, as well as bringing technical and economic advantages such as good controlability and high efficiency, is also associated with disadvantages, in particular due to grid voltage distortion. The effects are dependent upon the type of power semiconductor, the connection type, the operating state of the power converter and the prevailing grid parameters at the point of connection.

The operation of powerful frequency converters in connection with weak grids leads to an increase in harmonic voltages in the grid, which could have negative effects on the connected consumers. To prevent operating faults and system failures, the harmonic currents supplied by inverters must be kept low, preferably at the point of origin. Systems specifically designed to create low grid effects can be used for this purpose, or alternatively the harmonics (and voltage changes) that exist can be reduced by passive circuits and active intervention. Such devices are known as filters and grid support devices. The discussion that follows will be mainly limited to filters. These are connected in parallel at the grid supply point. Higher-frequency currents flow away via the filter unit and can thus be kept from the grid.

The design of the harmonic filter, as well as requiring precise knowledge of the grid structure and all its parameters, also calls for the specification of the permissible harmonic current at the point of connection. This so-called compatibility level is predetermined by the responsible power supply company.

4.5.1   Filters

Filters are generally made up of capacitors and inductors. Ohmic resistors have the task of reducing high impedances where filters have parallel resonances, or extending the frequency

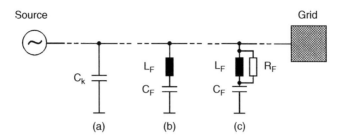

Figure 4.63 Equivalent circuit diagram of compensation and filter devices: (a) power factor correction capacitor, (b) harmonic absorber filter and (c) wide-bandpass filter

range in which filters exhibit low impedances. Only the main features will be considered in the following (Figure 4.63). Complex systems, e.g. double high-pass systems, will not be covered.

Even if capacitors are not used as harmonic filters, their behavior must still be assigned great importance. Capacitor banks consist of one or more capacitors connected in parallel. They have the task of compensating for the inductive reactive power of motors and generators (e.g. by the capacitive reactive power Q_c) and improving the power factor (Figure 4.61(a)). Due to their impedance behavior according to the simple equation

$$Z_c(f) = \frac{U_c^2}{Q_c} = \frac{1}{2\pi f C_k},\tag{4.40}$$

increasing frequency and rising capacitance C_k bring about lower impedances Z_c. This has the result that harmonic currents are led away but fundamental components are almost blocked. Thus power factor correction capacitors also have a filtering effect.

The variables mentioned in the intermediate equation characterize the reference conductor voltage U_c and the compensating reactive power Q_c. As networks usually display inductive behavior at high frequencies, the grid impedance increases with increasing frequency. Another point to be considered is that current harmonics provide an additional load on the capacitor banks. This results in warming and a reduction in insulating capacity if the capacitors are dimensioned for the fundamental power. This effect can be countered by designing the capacitors for around double the initial voltage.

Six-pulse inverters supply relatively high currents into the grid at 250 and 350 Hz. Due to their high impedance at low frequencies, a capacitor bank can only filter part of these harmonics. An increase in capacitor output would increase the filtering effect. However, the generator or inverter would be significantly overcompensated at its fundamental frequency, thus leading to a highly capacitive supply.

Better filtering can be achieved by the series connection of a capacitor and an inductor – a so-called harmonic absorber. These have infinitesimally small impedances close to their resonant frequency

$$f_{res} = \frac{1}{2\pi}\sqrt{\frac{1}{L_F C_F}}.\tag{4.41}$$

By the selection of filter inductance L_F and filter capacitance C_F as well as compensating output, the harmonic absorber can be dimensioned such that the currents near to the resonant frequency f_{res} (e.g. 250 Hz) can be absorbed. Below the resonant frequency, the filter device has capacitive behavior and, above it, inductive. Thus small compensating outputs lead to a steep resonance curve; high compensating outputs, on the other hand, bring about a shallow path.

Harmonic absorbers are best used where currents are to be filtered within a narrow frequency band. The fundamental component compensation output and the capacitance of the filter circuit can thus be kept low.

If, on the other hand, currents in a wide frequency band are to be removed from a grid, wide-bandpass filters are required. They consist of a harmonic absorber in which an ohmic resistor is connected in parallel to the inductor (Figure 4.63(c)). This parallel connection has the result that at high frequencies the inductance of the resistor is bypassed and the filter impedance is kept low over a broad frequency band. The resonant frequency of the broad-band filter (i.e. the frequency at which the filter exhibits the lowest impedance) is determined as for the harmonic absorber according to Equation (4.41). The behavior of the broadband filter is characterized by the capacitor at low frequencies and by the ohmic resistor at high frequencies.

4.5.2   Filter design

Taking the higher-voltage grid structure and the permissible compatibility level into account, a total impedance of the grid and filter unit must be achieved by filter design and the selection of components at which harmonics are kept within the range of the permissible compatibility level. We must take into account here the fact that critical operating ranges can occur due to parallel resonances between the filter unit and the grid and among the filters themselves. Moreover, filters must be robust and insensitive to parameter fluctuations such as changes to grid short-circuit power and load, and must always fulfil the necessary connection conditions.

Capacitors for reactive power compensation without an inductive unit can cause resonances and the associated voltage overshoot close to the harmonic source. In the grid areas that contain power converters, inductive-capacitors must therefore be used, whereby the resonant frequency must be selected below the first harmonic frequency.

Filter units are usually laid out in several stages. This makes it possible to tune individual filter levels to the same and different resonant frequencies. In principle, a filter circuit is laid out for the lowest harmonic frequency. Further filters for higher harmonics are dimensioned in order of increasing ordinal numbers. The resonant frequency is selected to be somewhat lower than the frequency of the harmonics to be removed, in order to avoid overvoltage and excessive capacitor currents. By using appropriate levels, the capacitive reactive power in the range of given power factors can be adapted to different load situations.

If filters with different resonant frequencies are used, a parallel resonance is created by their connection in parallel. This occurs at a frequency at which the inductive reactance component of the one filter and the capacitive reactance of the other are equal. Maximum impedance occurs at this point. Harmonic currents that occur in the region of an impedance peak can be reduced only slightly or not at all. If two filter systems (harmonic absorber, wide-bandpass filter, etc.) of different frequency designs are used, the parallel resonance must be positioned either between two harmonics or in a frequency range in which the supplied harmonic currents lie significantly below the permissible values.

It should also be noted that filter levels not only produce parallel resonances between themselves but also produce one or more parallel resonances with the higher-voltage grid, possibly resulting in excessive increases in impedance. This is the case if filter and grid impedance have the same magnitude but opposite signs, e.g. if the filter impedance is capacitive and the grid impedance has the same value but is inductive, or vice versa.

A precise analysis of current and voltage distribution at all occurring frequencies allows the actual loading of individual components in all switching states of a filter to be determined and the components to be designed accordingly. It should be noted here that the total r.m.s. values of voltage

$$U_{rms} = \sqrt{U_{1rms}^2 + U_{2rms}^2 + U_{3rms}^2 + \cdots} \qquad (4.42)$$

and current

$$I_{rms} = \sqrt{I_{1rms}^2 + I_{2rms}^2 + I_{3rms}^2 + \cdots} \qquad (4.43)$$

are made up of the sum of all the respective individual values. Thus relatively expensive simulation programmes are necessary for exact dimensioning. The voltage endurance of capacitors and the total r.m.s. current are of central importance for inductance. However, harmonic current loads, e.g. the fifth and seventh harmonics of capacitors, and the harmonic voltages of inductors must also be taken into account.

4.5.3   Function of harmonic absorber filters and compensation units

Figure 4.64 illustrates the function of the filter and compensation unit of a relatively small village supply system with a maximum output of approximately 70 kW. The frequency spectrum and voltage path (in phase L_1) are shown at around half the nominal load without a filter in Figure 4.64(a) and with a filter unit connected in Figure 4.64(b) [4.50, 4.57, 4.67]. It is not of primary importance here whether the harmonics are caused by the consumers, the photovoltaic system, the battery or the wind turbine.

All the uneven components in the spectral function shown in Figure 4.64(a) are reduced in Figure 4.64(b) to a few (six-pulse) harmonic components typical of a power converter, leaving a near-sinusoidal voltage path. This clearly illustrates the function of the filter unit.

Different conversion systems exhibit large differences with regard to harmonics, depending upon grid connection. In the case of asynchronous generators coupled directly to the grid, grid effects do not generally increase with the number of generators. On the contrary, harmonics and subharmonics that are already present in the grid are usually weakened, as shown by the measurements in Figure 4.49(a).

In the case of turbines with a frequency converter supply, by contrast, a higher harmonic component is supplied to the grid as the number of turbines increases or output rises (see Figure 4.49(b)). Six-pulse line-commutated inverters give rise to significantly greater grid effects, particularly at the fifth and seventh harmonics, than twelve-pulse inverters.

Similar effects are achieved by combining wind turbines that supply the grid via frequency converters with turbines fitted with asynchronous generators coupled directly to the grid. This is due to the increased short-circuit power and the filter function of generator and compensation reactance, which result in significantly lower grid effects (compare Figure 4.65(a) and (b)). The associated voltage paths are represented in Figure 4.48.

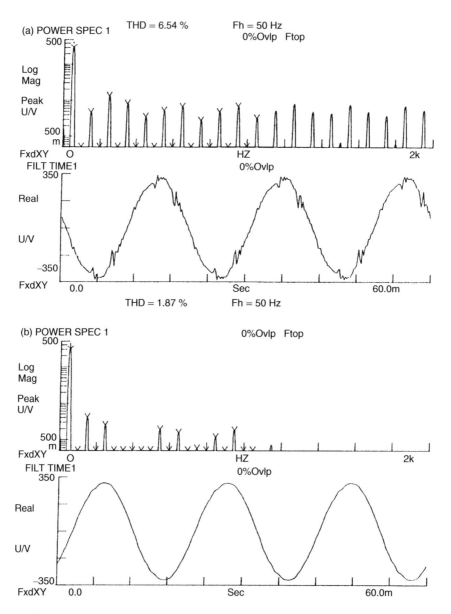

Figure 4.64 Frequency spectrum and voltage path of a village power supply system (70 kW): (a) without and (b) with a filter and compensation unit

4.5.4   Grid-specific filter layout

When large wind turbines supply weak grid areas, e.g. dead-end feeders in wind-rich coastal areas, powerful grid effects must be expected. This is particularly the case if a high-output wind turbine supplies its energy into the grid via a frequency converter with controlled rectifier, constant current d.c. link and a line-commutated six-pulse inverter in a bridge connection.

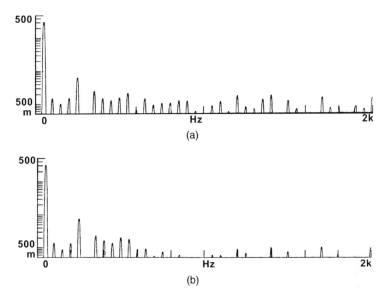

Figure 4.65 Frequency spectrum for (a) 275 kW input power via a six-pulse inverter and (b) 320 kW input power via a six-pulse inverter and 330 kW input power via an asynchronous generator coupled directly to the grid

When a 1.2 MW turbine was erected, the current grid situation at a coastal location gave rise to a new dimension of grid effects, which could be largely attributed to grid resonances (see Section 4.4) [4.68]. Test runs showed a high level of harmonics in the vicinity of the grid resonance point at the 25th ordinal number, despite the grid filter installed, which was designed according to conventional methods as a harmonic absorber for the fifth and seventh harmonics. This phenomenon cannot be attributed primarily to the effects of the wind turbine, but is basically caused by the grid configuration. The turbine could have been operated without any problems if it had been connected at a different grid connection point with greater grid short-circuit power. Such effects of grid resonance have been observed in other dead-end feeders recently, and will increase in importance with increasing grid loading. The descriptions thus clearly show the limits to the effectiveness of filters.

In order to meet the compatibility level stipulated by the electricity supply company at the grid power interchange, it is necessary to design filters specifically adapted for the entire electrical system. These must be adapted to take into account the characteristics of the power converter, the dead-end feeder and the higher-voltage grid structure. This requires precise knowledge of the following grid and turbine data:

- magnitude and angle of the network's impedance at all frequencies occurring at the point under consideration, which is generally the point of supply to the public grid;
- magnitude and angle of impedance at all frequencies of the network occurring at the infeed point of the harmonic source;
- level of supplied harmonic currents;
- level of permissible harmonic voltage at the point under consideration or at the grid point of supply and the permissible power factor range;
- impedances of other electrical equipment between the infeed point and the point of supply.

In filter design, the frequency path of filter impedance should be such that the permissible harmonic levels are not exceeded at the grid point of supply. These levels, in combination with grid impedance, determine the permissible currents. If a source of harmonics supplies large currents, the filter unit must accept the residual current. Thus, for every frequency that occurs a corresponding filter impedance can be calculated.

Figure 4.66 shows a sample excerpt from the calculated current–voltage distribution in the wind farm, obtained while all generators on the 165 kW turbines were in operation. The frequency-dependent current and voltage values at all nodes, components or connection points (e.g. marked by the arrow for node 54 at 850 Hz) can be determined from this for all operating states with and without the filtering unit. The levels can thus be evaluated and corrected by filter adjustment.

Based upon the above example, it was possible to demonstrate the functionality of the procedure using this type of filter layout. Figure 4.67 shows the harmonic voltage readings after the installation of the filter unit, compared with the levels calculated in the simulation and the permissible values. It is evident from this that the measured and calculated indicators match well for all ordinal numbers that are not divisible by three. Significant deviations can be observed at multiples of the third harmonic. These are due to the assumption of mean grid harmonics, which clearly has only limited validity in this grid branch. If precise knowledge exists about these grid distortions, or they have been previously determined, all the inconsistencies shown here can be prevented and values exceeding the limit values avoided.

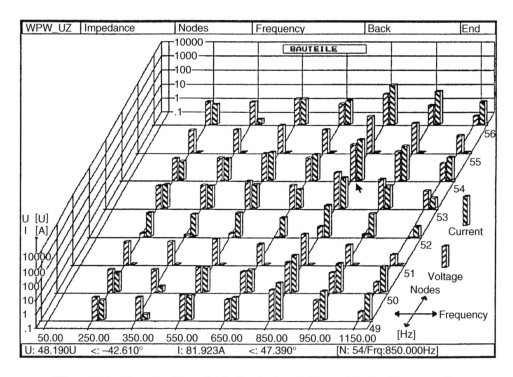

Figure 4.66 Current–voltage distribution in the wind farm with all turbines operating

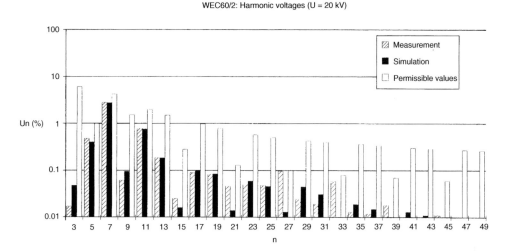

Figure 4.67   Measured, calculated and permissible relative harmonic voltages for a 1.2 MW inverter output with a grid-specific filter (20 kV levels)

For grid connections that are susceptible to grid resonances, filter systems can be designed using computerized simulations that allow the local grid connection conditions to be fulfilled. Thus low filter unit costs can be achieved and high grid loads attained.

4.5.5   Utilizing compensating effects

The discussions in Sections 4.3.3, 4.3.4 and 4.5.3 show that low grid effects can be achieved by a favourable combination of differently designed wind energy converters [4.60]. Firstly, asynchronous generators with inverter systems, which produce a filtering effect and increase the short-circuit power, can be used to reduce harmonics. Secondly, the grid-supporting characteristics of units with frequency converter inputs can help reduce the voltage changes or other grid effects caused by turbines connected rigidly to the grid.

A further method of reducing grid effects is possible in the case of the installation of several wind turbines with a frequency converter supply by adjusting their inverters so that they function with different trigger angles in relation to the grid. By an appropriate selection of d.c. link voltage or transformer voltage for two turbines, the trigger angle of the two inverters (Figure 4.68(a)) can be adjusted so that the critical harmonic (here, for example, the 29th) of both systems occurs half a cycle apart, and is thus largely cancelled out (see Figure 4.68(b)) [4.69].

For inverter 1 the inverter stability limit is assumed to lie at the point where the d.c. link voltage remains the same during a voltage dip of 15%, which corresponds to a nominal trigger delay angle of $\alpha_{1N} = 138°$. According to Figure 4.67, a nominal trigger delay angle of $\alpha_{2N} = 131.79°$ is found for inverter 2 for the minimum amplitude of the 29th harmonic. The amplitude spectrum of the superposed grid current for a 1.2 MW plant (value calculated according to Figure 4.70) illustrates the significant reduction in all harmonics between the

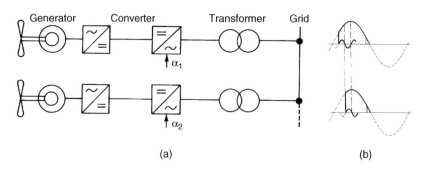

Figure 4.68 (a) Block diagram and (b) harmonic cancellation of two wind turbines with different inverter trigger angles

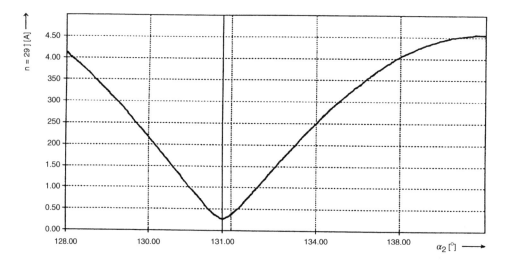

Figure 4.69 Current amplitude of the 29th harmonic as a function of the trigger delay angle α_2 (simulation result)

11th and 49th ordinal number. Therefore, by using an appropriate system layout, grid effects can be significantly reduced using this process, both for operation at nominal ratings and at partial load, without incurring additional construction costs. Laboratory investigations with line-commutated inverters confirm the effectiveness of this process (Figure 4.71).

The available grid capacity can be utilized well by the formation of a turbine cluster if the turbines are designed with a suitable generator, grid connection and power regulation system, and different turbines are combined in a manner appropriate to the location.

In low-power grids with a proportion of the power supplied by wind power, particularly large system perturbation can be expected. Practical investigations have shown, however, that by using an appropriate layout of the grid and its components, wind power can contribute up to 100% of the supply. Any system disturbance can be avoided by the use of devices to influence the grid characteristic in connection with static or rotating phase shifters, batteries with

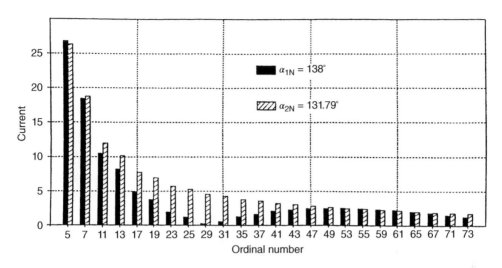

Figure 4.70   Amplitude spectrum of the grid current at the same (striped) and different (black) power converter trigger delay angles (simulation results)

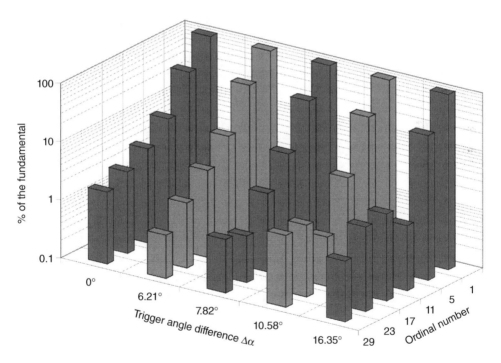

Figure 4.71   Amplitude spectrum of the grid current (measurements) as a percentage of the fundamental for cophasal supply, $\Delta\alpha = 0°$, and trigger difference, $\Delta\alpha$, for the reduction of the 29th harmonic, $\Delta\alpha = 6.21°$, the 23rd harmonic, $\Delta\alpha = 7.82°$, the 17th harmonic, $\Delta\alpha = 10.58°$, and the 11th harmonic, $\Delta\alpha = 16.35°$

reversible converters, grid filters, compensation units and grid regulators. In this manner, any existing grid effects plus those that can be expected as a result of the installation of additional turbines can be reduced. Expensive measures to reinforce or extend the grid can therefore be largely avoided.

4.6   Grid Control and Protection

Power station turbines and generators are generally fitted with proportional speed regulators. As shown in Figure 4.70, the speed–power curves of individual generators can be used to derive a steady-state frequency–power characteristic $\omega = f(P_E)$ for the total generator active power P_E of a grid, and this characteristic has a much flatter path than the characteristic of the largest individual machine. For the load P_L, including the power station's own consumption and the loss in the grid, an approximately linear steady-state frequency–output characteristic $\omega = f(P_L)$ is found, the gradient of which is determined by the total characteristic of the consumers. The steady-state operating point in the grid is found at the intersection of the two characteristics at frequency ω_0, where the supplier output just covers the consumer component and the grid losses, i.e. $P_E(\omega_0) - P_L(\omega_0) = 0$. Due to the flat characteristics, each individual generator works with a virtually fixed-frequency grid.

If a load-specific displacement of grid frequency occurs, the producer with a frequency-dependent power characteristic is called upon to maintain the frequency according to the drooping characteristic by output changes. As well as this so-called primary control, frequency and coupling output are monitored and adjusted by so-called secondary control and load management, which will not be considered further here.

4.6.1   Supply by wind turbines

In contrast to conventional power stations, wind turbines, like all renewable power generation systems, usually supply the entire energy offered by the wind into the grid, regardless of the grid state. Moreover, starting and stopping procedures are not usually centrally coordinated. In low-power sections under 10% this does not cause any serious effects. In Schleswig-Holstein roughly 20% of the electricity supply is already covered by wind power.

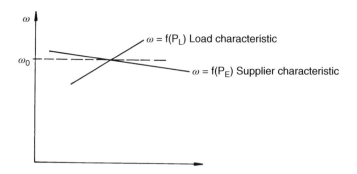

Figure 4.72   Supplier and consumer characteristics of a grid [Chapter 3, Reference 13]

In the near future, however, it is expected that energy contributions from wind power will reach, or even exceed, power consumption values at low-load times in the North German coastal and inland region and low mountain ranges, particularly in windy conditions. Grid effects and measures necessary for grid regulation can take on considerable importance here in an uncontrolled wind energy supply.

Up until now, grid management has taken place based upon the assumption that small power plants, and in particular wind turbines, contribute nothing to the maintenance of grid parameters. They are largely viewed as being only negative consumers, who can be connected as desired. To prevent the formation of asynchronous islands in the grid that are difficult to control, decentralized installations are separated from the grid early, when large suppliers (conventional power stations) fail. Therefore even higher proportions are lost, which could support the grid if there were enough wind.

Initial investigations of output fluctuations [4.69] and interchange power [4.70] in wind farms and grids [4.71] were based upon computerized simulations using the least favourable assumptions. The results of measurements on an existing wind farm with a spatially offset arrangement of individual wind turbines, as discussed in Section 4.3.2, lead us to expect only small output gradients for total output owing to the time sequence of wind and output events. Changes to the grid load can be divided into variations due to the time of day, which usually occur slowly and are for the most part predictable, and changes that occur by chance.

There are fears that the extreme fluctuations in the supplied wind power will lead to problems in grid control or necessitate expensive compensatory measures. Research results from the scientific measurement and evaluation programme, however, refute this. According to reference [4.72] the greatest occurring power fluctuation from one hour to the next in the supply areas of E.ON and RWE amounted to around 20% of the total installed nominal power, with a probability of occurrence of 0.01%, i.e. less than once a year.

Even with today's level of meteorology, local wind speeds and directions and their gradients can be reliably predicted to a large degree. This forecasting could be significantly refined and improved in the future to meet the needs of the electricity generation sector. Therefore, with a more precise knowledge of wind turbines and their behavior and location, more reliable predictions could be made about output over time (see Section 4.3.2.3). With such predictions, methods could be developed in the future that permit long-term grid supporting measures to be taken, replacement turbines to be put in operation in anticipation or consumers that are not at the time absolutely necessary to be switched off.

4.6.2   Grid support and grid control with wind turbines and other renewable systems

The value of wind power can be decisively increased if it is capable of contributing to grid support. Thus wind power could be transformed from a negative consumer into a grid-supporting quantity. The previously somewhat negative image of this environmentally friendly power source will be significantly improved and its long-term prospects guaranteed.

In wind turbines and other small power stations, synchronous or double-fed asynchronous generators with a grid connection via self-commutated pulse-controlled converters are being increasingly used as the mechanical–electrical conversion system. Such electronic power conditioning systems, which have also become standard in photovoltaic systems, offer diverse

conventional power stations similar intervention options with regard to the transmission of power to the grid, such as the adjustability of voltage and reactive power and the control of the supply of active power. These technical possibilities, however, remain largely unused to date.

Within the framework of a new research and development project, a concept for grid regulation has been developed and realized in hardware and software. The practical investigations were carried out at various locations in Germany and on the Canary Islands. Thus different grid configurations in combined and separate grids and different meteorological and geographic conditions could be taken into account.

It is the task of a grid regulation unit to match the operating behavior of renewable, decentralized generation plants (wind power and photovoltaic generators, etc.) with that of conventional power stations and to ensure that they participate actively in grid support. Figure 4.73 shows the configuration and the components for grid regulation. Before practical realization, comprehensive preliminary studies were carried out. To this end, the structure of the subnetworks relevant at the time were recorded and comprehensive simulation calculations (load flow, voltage changes, etc.) were performed, in order to be able to estimate possibilities for control interventions and their effects upon the grid.

The effects of five supply systems of the 500 kW class upon an approximately 23 km long 20 kV grid, connected to the 110 kV level via a 40 MVA transformer with a short-circuit power of around 320 MVA, will be illustrated by the following descriptions, which first of all consider the situation without control interventions. Figure 4.74 shows the differences of voltage paths along the line without (a) and with (b) maximum supply. This shows that in all operating ranges the voltage deviations at the point of common coupling remain below 2%. However, the voltage

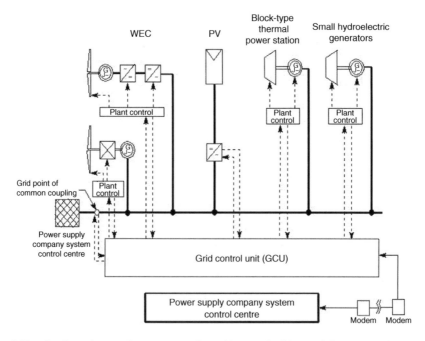

Figure 4.73   Configurations and components for grid control with (spatially separated) regenerative power supply systems

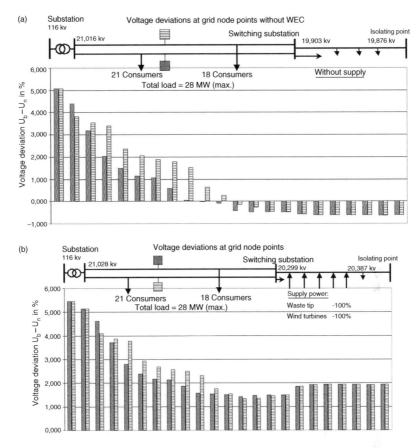

Figure 4.74 Voltage graphs in a 20 kV grid branch of the combined grid (a) without and (b) with maximum supply by renewable systems in the dead-end feeders

change in the grid branch without supply amounts to almost 6% and at maximum supply only around 3%. Furthermore, Figure 4.75 also shows the differences in the flow of active and reactive power and in the utilization of the transmission system. Whereas in the grid range under consideration the active and reactive power always flows from the transformer substation to the grid disconnection point when there is no supply (a), the power flows are sometimes reversed if there is supply (b) depending upon the load case, with active and reactive power sometimes flowing in opposite directions to each other. By control interventions, however, it is possible to achieve the desired power flows and favourable grid loads. Furthermore, it is clear that even where supply is 10% of the consumed power, load is reduced by a good 10%, particularly in those grid areas with high utilization. As a result the transmission losses (which will not be considered further here) can be correspondingly reduced.

Suitable measurement, conversion and control systems are necessary for control interventions for the support of grid voltage or the adjustment of reactive power. We will not go into further details at this point regarding the various design options that are usually selected specifically to suit the power supply company and component and turbine manufacturer.

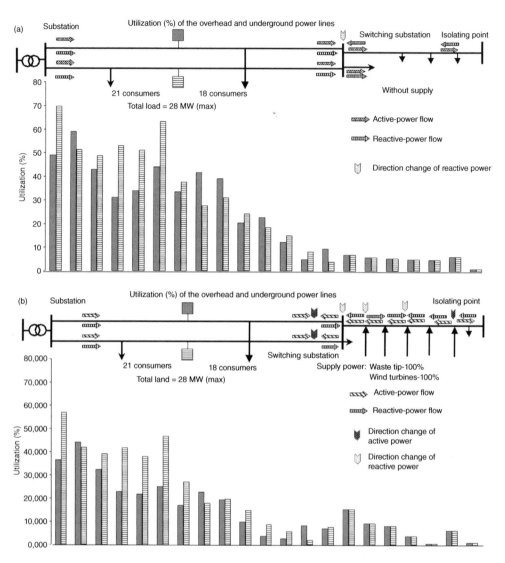

Figure 4.75 Capacity utilization of the transmission systems with active and reactive power flow (a) with no supply and (b) with maximum supply by renewable generation systems

Detailed descriptions on this subject based upon the above-mentioned EU project are given as examples in references [4.73–4.76].

The results of measurements on the grid branch sketched out at the start will be briefly characterized. Figure 4.76 shows, at an approximately constant supply of active feed of 250 kW, the voltage change (with approximately 1 V superimposed voltage fluctuation) from around 402 to 398 V, i.e. a reduction of approximately 1% (in relation to the nominal value), caused by ramp-shaped reactive power variation of 150 kvar (overexcited, inductive supply) to −25 kvar (underexcited, capacitive supply). Even when the reactive power increases, the voltage rises

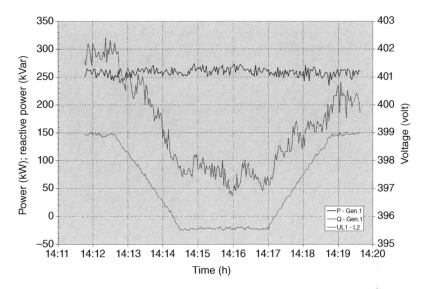

Figure 4.76   Variation in reactive power of a 0.5 MVA power supply system and effects upon the voltage of a weak grid branch

back to almost its original value. As the measurements show, the voltage changes at the point of common coupling in the grid are approximately half the level of those at the point of supply.

At roughly the same reactive power variations, significantly greater effects were found in separate networks. For example, a voltage change of around 3% was measured in a separate network with around 500 MW installed power, and a voltage change of around 6% was measured in a 100 MW grid [4.68]. These results illustrate the particular effectiveness of the procedure in weak grids. However, it is also clear that individual supply systems of the 0.5 MW class are not capable of holding branches in the combined grid at a constant voltage.

The grid supporting function of the grid control unit (GCU) shown in Figure 4.77 is illustrated on the basis of Figure 4.78. In a selected voltage range, e.g. between 20.3 and 20.7 kV, the control unit does not intervene (0900 to 1200 hours). If the voltage falls the reactive power is increased and if the voltage rises the reactive power falls, or is run in the capacitive feed range. The installed wind power in Germany already amounts to approximately 30 GW and is increasing further. Up to 2009, however, there was no obligation for wind turbines or wind farms to participate in the guaranty of a safe grid operation.

After the System Service ordinance for wind turbines enacted in July 2009, all plants that were to be connected to the high-voltage grid after 30.6.2010 had to participate in the system operation and to fulfil the connection criteria of the Transmission Code 2007 [4.77].

4.6.3   Central reactive power control

Also, wind farms connected to medium or high voltage are able to support the grid by means of specific reactive power controls. In this, the reactive power controls are operated by a central

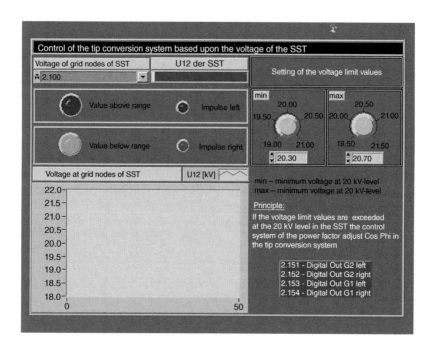

Figure 4.77   Three-Grid control unit (GCU)

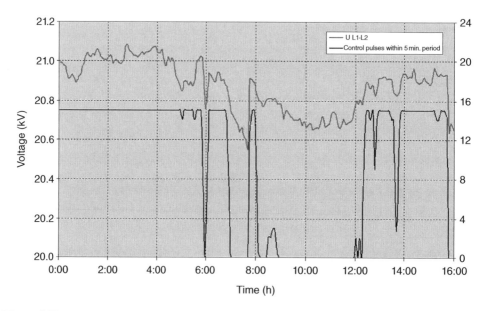

Figure 4.78   Graph showing grid voltage and reactive power control interventions by the grid control unit for voltage stabilization

wind farm controller that controls the reactive power at the GCP. Depending on the requirement of the grid operator, such wind farm controllers are usually able to make a constant reactive power Q_{ref}, a voltage-independent reactive power $Q_{ref}(U)$ or a constant power factor $Q_{ref}(U)$ available. In the case of the $Q_{ref}(U)$ static, the reactive power control depends on the condition of the grid, whereas in the other two methods, the wind farm feeds independently of the grid condition with constant reactive power or constant power factor.

Independently of the prescribed rated value, the action control value is always the reactive power as only this can be directly influenced. For this reason in the $Q_{ref}(U)$- or cos $(\varphi)_{ref}$-mode, the nominal value Q_{ref} must first be calculated from the prescription of the wind farm controller at the GCP (see also Figure 4.79). For the voltage-dependent reactive power nominal value $Q_{ref}(U)$, this results directly from the measuring value of the grid voltage with the $Q(U)$-static stored in the controller. For a given power factor cos $(\varphi)_{ref}$, the nominal value Q_{ref} is calculated from the power factor and the measurement value for the efficiency at the GCP.

The reactive power feed at the grid connection is now controlled by the nominal value that has been obtained, in which case, for instance, a PI controller can be used. The actuator size of the central wind farm controller is distributed among all the available plants in the farm. The converters set the prescribed nominal values for the reactive power feed and thus act as control devices of the reactive power circuit. The reactive currents fed in by the converter lead to a reduction in voltage over the reactances in the wind farm and the grid and thus to a reactive power feed at the GCP that again is measured by the wind farm controller (see also Figure 4.80).

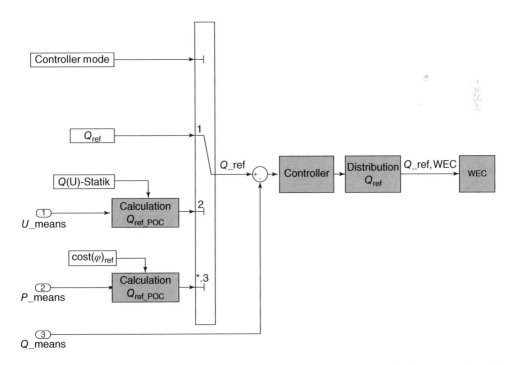

Figure 4.79 Principle of function of a central reactive power controller with the three control modi, reactive power, $Q(U)$-static voltage drop and power factor [4.78]

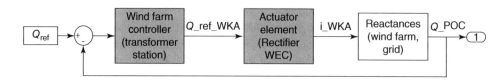

Figure 4.80   Reactive power circuit of a wind farm with central wind farm controller [4.78]

As the rated values of the central reactive power controller need to be transmitted to the individual plants, the communication arrangements play an important role. Therefore, the communication delay between the farm controller and the plants act as dead time in the reactive power circuit. The properties of the communication arrangements thus influence the dynamic behavior and the stability of the control circuit that must be correspondingly taken into account in the design of the controls.

4.6.4   System services and operation

The transmission grid operator (TGO) is responsible for safe resource-saving and economical operation of the transmission grid as well as for a reliable supply to the users of electrical energy. For this reason, besides transmission and distribution of the energy, system services must also be provided in order to satisfy this requirement. According to the Transmission Code 2007, the following are included in the system services:

- Maintaining the frequency
- Maintaining the voltage
- Re-establishing the supply
- Operating the system (Figure 4.81)

4.6.4.1   Maintaining the frequency

According to Equation (4.44), an equilibrium must exist in the electrical grid for each moment between generated power P_G on the one side and the losses P_V and utilization P_L on the other side according to the following equation.

$$P_G - (P_V + P_L) = 0 \tag{4.44}$$

Even the smallest deviation leads to a change in the grid frequency and must be equalized by the use of control power. Such balance disturbances can be caused by load ramps, load swings, load noise or load jumps. The first two cases depict changes in hourly or minute range and can be controlled by a corresponding adaptation of the mode of operation. In the last two cases, that can occur, for instance, due to the failure of a power station, the deviation of frequency must be

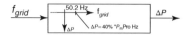

Figure 4.81   Active power reduction for overfrequency of generating plants with regenerative sources of energy [4.63]

caught up by the use of control measures. This occurs in the first milliseconds by the electricity stored in the capacitors and inductivities of the grid elements and the magnetic energy. Also, the kinetic energy stored in the rotating masses of the grid can be used to balance the deficit. Only then, with the use of primary control power (in the range of 5 to 30 seconds), secondary control power (5 to 10 minutes) and minute reserves (>30 minutes) can the power equilibrium re-established. The control power is prescribed by the TGO and provided by the generating plants. In order to take part in the provision of control power, the power stations must fulfil technical conditions that are defined in the pre-qualification requirements determined by the TGO.

4.6.4.2 Maintaining the voltage

In contrast to the frequency, the voltage within a grid is not constant but varies depending on the load. Motorized loads as also static grid elements (lines, transformers) take up reactive power and lead to changes in voltage. The task of the TGO is to ensure a balanced reactive power and in this way the maintenance of the voltage in as tight a range as possible. For physical reasons, therefore, reactive power must be made available near the site in order to prevent overloading of the lines in the grid due to increased reactive power flows. The control of the reactive power flows and the voltage can occur in several ways [4.79].

The control of the voltage with the aid of **generators** occurs with the setting of the voltage-rated value of the voltage controller or the excitation and thus the output reactive power of the generator. In this, every connected generating plant must be able to provide reactive power inside a range prescribed by the TGO.

If the provision of the reactive power by the generators is insufficient, then **reactive power compensation plants** can provide specific inductive or capacitive reactive power and thus influence the voltage in the grid. Besides this, they can supply users directly with the required reactive power and can be used for setting a certain power factor.

Also, the **transformers** used for coupling grid components of different voltages can also influence the power flow in the grid by altering the stepping levels. In this, the control of efficiency, reactive power or both is possible depending on the amount and phase length.

An influence on the power flows and the voltages at the grid nodes can also be achieved by a change in the **grid topology**

4.6.4.3 Re-establishing the supply

In the case of a large disturbance, the TGO together with neighbouring TGOs, the subordinate distribution grids and power station operators must see to the re-establishment of the power supply. In this, the capabilities of power stations for island operation (operation of asynchronous part-grids) as well as for blackstart (starting the power station from the stoppage) are of decisive importance.

4.6.4.4 Operating the system

The TGO, within the scope of the operation, must ensure that with the use of the momentary available operating possibilities and operating means, all users can at any time be supplied with qualitatively sufficient electrical energy (voltage, frequency) and malfunctions can be mastered and limited.

Operating the system in this sense means operating the grid and the power station coordination and individually [VDN 2007]:

- Controlling the switching condition
- Monitoring the adherence to the operating limits of current and voltage
- Monitoring the $(n - 1)$ safety
- Using control power for imbalances between generation and use
- Controlling the use of reactive power for maintaining the voltage
- Implementing the operating management
- Bottleneck prognosis and bottleneck management
- Carrying out topological measures
- Re-establishing the grid after malfunctions.

A difference is made between normal operation, malfunction operation and endangered operation of a grid. In normal operation, all operational limiting values (voltage, flows, grid short circuits) are adhered to; the reactive power use is optimized. The $(n - 1)$ criterion is fulfilled; the agreed to voltage band is adhered to by the measures described above. In endangered or malfunctioning operation (e.g. in the case of a failure of an operating means or a power station), the TGO is responsible for the quickest possible resumption of normal operations and the $(n - 1)$ safety. In endangered operation, measures are to be divided into hour/minute regions segments, whereas in malfunction operation, partly automatic measures (e.g. load-shedding) are carried out.

4.6.5   Connection of wind turbine to the transmission grid

For wind turbines as well as for conventional power stations, there are minimum requirements for connection to the high and highest voltage grids that are prepared by each TGO on the basis of the Transmission Code. Decisive for the system services are the criteria of the active power output or maintenance of the frequency, the provision of reactive power and the rebuilding of the supply that is mentioned with regard to the wind turbines (WTs)

4.6.5.1   Active power output and upholding the frequency

An imbalance in generation and use of active power leads to swings in the grid frequency thus endangering the operation of the grid. For this reason, the active power output of the generating plant must be controllable in accordance with the requirements of the TGO. For EEG-generating plants, it is valid that at a frequency of more than the 50.2 Hz, the momentary active power is decreased with a gradient of 40% of the momentary available power of the generator per Hertz (Transmission Code). In the region of 47.5 to 50 Hz, EEG plants experience no limitations but for a frequency of more than 51.5 Hz or less than 47.5 Hz, there is a separation from the grid (Figure 4.81)

Wind turbines or wind farms are freed from participation in the primary controls [4.77]. For preparation of secondary control power and minute reserve, they are entitled when they fulfil the pre-qualification requirements of the respective TGO.

4.6.5.2   Provision of reactive power and maintenance of the voltage

With respect to the maintenance of the voltage in the grid, each generating plant must be equipped such that rated value determinations for reactive power and voltage (power factor,

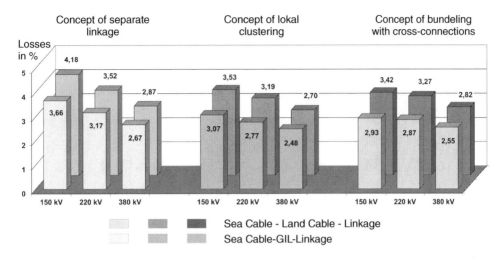

Figure 4.82   Investment per annum for diffreat concepts (calculation of 2008)

reactive power value or voltage value with tolerance band) can be implemented by the TGO in the case of rated power output within a determined framework. This framework is determined by three defined voltage/reactive power-operating diagrams (U/Q diagrams) in the Transmission Code from which the TGO selects one variant on the basis of the grid requirements. The agreed to reactive power range must able to pass through by the WT within 4 minutes. The voltage-/reactive power control must be possible either by the generating plant itself or by means of additional arrangements. Figure 4.82 shows one of the three variants of the Transmission Code that has also been defined for the grid connection of the E.ON Netz TGO for the Alha Ventus wind farm in its grid connection rules.

For operation in the part-load region, separate power/reactive power diagrams (P/Q diagrams) are prescribed for WTs [4.77] that are allocated to the U/Q diagrams and must be reachable inside each possible operating point.

4.6.5.3   Re-establishing the supply

In the case of a separation from the grid, a generating plant (own requirements case) as well as an island formation must be in a position to maintain this operating condition for several hours. The black start capability of generating plants is no minimum condition but must nevertheless be able to be carried out by WTs – depending on the linkage variant. A grid re-establishment of supply is decided in cooperation with the TGO.

4.7   Grid Connection Rules

When operating wind turbines in parallel with the public supply grid – as mentioned in the introduction to Chapter 4 (Table 4.1) – the valid international and national standards and directives apply. The grid quality characteristics and limit values in the low-voltage and medium-voltage grid according to Table 4.7 from reference [4.8] that customers and suppliers on the low- and medium-voltage transfer point of the grid may expect can be assumed here.

Table 4.7 Grid quality characteristics and limit values in accordance with DIN EN 50 160 and the VDEW directives for in-plant generation. From reference [4.81]

Grid quality characteristic	Averaging time	Reference value or limit value			
		EN 51060 LV grid	EN 50160 MV grid	VDEW: EEA at the LV grid	VDEW: EEA at the MV grid
Grid frequency (at synchronous connection to interconnected grid)	10 s	50 Hz + 4–6% (100% value) 50 Hz ± 1% (99.5% value)	50 Hz + 4–6% (100% value) 50 Hz ± 1% (99.5% value)	–	–
Grid frequency (without synchronous connection to interconnected grid)	10 s	50 Hz ± 15% (100% value) 50 Hz ± 2% (95% value)	50 Hz ± 15% (100% value) 50 Hz ± 2% (95% value)	–	–
Slow, quasi-stationary voltage changes	10 min	U_n ± 10–15% (100% value) U_n ± 10% (95% value)	– U_c ± 10% (95%value)	$\Delta u_a = 2\%$ –	$\Delta u_a = 2\%$ –
Long-time flicker intensity	120 min	$P_{lt} = 1$ (95% value)	$P_{lt} = 1$ (95% value)	$P_{lt} = 0.46$	$P_{lt} = 0.46$
Voltage asymmetries	10 min	$U_g/U_m < 2\%(3\%)$ (95% value)	$U_g/U_m < 2\%(3\%)$ (95% value)	–	–
Total harmonic distortion ($v = 1–40$)	10 min	THD = 8.0%	THD = 8.0%	–	–
Harmonics	10 min	Harmonic voltage (95% value)			

Permissible harmonic current related to S_{kV} (A/MVA)

Uneven, not multiple of 3		6.0% of U_n	6.0% of U_c	10 kV	20 kV	
$v = 5$	10 min	5.0%	5.0%	2.5	0.115	0.058
$v = 7$	10 min	3.5%	3.5%	2.0	0.082	0.041
$v = 11$	10 min	3.0%	3.0%	1.3	0.052	0.026
$v = 13$	10 min			1.0	0.038	0.019

(continued overleaf)

Table 4.7 (*continued*)

	Interval	10 kV			20 kV	
Uneven, not multiple of 3						
$v = 17$	10 min	2.0%	2.0%	0.55	0.022	0.011
$v = 19$	10 min	1.5%	1.5%	0.45	0.018	0.009
$v = 23$	10 min	1.5%	1.5%	0.3	0.012	0.006
$v = 25$	10 min	1.5%	1.5%	0.25	0.010	0.005
Uneven, multiple of 3						
$v = 3$	10 min	5.0%	5.0%	4.0	–	–
$v = 9$	10 min	1.5%	1.5%	0.7	–	–
$v = 15, 21$	10 min	0.5%	0.5%	–	–	–
Uneven, $v > 25$	10 min	–	–	$0.25 \times 25/v$	$0.06/v$	$0.03/v$
Even harmonics						
$v = 2$	10 min	2.0%	2.0%	$1.5/v$	$0.06/v$	$0.03/v$
$v = 4$	10 min	1.0%	1.0%	$1.5/v$	$0.06/v$	$0.03/v$
$v = 6\text{–}24/6\text{–}38$	10 min	0.5%	0.5%	$1.5/v$	$0.06/v$	$0.03/v$
Subharmonics						
$\mu < 40$		–	–	$1.5/\mu$	$0.06/\mu$	$0.03/\mu$
$\mu > 40$		–	–	$4.5/\mu$	$0.18/\mu$	$0.09/\mu$
Voltage-related events (criterion is the number of limit violations)						
Rapid voltage changes (e.g. due to switching operations)		$U_\mathrm{n} \pm 5\%$ (in general)		$U_\mathrm{c} \pm 4\%$ (in general)	$\Delta u_\mathrm{max} = 3\%$ (<1/5 min) –	$\Delta u_\mathrm{max} = 2\%$ (<1/1.5 min) –
Voltage dips ($t < s$; $U > 0.01\,U_\mathrm{n}$ and $< 0.90\,U_\mathrm{n}$)		$U_\mathrm{n} \pm 10\%$ (several times per day) $10/a\text{–}1000/a$		$U_\mathrm{c} \pm 6\%$ (several times per day) $10/a\text{–}1000/a$	–	–
Short-time interruptions ($t < 3$ min; $U < 0.01\,U_\mathrm{n}$)		$10/a\text{–}300/a$		$10/a\text{–}300/a$	–	–
Long-time interruptions ($t < 3$ min; $U < 0.01\,U_\mathrm{n}$)		$10/a\text{–}50/a$		$10/a\text{–}50/a$	–	–
Grid frequency overvoltages		$U_\mathrm{eff} < 1.5$ kV		$U_\mathrm{eff} < 1.7\text{–}2\,U_\mathrm{c}$	–	–
Transient overvoltages		$U_\mathrm{eff} < 6$ kV			–	–

The quality characteristics of the grid are usually evaluated statically and specified for the frequency values of 100 and 95%. A differentiation can be made here between relatively slow frequency and voltage changes, flicker, voltage asymmetries and harmonics.

Furthermore, voltage-related events should be taken into account. These are determined on the basis of the number of times limit values are exceeded during the monitoring period. As shown in Table 4.7, these cover rapid switching-related voltage changes, voltage dips, short-time interruptions, long-time interruptions and grid-frequency overvoltages.

Electromagnetic compatibility is dealt with in IEC 61 000. A differentiation is made here between conducted disturbances (which occur up to several 10 kHz) and nonconducted disturbances that dominate at higher frequencies (DIN VDE 0 838 to 0 847), which we will not deal with further. Table 4.8 gives a brief overview of the international (IEC) and national classifications (DIN, VDE) for conducted disturbances. Wind turbines are dealt with internationally in IEC 61 400 and nationally classified (in Germany according to VDE 0 127). These standards and directives include the classification according to Table 4.9. Since 1992, the Fördergesellschaft Windenergie (FGW) has issued directives specifying, for example, clear measuring procedures with which comparable data can be determined for different wind turbines. As shown in Table 4.10, they cover the evaluation of the energy yield calculations, grid integration and noise emissions.

According to the Energie-Wirtschafts-Gesetz (Energy-Economy-Laws) (EnWG) of 1998 the operator of the transmission grid is responsible for the organization and operation of the interconnected system. The operator of the transmission grid was therefore made responsible for the safety and reliability of the transmission system. Furthermore, they have to guarantee the quality of the power supply. Moreover, all grid users and participants in the deregulated electricity market must be allowed nondiscriminatory access to the transmission grids and their use. To this end, all grid users must adhere to minimum technical requirements described in the *GridCode 2000* directive. The rules compiled by the Verband der Netzbetrieber (Association of Grid Operators) (VDN) are oriented towards the requirements for uninterrupted interconnected operation in the national framework and take into account the international specifications for grid operation of the Union für die Koordinierung des Transportes elektrischer Energie (ENTSO-E).

Table 4.8 National and international standards for EMC including from electricity supply systems

IEC-standard		DIN-No.	VDE-reference	Title/content
IEC$_x$	61000-1-			General
IEC	61000-2-	DIN EN 61 000-2-x	VDE 0839 Part 2	Building and civil engineering
IEC	61000-3-	DIN EN 61 000-3-x	VDE 0838	Limiting values
IEC	61000-4-	DIN EN 61 000-4-x	VDE 0847 Part 4	Test and measuring methods for evaluating the electro-magnetic compatibility
IEC	61000-5-	DIN EN 61 000-5-x	VDE 0847 Part 5	Installation guidelines and aid measures
IEC	61000-6-	DIN EN 61 000-6-x	VDE 0839 Part 6	Technical standards
IEC	61000-9-			Miscellaneous

Table 4.9   National and international standards for wind turbines

IEC standard	DIN No.	VDE reference	Title/content
IEC 61400-1	DIN EN 61400-1	VDE 0127 Part 1	Wind turbines – Part 1: Design requirements
IEC 61400-2	DIN EN 61400-2	VDE 0127 Part 2	Wind turbines – Part 2: Requirements for small wind turbines
IEC 61400-3	DIN EN 61400-3	VDE 0127 Part 3	Wind turbines – Part 3: Design requirements for wind turbines on open sea
IEC 61400-4	DIN EN 61400-4	VDE 0127 Part 4	Wind turbines – Part 4: Design requirements for drives for wind turbines
IEC 61400-5			Wind turbines – Part 5: Rotor blades
IEC 61400-11	DIN EN 61400-11	VDE 0127 Part 11	Wind turbines – Part 11: Noise measurement procedure
IEC 61 400-12-x	DIN EN 61 400-12-x	VDE 0127 Part 12x	Wind turbines – Measurement of the power behavior
IEC 61400-13			Wind turbines – Measurement of mechanical loads
IEC 61400-14			Wind turbines – Information of emission-relevant acoustic power level and the tonality
IEC 61400-21	DIN EN 61400-21	VDE 0127 Part 21	Wind turbines – Part 21: Measurement and evaluation of the grid compatibility of grid-coupled wind turbines
IEC 61400-22	DIN EN 61400-22	VDE 0127 Part 22	Wind turbines – Part 22: Conformity testing and certification
IEC 61400-23	DIN EN 61400-23	VDE 0127 Part 23	Wind turbines – Part 23: Rotor blades – experimental structure testing
IEC 61400-24	DIN EN 61400-24	VDE 0127 Part 24	Wind turbines – Part 24: Lightning protection
IEC 61 400-25-x	DIN EN 61 400-25-x		Wind turbines – communication protocols
IEC 61 400-27-x	DIN EN 61 400-27-x	VDE0127 Part 27-x	Wind turbines – electrical simulation models

Table 4.10 Overview of the current FGW regulations [4.83]

Part	Revision	Date	Title
1	18	01.02.2008	Determination of the noise emission values
2	16	28.01.2010	Determination of the power curve and standardized energy yields
3	22	01.07.2011	Determination of the electrical properties
4	5	22.03.2010	Requirements of the modeling and validation of simulation models
5	4	01.06.2008	Determination and application of reference yields
6	8	19.05.2011	Determination of wind potential and energy yields
7	1	15.10.2010	Maintenance of power stations for renewable energies
8	5	01.07.2011	Certification of the electrical properties of generating units and plants

In the framework of the GridCode [4.80]:

- the *MeteringCode 2004* VDN gives directives and change procedures,
- the *TransmissionCode 2003* gives grid and system rules of the German transmission grid operators and associated appendices and sample data,
- the *DistributionCode 2003* gives rules for access to distribution grids and
- the *GridCode* gives cooperation rules for the German transmission grid operator.

Since 1.08.2003, new regulations for connection to the transmission grid have come into force for the German EON grid operator (now TENET TSO). These rules set down minimum technical requirements for connections to the high and extremely high voltage grid. Changes relate particularly to wind turbines. The massive expansion of wind farms makes changes necessary so that the stability and availability of the transmission grid is assured when higher levels of wind power are supplied. Directives of other grid operators (RWE, Vattenfall, EnBW) are awaited.

Up until now it has been stipulated that wind turbines must be disconnected from the grid within 100 to 200 ms in the event of grid voltage and frequency values exceeding or falling below limit values. This immediate disconnection of the wind turbines causes problems for the transmission grid operators with regard to the adherence to ENTSO-E criteria for the primary regulation of frequency. For example, in the event of a severe grid fault in North Germany, the grid voltage could fall so far that 3000 MW of wind power could fail suddenly. This power deficit, which makes up around 1% of the ENTSO-E peak load, would disrupt the grid frequency of the entire mid-European interconnected system in sympathy. In view of the politically-desired expansion of onshore and offshore wind energy exploitation in Germany and Europe, E.ON, together with the other transmission grid operators, has been asked to compile rules [4.80, 4.82] that draw upon wind turbines to support the grid. It has been specified that wind turbines have to continue operating in a considerably wider voltage and frequency range than previously. Furthermore, the reactive power output from wind turbines or wind farms must be adjustable on-line by the specification of a target value for power factor or voltage, as described in Section 4.6.

Furthermore, in the event of faults in the grid, very high requirements should be imposed upon the stationary and dynamic behavior of wind turbines and wind farms. In the event of voltage dips, wind turbines may not be disconnected from the grid in a wide voltage–time graph (up to 3 seconds). In the event of voltage dips at 15 to 60% of the nominal voltage, they should supply the greatest possible apparent power (in overexcited generator mode) to support the grid. Further specifications that are in line with the large-scale use of wind energy will be provided in the framework of these rules regarding switching on, active power transmission and the disconnection of the wind farm from the grid. These will not be described further here. In the coming years these will shift from the current land-based to the increasingly sea-based installations of the future. Grid connections for these wind farms, which are generally very large, will be in the high and extremely high voltage levels. Therefore, comparable directives are required to those that have already been described for low- and medium-voltage grids [4.22, 4.23].

4.8   Grid Connection in the Offshore Region

In the 1980s and start of the 1990s, WTs of 10 to 100 kW size dominated the market. The grid connection of WTs and wind farms was limited at the time to some hundreds of kilowatts to a few megawatts. From the middle to the end of the 1990s, wind farms of 100 MW were erected. Today, the individual plants are already in the 6 MW class and 8 to 10 MW units are expected in a few years [4.84]. The safe operation of plants of this size is an important condition for offshore wind energy utilization.

For reasons of costs (especially of foundations and grid connections), only large units (e.g. from 5 MW) will be used for installation of turbines at sea. Planning is being carried out for offshore wind farms in the 100 MW to several gigawatt regions [4.85–4.89]. Their size thus approaches that of today's power stations on land. These grids at sea must be selected and designed with comparable features and equivalent connections. Important conditions for these are that wind farms are equipped with power station properties. These properties will be mentioned in the following. In addition, possibilities for linking approximately 25-GW offshore WT capacity in the German North Sea area will be discussed.

4.8.1   Offshore wind farm properties

The large capacity of offshore wind farms (mostly in the GW region) requires a coordinated grid feed. For this purpose within the framework of a research project, investigations were carried out on cluster management systems for wind farms on land [4.90] that were transferred and adapted to wind farms at sea [4.91, 4.92]. In this way, primarily the economic use of wind energy was to be increased. In order to achieve this, system services are required. Analogous to conventional power stations, these include [4.93–4.95]

- maintenance of the voltage;
- control of the frequency;
- control reserve planning; and
- dynamic behavior for short circuits (fault ride-through).

Modern WTs of the 5 MW class are easily able to fulfil these requirements. In addition, they offer additional possibilities that modern power stations often cannot provide. Here, in offshore applications, differences are made between WTs with the following.

- Electrical or PMSGs as well as
- Asynchronous generators with short-circuit rotors that are operated via fully rated converters or
- Double-feed asynchronous generators that are operated via part-converters at the grid.

Further new variants are the *Windrive systems* that, by means of hydrodynamic power couplings with direct grid-coupled rigid rotation synchronous generators, permit a stable and vibration-free grid operation with WT speed variations similar to the abovementioned systems.

WTs with synchronous or short-circuit rotor asynchronous generators are operated via a fully rated converter, meaning that the whole electrical energy of the generator is fed into the grid by means of a converter. In this way, with corresponding intermediate circuit design, the whole of the converter power for

- reactive power provision as well as for
- voltage control or support

can be used in stationary operation. In addition, because of the very quick access possibilities (in the microsecond region) and with the aid of pulsed converters in the grid

- Harmonics as well as
- Flicker and
- Voltage collapses

can be partly or completely balanced with the whole converter power [4.96–4.98]. In grid operation, many required short-circuit performances for operating safety elements can be provided by generators but converters must be protected from short circuits very quickly and effectively. Thus, WTs with synchronous or asynchronous generators and fully rated converters are not in a position to output short circuits to the grid.

Wind turbines with double-feed asynchronous generators possess a power branching. The stator of the generator is directly connected to the grid. The rotor of the generator, however, is coupled to the grid via a converter only to the extent of 30 to 40% of the generator power. In this way, the generator that is directly switched to the grid, can – if required – feed short circuit power controllably to the grid. Due to the smaller power capacity of the converter, the reactive power provision, as well as the voltage occupation or support as also the balance possibilities in the case of harmonics, the flicker and voltage collapses, are limited to power equivalent changes.

WTs with rotation variable drives and direct grid-coupled synchronous generators can control or support short circuit and reactive power provision and grid voltage in the power capacity in the machine. Because of the contribution of the synchronous machine to the increase of the short circuit, the harmonics, flicker and voltage collapses are reduced. Yet, their quick and targeted control correction – as with converter systems – is not easily possible.

4.8.2   Stationary and dynamic behavior of offshore wind farms

From a technical point of view, the large installed capacity (several gigawatt clusters) and the long transmission lines (up to approximately 150 km) from wind farm to connection point of the (for instance, German) grid are the particular features of offshore wind farms. For this reason, the properties of the transmission cable and the operating media as well as the selected transmission concepts have a decisive effect on the system behavior. Basic knowledge and experience on this are expected in two to three years from research work [4.99]. For this purpose, and on the basis of measurements of the stationary behavior of wind farms, clusters and grid regions are investigated as regards occurring power peaks, power fluctuations, harmonics, voltage swings and flicker as well as reactive power changes and reactive power controls. In addition, account is also taken of the transient behavior of internal wind farm grids in the case of malfunction.

   Control algorithms to be developed [4.100, 4.101] for wind farms and their converter systems and also for wind farms and clusters can increase the system efficiency with the provision of grid services that support grid formation and grid operation. Therefore, the behavior of internal and external symmetrical and asymmetrical malfunctions are of particular importance. Furthermore, the applicability and effect of normal protection concepts in the offshore region are tested. New application-specific protection concepts and equipment must also be developed and tested. The function of the safety systems, e.g. for emergency power supply, the structure of the grid also with black starting features and the grid behavior in extreme situations must also be investigated and ensured.

4.8.3   Wind farm and cluster formation at sea and grid connection

By 2020, it can be expected that offshore wind farms with approximately 25 GW overall power will be installed. In order to transmit this enormous power, equal to around 20% of the power installed in Germany, over distances of up to 140 km to the German grid use is made of three-phase and direct current transmission methods in the highest voltage ranges. In this, the advantages and disadvantages of both methods of transmission are weighted and evaluated very differently. Up to now, three-phase transmission has been used in preference – especially in view of the long-term experience, the high state of the technology and the excellent operating safety. For this reason, the following discussion will concentrate on these.

   Up to now, the practical sequences of wind farm plants have been decided mostly by the individual project management. This means that the operator of each offshore wind farm designs the structure and grid connection with their own cable route and without coordination with other operators. In this way, however, no overall economical, ecological contractual and sustainable offshore WT utilization can be achieved. It will be more difficult to achieve longer overall construction phase for the completion of all projects lasts.

   In contrast to individual project implementation, substantial advantage can be gained with coordinated wind farm installations. In this, for instance, more or less spatially neighbouring wind farms can form themselves into clusters that can feed their electrical energy into common grid nodes at sea or also on land.

   From these nodes, the electricity can be transmitted to the grid and possibly fed into load centres. For this, it is necessary for the individual wind farm projects to agree and coordinate their grid connections in their time sequences and spatial linkages. Thus, all wind farm operators can make use of a common high-voltage transmission system at sea. Besides ecological and economical advantages, this system permits new perspectives to open up for control and implementation in normal operation and in the case of grid malfunctions.

The following discussion based on the research project 'Offshore Integration großer Offshore-Windparks in elektrische Versorgungssysteme' [offshore integration of large offshore wind farms in electrical supply systems] (BMWI, BMU) [4.102–4.104] and summarized in references [4.105] are examples for three construction steps with approximately 3500, 4800 and 5500 WTs in 14, 18 and 24 wind farms with approximately 16, 22 and 25 GW. Here, the technical implementation with individual connections of the offshore wind farms with local cluster formation or bundling the connections and cross-connections will be viewed in greater detail. Further discussions can be found in references [4.106–4.115].

4.8.3.1 Separate linkage of the offshore wind farms

In the connection of the individual wind farms via respective own high-voltage routes according to Figure 4.83, a different number of cable connections are necessary for the three voltage levels mentioned above. At the 150 kV level, 157 cables are required at sea and 136 on land. The 220-kV connection requires 110 cables at sea and 94 cables on land. With a 380 kV connection, one can count on 62 cables at sea and 53 on land. Here, it is assumed that each wind farm is directly connected at sea with separate cables. These are combined at respective GCPs at the coast and together integrated at a transformer station into the German grid. For this, 14, 18 or 24 sea platforms are required in the three time horizons. However, an equalized loading of the transformer stations and the connection points in the grid cannot be achieved.

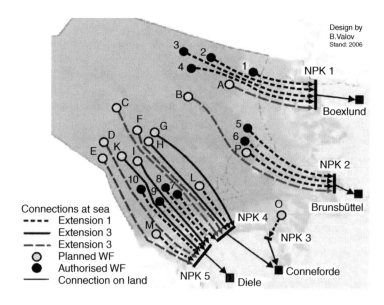

Figure 4.83 Concept of separate connections of offshore wind farms

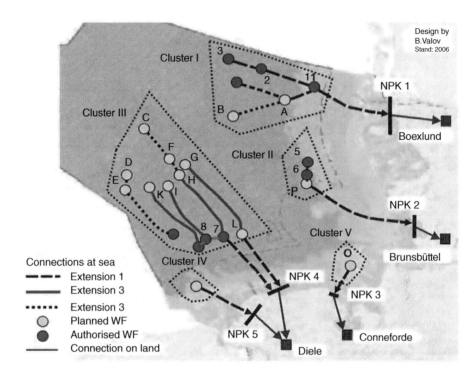

Figure 4.84   Concept for local wind farm cluster formation

4.8.3.2   Local wind farm cluster formation

In areas of geographic vicinities with similar time horizons of planning, wind farms can be combined into clusters according to Figures 4.83 and 4.84. In this way, in comparison to the separate connections, the number of sea platforms can be reduced to 13 (in expansion step 1) 17 (extension step 2) and 19 (extension step 3).

There is then the further possibility of using plants in common for reactive power compensation, communication, control, regulation, emergency power supply, etc. In designing the cable routes for the final extension power, the space requirements and access at sea are substantially reduced and the costs much lowered.

Figure 4.84 shows that in this way, six cable routes are required at sea (instead of 24 individual linkages in Figure 4.83). The numbers of cables are reduced for 150 kV to 129, for 220 kV to 88 and for 380 kV linkages to 54 at sea as well as on land. In addition, economical grid loads are achieved.

4.8.3.3   Combining the linkages with cross-connections between the clusters

The combination with cross-connections of the clusters according to Figure 4.85 permits cable laying and connection to be much more economical than separate connections and transformer works and linkages to the grid to be mostly evenly loaded. In addition, during the first phase in the erection of the first 16 GW, the coastal offshore wind farms can be combined to clusters and connected to the grid through common routes.

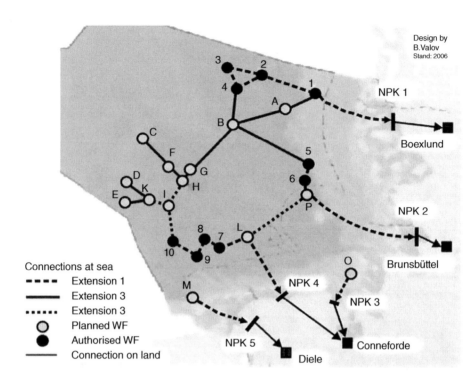

Figure 4.85 Concept of combining and linking the offshore wind farm cluster by means of cross-connections

In the following phases the further wind farms are added in steps and the cross-connections are added (see Figure 4.85). These results, similar to the local wind farm cluster formation, in 13 (extension 1), 17 (step 2) and 19 sea platforms in step 3 being required.

Further, for the 150 kV connection, there are 126 for 220 kV transmission, there are 90 and for 380 kV connection there are 59 cables at sea as well as on land. This concept permits a high degree of flexibility in operational management and energy transmissions. The additional costs of cable for the cross-connections at sea will be lower than the costs for grid-strengthening measures in the grid. In addition, there are possibilities for mutual support of wind farms in the data transfer and black start as well as for emergency supply, etc. Also, with the large area connections, balancing effects of power and voltage gradients can be expected.

Thus, the reliability and availability, as well as the operational safety are enhanced in comparison to other concepts and overall balancing behavior in high-voltage energy transmission systems and the overall electrical supply at reduced costs can be achieved. These will be briefly discussed in the following.

4.8.3.4 Costs and losses of the linkages

The project costs for the respective concepts are divided into the costs for the WTs with transformer and other investments at sea. The latter include the sea platforms, the medium-voltage cables including the laying in the wind farm, power transformers, reactive power compensation installations and high-voltage sea cables with laying for the three desired voltage levels of 150, 220 and 380 kV. Further, all wind farm grids were assumed to be 30 kV rated voltage. Unified sea cable types were selected for internal medium-voltage connections between the plant in the wind farm and for both high or highest voltage cables between the wind farms and the grid connection at the coast.

As the cost information for plants and construction work, for instance, between references [4.107, 4.111, 4.112] and Ref [4.116] differ strongly, it would be more realistic to use average values. Finally, the operating safety was checked on the basis of grid calculations and thus the feasibility of the selected transmission system. A difference was made between land cable and gas-insulated lines (GIL) for the linkage between the GCPs at the coast (ECP) and the connection point to the grid.

With even distribution of the costs during the overall time period, the highest investments (Figure 4.86) and greatest losses (Figures 4.86 and 4.87) resulted in the concept of the separate linkages with the 150 kV cable connections. Lower investment costs can be expected from local clustering and the combination or bundling with cross-connections. An increase in the operating voltage especially to the 380 kV level brings with it further cost savings. In comparison to land cables, the use of GIL leads, although leading to lower investment costs, also involves greater losses.

4.8.4 Electrical energy transmission to the mainland

Three-phase as well as direct current transmission concepts can be used for the transmission of electrical energy from large offshore wind farms over great distances (also more than 100 km) from the mainland or for further transmission into the highest voltage grids of the German gird.

The high transmission power and the large distances to the connection point on the mainland require special design and implementation of offshore operating media. Conventional three-phase systems with a working frequency of 50 Hz correspond to the state of the

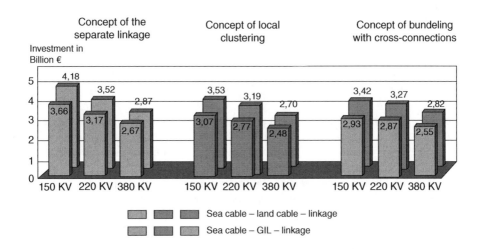

Figure 4.86 Investments per annum for different concepts (calculation of 2008)

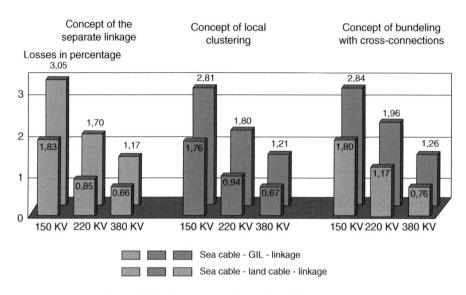

Figure 4.87   Transmission losses for different concepts

technology and are thus available as mature systems with a high standard of safety. In the formation of grids, they can also be carried out in a multi-point manner [4.117, 4.118]. New concepts for this open up extreme power regions. Direct current transmission systems have not yet included the necessary power region for the planned offshore wind farms. The offshore transmission variants offered by well-known manufacturers can only be used for individual wind farms (to approximately 1000 MW). For the proposed power of wind farms at sea of 15 000 MW and more, new transmission concepts are required, which also cannot include traditional working frequencies (e.g. 16.67 Hz) in order to achieve substantial power increases and larger transmission distances with three-phase systems [4.119].

In the grid integration of larger offshore wind farm far from the coast (100 to 150 km), use is also made of

- high-voltage three-phase transmission (HVT) as well as
- three-phase transmissions with reduced frequencies (e.g.10 to 20 Hz)
- three-phase transmissions with full compensation of the capacitive cable coverings.

In both cases, the wind farm frequency is uncoupled from the grid frequency on the mainland (ENTSO-E or German grid). In the use of dc link converters, the frequency in the wind farm can be set in a sliding manner depending on the currently available wind velocity or the power capacity of the wind farm. With this additional degree of freedom, the design and the method of operation of the WT can be optimized in that, among others, the controllers of the system are designed for high stability and maximum energy yields.

A further increase in the WT capacity and the transmission distance is of great interest for offshore wind farms. The operating media required for this will overstep the currently achieved limiting values of the masses and dimensions of mechanical and electrical systems in the onshore technology and reach the limits in production and logistics.

A conversion to higher working frequencies in the wind farm grid can (compared to 50 or 16.67 Hz) substantially reduce the amount and thus the costs of grid operating media and solve possible transport problems. Low working frequencies in the offshore region, in contrast, bring with them substantial reductions in the transmission losses and the compensation requirement of the reactive power. Furthermore, the maximum sea cable lengths can be extended from approximately 100 to 150 km for the conventional 50 Hz frequency to 300 to 400 m for the so-called 'Railway frequency' of 16.67 Hz. With direct generation of these low working frequencies by the WTs, it is not necessary – compared to direct current technologies – to install a converter station on the offshore platform.

Even lower working frequencies can bring further advantages with them. However, the frequency reduction has the disadvantage that the mass and thus the dimensions of the sea platforms for operating media to be installed (transformers, reactive power compensation installations, etc.) rise according to the frequency reduction and the duration of transient processes (e.g. for short circuit) becomes longer.

For a large-scale use of wind energy, there is a possibility of opening up greater perspectives of this technology using technical and economical optimum working frequencies. For this, it is necessary to keep the design of system components (generator converter, transformer, sea cable, land cable, compensation plants, switches, etc.) and safety equipment as economical as possible, considering the special requirements (e.g. at sea). However, extensive fundamental and application investigations in this area will be needed.

In Germany, in the future, the grid operators should consider being able to use so-called 'socket boxes at sea' [4.120]. For this, the grid operators must optimize and strengthen the lines on land in order to integrate the strongly growing wind power and distribute it in the grid.

4.8.5 Reactive power requirement and reactive power provision in the offshore grid

The electrical grid of an offshore wind farm consists of the WTs with machine transformers, power transformers, offshore and onshore transformer stations, internal cabling, the grid connection at sea and on land and the compensation installations. Transformers, cables and compensation coils draw inductive or capacitive reactive power from the grid, so that the elements, besides the active power to be transmitted is also loaded with reactive power. It reduces the transmission capacity and influences the voltage relationship in the gird. For this reason, compensation installations are necessary that relieve the load on the GCP of the wind farm and stabilize the node voltage of the grid.

4.8.5.1 Reactive power requirement of sea cables

The equivalent circuit diagram of a power element according to Equation (4.12) contains resistance (R', G') and reactance (L', C') portions. In the resistance, R' ohmic losses occur; the conductance G' leads to dielectrical losses due to the insulation of the cable. The inductivity L' and the capacity C' result from the geometry of the conductor arrangement and the materials used. Here, the free lines that can only be installed economically on land prevail due to the larger distances of the conductors of the inductive portions. The capacity is distributed over the whole length of the cable so that it acts as a large capacitor that is constantly loaded and unloaded with alternating current in operation. In this way during idling, reactive power is taken up via the capacities of the cable. A capacitive loading current is flowing. In the loading of the cable by means of an active power, inductive reactive power is additionally converted in the inductivity of the cable. The reactive power requirement of a cable Q_{Cable} is therefore made up of an inductive Q_{Ind} and a capacitive Q_{Cap} amount (Equation (4.45)).

$$Q_{Cable} = Q_{Ind} - Q_{Cap} \tag{4.45}$$

where

$$Q_{ind} = X_{ind} \cdot I^2 = 3 \cdot \omega L' l I^2 \tag{4.46}$$

$$QCap = \frac{U^2}{X_{Cap}} = 3 \cdot \omega C' I U^2 \tag{4.47}$$

and $C' = b$ capacitive cover, single phase

The result of the reactive power drawn from the cable in dependence of the apparent power S (in proportion to the natural power P_{nat})

$$\frac{Q_{Cable}}{Q_{Cap.\,max}} = \frac{Q_{Cap}}{Q_{Cap.\,max}} \cdot \left[\left(\frac{S}{P_{nat}} \right)^2 - 1 \right] \tag{4.48}$$

Determining factors for the reactive power take-up are the length of the cable and the amount of the rated voltage. The larger both are the more capacitive reactive power the cable takes up. Figure 4.88 shows in graphic form the reactive power taken up in dependence of the length and the transmitted power.

Cables can only be operated below the natural power, and the thermal loading capability is far below the natural power. The capacitive reactive power, for instance, can be partly compensated by the installation of compensating throttling coils.

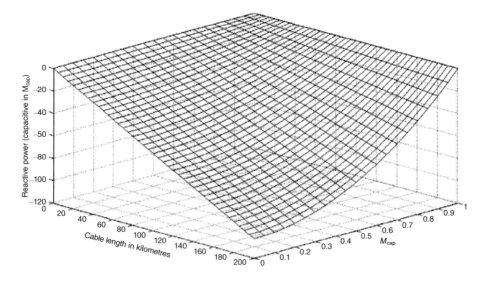

Figure 4.88   Reactive power of a cable in dependence of the length and the transmitted power

4.8.5.2 Transformers

As the single-phase equivalent circuit diagram 4.12 shows, transformers consist of an arrangement of coils and resistances and thus represent inductive users. They require inductive reactive power in order to build up the magnetic field. In this, R_1 represents the ohmic resistance of the primary or reference side of the transformer and R_2 the converted dimension of the reference side. The main reactance X_h characterizes the complete magnetic coupled portions between primary and secondary sides of the transformer. $X_{1\sigma}$ and $X_{2\sigma}$ represent the coupling, reduced due to magnetic scatter, between primary and secondary windings, whereby $X'_{2\sigma}$ is again converted on the reference side. The resistance R_0 *represents* mostly the resistance characterized by the iron losses.

If primary and secondary-side voltages U_1 and U_2 are divided by their rated sizes (e.g. 400 V, 20 kV) as given sizes, they can also be seen, for instance, as changes and influences beyond their voltage levels.

Here, too, the give reactive power is divided into a part Q caused by main reactance X_h *and thus* dependent only on the voltage and part Q_u dependent on the load of the transformer. For determining the impedance X_h and X_{Tn}, there are, besides the dimension apparent power S_i and the rated voltage U_{rl}, the rated sizes P_0 (idling losses) and I_0 (idling current) from the idling test as well as the rated sizes P_k (short circuit losses) and the u_k (short circuit voltage) necessary from the short circuit test.

$$\frac{Q_{\text{Trafo}}}{Q} = \frac{Q_u}{Q} + \frac{Q_i}{Q} = \frac{U_1^2}{X_h} + X_{\text{Tr}} \cdot I_1^2 \quad \text{Trafo} - \text{Trans} \tag{4.49}$$

where

$$X_h = \frac{U_{rl}}{\sqrt{3 \cdot \left[\left(I_0 \frac{S_r}{\sqrt{3} \cdot U_{rl}} \right)^2 - \left(\frac{U_{rl}}{\sqrt{3} \cdot R_0} \right)^2 \right]}} \tag{4.50}$$

$$R_0 = \frac{U_{rl}^2}{P_0} \tag{4.51}$$

$$X_{\text{Tr}} = \sqrt{\left(\frac{U_{rl}^2}{S_r} \right) \left(u_k^2 - \frac{P_k^2}{S_r^2} \right)} \tag{4.52}$$

4.8.5.3 Influence of the compensation installation on the transmission capacity of the grid linkage

The transmission capacity S_{max} of a cable is determined to be the maximum permissible temperature that depends on its part on the thermal stability of the insulating medium. For VPE cables, this limit is 90 °C [4.121] and exceeding this temperature leads to premature ageing of the insulation.

Decisive for the loading of a line is the amount of the current or the apparent power that is made up of the transmitted active power and the reactive power of the line.

$$S = \sqrt{P^2 + Q^2} \leq S_{\text{max}} \tag{4.53}$$

This results in the maximum transmittable active power:

$$P_{max} = \sqrt{S_{max}^2 - Q^2} \tag{4.54}$$

For an uncompensated cable, the transmittable active power is determined only by the capacitive referenced reactive power (see also Figure 4.89) – when ignoring the inductive part of the reactive power

$$P_{max} = \sqrt{S_{max}^2 - Q_{kap}^2} = \sqrt{S_{max}^2 - 3 \cdot \omega C' l U^2} \quad \text{kap} = \text{Cap} \tag{4.55}$$

In order to compensate for the high capacitive reactive power that a cable draws from the grid, static throttling coils can be installed at the ends of the cable that take up the inductive reactive power and can thus cover the capacitive reactive power of the cable. In this, it is possible to compensate on only one side of the cable or on both sides, whereby in this case always half the required compensating reactive power is taken up on each side.

In the ideal case, compensation coils would be distributed along the line at regular intervals and in this way to achieve an even cover of the reactive power requirement. In the determination of the loading capacity of the line, it must be ascertained that the compensation power also loads the cable (however only the offshore side of the cable; the coil on the grid side is only required for unloading the primary grid in the provision of the reactive power). It is now decisive whether the loading of the GCP is greater through the reference of the capacitive reactive power or the sea side connection due to the reference of the compensation reactive power. This means that

$$3 \cdot \omega C l U^2 < Q_{Comp} \tag{4.56}$$

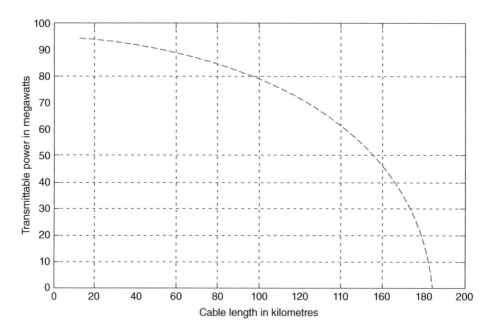

Figure 4.89 Free active power capacity of an uncompensated cable

Figure 4.90 Auxiliary switching for clarification of the design of the compensation installations

determines the (fixed) compensation reactive power Q_{Comp} the maximum transmission power, in the other case the reference capacitive reactive power of the cable Q_{Cap}. Simplified, an equivalent circuit is viewed in which a 70-km long sea cable is operated in idling mode and is directly connected to the grid without transformers (Figure 4.91). From this, there results the clear depiction of Figure 4.90 because of the different types of compensation, the courses of the current amount on the line change and thus influence the choice of the compensation amount. The cases without compensation and with full compensation onshore are covered in the same way and in both cases the line has the greatest load on the grid side due to the fed-in capacitive current. With full compensation offshore, the cable has the highest load at this point due to the inductive power. It is more advantageous when the compensation power is divided, whereby a compensation along the line represents the optimum case (the current is very low in the middle). However, for offshore wind farms, this is accompanied by great effort. In the division to the ends of the cable, the distribution with each half power is the optimum case; in all other cases, the line would be more heavily loaded on one side which limits the transmission capability. These statements are verified also by the voltage sequences that are the most advantageous for a distributed compensation with regard to a voltage increase.

A further type of compensation for the offshore application is an inductivity in the middle of the cable. In this way, the two installations at the end of the cable fall away and thus on the offshore side, a smaller platform is needed. At the same time, however, this also means the erection of an additional platform for the compensation installation. The coupling of further wind farms on long transport paths could also make possible further reactive power support points and result in additional platforms.

From an electrical point of view, the installation of the full compensating power in the middle of the sea cable leads to a full easing at both end nodes of the cable. The capacitive reactive power ($Q_{Cap} = 0.5 \cdot \omega C' l U^2$) for the second half can be fully covered by the compensation throttling coil. But as this is designed to be twice as large it means at the same time, a complete compensation of the first part of the line so that no reactive power is drawn off at the GCP. Alternatively, it is possible to install only half of the required compensation power in the middle of the cable but then only the second part of the line is covered, whereby the first part of the required capacitive power is drawn from the grid. Thus, the additional installation of half the compensation power at the coast is required for unloading the grid node.

Besides the fixed compensation installations mentioned here that were already mentioned in Section 3.4 for asynchronous generators and in Sections 4.5.1 to 4.5.3 within the scope of the filter design, it is also possible for flexible and controllable compensation system to be used for reactive power compensation on the onshore as well as the offshore sides.

The use of flexible or three-phase transmission systems (flexible AC transmission systems, FACTS) for reactive power compensation.

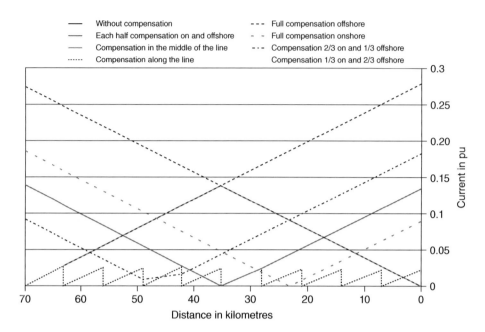

Figure 4.91 Current occupation sequence of a sea cable for various types of compensation (idling, values referenced to rated value)

4.8.6   Flexible AC transmission systems (FACTS)

Besides the grid-strengthening measures such as grid extension, introduction of higher voltage levels, addition of new routes, etc., flexible transmission systems also offer the possibility of improving the capabilities of existing grids. This permits quick control of the compensation installation by means of power electronic structural elements, which have fundamental advantages in the control of the voltage and power. Flexible transmission systems can be divided into static compensators (SVC, STATCOM), series compensators (TCSC, SSSC) and universal power flow controllers (UPFC)

4.8.6.1   Static reactive power compensator (SVC)

Static reactive power compensators (static voltage control – SVC) are divided into

- thyristor-controlled inductivity (TCR, thyristor-controlled reactor);
- thyristor-switched capacity (TSC, thyristor-switched capacitor);
- thyristor-switched capacity/thyristor-controlled inductivity (TSC/TCR;
- synchronous static compensator (STATCOM, static synchronous compensator);
- SVC PLUS (static var compensation of the Siemens Company).
- SVC-Light (static var compensation of the ABB Company with storage possibility).

SVC are normally designed so that the grid voltage can be changed by at least ±5%. Their positions are mainly selected to be near the centres of loads such as large city areas, in critical, normally far distant sub-stations and at the feed-in points of large industrial or railway users. In comparison to conventional solutions with switched capacitors and throttles, in the SVC, the mechanical switches are replaced by thyristors whose switching speed is determined by the grid frequency. TCR and TSC are normally possible in the medium component range (some kilovolts to 36 kV) via transformers at an existing tertiary winding of a grid transformer [4.122].

The **thyristor-controlled inductivity (TCR)** is an inductivity (iron-free throttling coil) that is switched through thyristors and the reactive power is continuously changed by the modulation of the thyristors so that the throttling current can be steplessly controlled [4.122]. In this way, a continuous adjustment region is possible without transient influences. However, harmonics are introduced. To prevent them, filter circuits can be installed in parallel. The harmonic influences can be reduced already with the installation of a capacitor.

Thyristor-switched capacities (TSCs) are switched on and off in a branched manner by means of anti-parallel thyristors. In order to prevent transient occurrences, the thyristor are ignited such that the capacity can be switched on and off with the smallest possible balancing method. In order to limit the switch-on current steepness, the capacity branches must be provided with a protective throttle. By means of thyristor-switched capacitors, a stepwise change in the capacitive region is possible [4.122, 4.123]. In this way, a stepwise control without transient influencing is possible, whereby harmonics are prevented and the losses are kept low.

The combination of thyristor-switched **capacitors and thyristor-controlled inductivities (TSC/TCR)** is the optimum solution in most cases. A compensator permits the use of the whole reactive power adjustment region. The continuous reactive power output is achieved by the change in the TCR ignition angle. As soon as the TSC region is compensated by the TCR, the capacitive region is switched off and the compensator works as an inductivity [4.122]. With this combination, continuous adjustment regions without transient influences are given, whereby the harmonic generation and the losses are kept low.

In the **synchronous static compensator (STATCOM)**, a buffer capacitor is operated via a converter with direct current intermediate circuit. In this, the direct current is converted to three controllable alternating currents with the frequency of the three-phase grid. Each generated voltage is in phase with the corresponding grid voltage. The voltages are connected to the grid by means of a transformer. By means of a change in the amplitude of the generated voltage, it is possible to exchange reactive power between STATCOM and the grid. In the case that the amplitude of the STATCOM output voltage is greater than the grid voltage is set for, then a leading current of the grid voltage flows over the reactance X into the grid and the converter makes provision for inductive reactive power. If the amplitude of the generated voltage is less than the grid voltage, then the converter takes up inductive reactive power. In the case of the same voltage amplitude, there is no exchange of reactive power. Phase and frequency can be influenced by means of a control so that the voltage intermediate circuit of the converter can be seen as a controllable voltage source. The reaction times of these systems are much less than with SVC, and the reactive power take-up can be controlled independently of the grid voltage. The STATCOM can be used for feeding reactive power in transmission grids, where the STATCOM gives up its reactive power for node voltage ($Q \sim U$) and for compensation of voltage collapses at user nodes in distribution grids of the medium voltage level.

SVC PLUS (static var compensation) is an advanced STATCOM with modular multilevel converter (MMC) technology. The use of this technology has additional advantages in comparison to the normal solutions for reactive power compensation and, because of its compact structure, is an ideal solution for limited space. The MMC solution permits an almost ideal sine-shaped wave form for the a.c. side. For this reason, only few harmonics occur and there is no need for harmonic filtering [4.124]. A transformer-less grid connection up to 36 kV and ±100 Mvar is possible. Three standardized prefabricated configurations for ±25, ±35 or ±50 Mvar are available as container solutions and can be configured for up to four units as a system, whereby high dynamic power, very short reaction times and highly efficient voltage support are given.

SVC Light with energy storage is a STATCOM based on the VSC (voltage source converter) technology and is equipped with IGBTs. These systems are connected to the grid by means of a string throttle coil and a power transformer. The SVC Light offers a high degree of reactive power storage also for low grid voltages. As the system possesses several rows of parallel batteries, it is able to also control active power. From a practical point of view, SVC Light also has further advantages such as very short reaction time, provision of system services such as frequency regulation, dynamic voltage regulation and reduction of harmonics as well as modular, work-erected units, whereby place, start-up time and costs are saved.

The **static row capacitor** permits the control of the reactive and active power flows by means of changing the line impedance or by the insertion of a series voltage. The main types of these technologies are the thyristor-controlled series capacitor (TCSC) and the synchronous static series compensator (SSSC) [4.125]. With the application of the thyristor-controlled series compensator, it is possible to improve the stability and transmission capability of the high-voltage grids. The series capacitor feeds capacitive reactive power in order to compensate the inductive voltage drop of the line, i.e. it reduces the effective reactances of the transmission line. The TCSC control consists of thyristor-switched throttling coils that are switched in parallel by segments of a capacitor battery. In practical applications, several systems are switched in series in order to achieve the operating properties for the corresponding rated voltage. A varistor is parallel-switched additionally to the throttling coil in order to prevent excess voltages.

The TCSC can work in three operating modes.

In lock-out operation when the thyristor valves are not conducting, the current flows only into the capacitor battery. In this operating mode, the TCSC behaves as a passive capacitor bank.

In the circulating mode when the thyristor valves are constantly ignited, they remain conducting the whole time and the TCSC behaves as a parallel switching of capacitor battery and throttling coil. This type of operation is used in order to reduce the load of the capacitor battery in case of short circuits.

In the capacitive enhancing operation, an ignition impulse is applied to the thyristor with zero flow of the capacitor voltage and in the parallel inductive branch, a capacitor-unloading impulse takes place. This current impulse adds itself to the grid current and causes an additional voltage drop at the capacitor that is added to the grid voltage. For this reason, the capacitor voltage is increased. This type of operation of the TCSC acts as an increase of the series capacity for the transmission path.

An important possibility of application for the TCSC is in the prevention of power swings with coupled electro-energy systems and the prevention of synchronous resonance appearances for series compensation that can lead to damage of the turbine–generator shaft with series compensation [4.126]. With this system, an improvement of the behavior of the transmission system with respect to the voltage and angular stability is achieved [4.122].

The **SSSC** is an alternative to the TCSC. In this, the difference is that the TCSC lies on high-voltage potential and must be controlled by means of fibre optics, whereas the SSSC lies on earth potential [4.127]. The SSSC possesses an internal voltage source that calls up the desired voltage with a view to angle and amplitude of conductor current. With this, active and reactive power can be controlled. Thus, a SSSC can behave as a series capacitor or a controllable series throttling coil. The difference is that the set voltage is not referenced to the grid current but can be controlled independently of it [4.122]. Up to now, the SSSC is included only as a component of a UPFC in two installations in electro-energy systems [4.126]. If more parallel transmission lines are equipped with SSSCs, then active and reactive power can be exchanged between the lines as desired. This installation is called 'IPFC (interphase power flow controller) or IPC (interphase power controller)'.

The **UPFC** consists of two rectifiers that are linked by a common direct current intermediate circuit. The collaboration of two converters with the use of the common direct current intermediate circuit unifies the function of an SSSC and a STATCOM. A power converter feeds the required active and reactive powers into the intermediate circuit (SSSR), in that the direct current is changed. The second power converter carries out the main function of the UPFC, in that it feeds an alternating current via a series transformer in series with the transmission line so that the active power current can be controlled. The UPFC possesses three control sizes and can be operated in various conditions. The parallel-switched first power converter regulates the voltage at the collector rail and the series-switched second power converter permits the active and reactive powers over the line. With the combination of these properties, the regulation of the voltage profile in the grid and the control of the power flows can be undertaken as desired. In this way, an improvement of the transient stability and the voltage stability can be achieved. The result is increased reliability and availability of the overall system.

4.8.6.2 Connection of compensation plants

The compensation plants typically used with a mechanical switch, such as the MSC capacitor bank are often used in combination with a transformer with step adjustment. However, the resulting frequent on- and off-switching due to power swings of the wind farm degrades the grid quality and reduces the life of the plant. For this reason, connection variants that are exactly tailored to the requirements of the wind farm should be selected.

4.9 Integration of the Wind Energy into the Grid and Provision of Energy

At present, four German States cover their power requirements with about 50% from wind energy. It can be expected that already by 2020, the installed WT power will be about doubled. However, the increasingly larger new installations will permit over-proportionally much larger energy feeds or power. In addition, with WTs installed at sea, about double as much energy yield can be expected as is generated by plants situated on land. An electrical energy transmission and distribution of the wind power will thus – especially in North Germany – increasingly spread to neighbouring regions such as abutting German States and abutting foreign countries.

4.9.1 Grid extension

Besides national grid extension possibilities (DENA studies I and II, [4.111, 4.112], etc.), 'a single grid for Europe' [4.128] could open new perspectives and permit a functioning competition in the European energy market. However, at present, the highest voltage grids of the European Member States are only relatively weakly interconnected. The stronger these links are the easier, it is for large quantities of regenerative energies to be integrated and a Europe-wide balance for wind energy can be achieved.

In the European ENTSO-E UCTE Association (Union for the Coordination of the Transport of Electrical Energy) of 34 TGOs from 23 countries, the transnational current flows have almost tripled since the start of the 1990s. However, the capacities for the exchange and trade between the countries have meanwhile reached their limits. The removal of these bottlenecks must now be given special attention.

The electrical energy supply in Germany and Europe up to now has mostly been oriented to the requirements of the times between the middle and end of the twentieth century. This means that power stations and regional and supra-regional grids were built in the vicinities of industry and population and thus load centres. In this way, it was possible – for that time period – to erect and safely operate a technical and commercially feasible electricity supply structure. In the meantime, on the one hand, the economic structures have fundamentally changed and, on the other hand, renewable energies up to now structurally weak generating centres are becoming increasingly important. As a result grid, expansion plans must take both the change in the economy and the expansion plans of regenerative energies into account. An 'offshore super grid' [4.128] in the form of high-voltage direct current transmission lines (HVDC) or a special wind energy transport via HVDC [4.129] could form the first milestone of a European linkage.

In Germany, measures for extension in the highest voltage transmission grid are necessary. For this purpose, the strengthening of existing current routes, the erection of highest voltage routes, the erection of cross regulators for targeted control of the load flow and the construction of plants for provision of reactive power are required. According to reference [4.107], by 2015, the existing approximately 18 000 km highest voltage routes must be extended by 850 and 400 km of the existing ones must be strengthened. The costs of these grid extensions are around 1.1 billion euro. Further 1050 km of new construction and 450 km of strengthening of lines are required by 2020. The transport of power – also over large distances of wind farm in the North and the Baltic seas for GCPs to land is further evaluated by reference [4.107] as technically feasible.

A timely implementation of the measures for extension and strengthening of the grids is required in order to ensure safe grid operation also in the future. With the utilization of variable-speed plants that feed into the grid via pulse-width-modulated inverters, permissible limiting values of harmonics can even be eliminated by means of active filtering. For, instance, by regulating the power factor or the current feed angle, the voltage at the GCP of a plant linkage can be adjusted or kept to a defined value (Section 4.6.2). In this way, existing grids can be relatively well loaded and costs for grid-strengthening measures can be saved [4.96, 4.130–4.132].

Especially, large feedbacks can be expected with weak grids with a high proportion of wind power. Practical investigations at island grids, however, have shown that with a targeted design of the grid and its components, even wind power proportions of 100% are possible. By means of measures for influencing the grid characteristics, possible grid disturbances can be prevented in linkage operation with the inclusion of phase displacement, battery storages and reverse flow devices, grid filters, compensation units and grid controllers [4.133–4.135].

These types of concepts provide information for the operation of grids and the power or energy provision from conventional and regenerative 'Generating systems' in combination with storage units.

4.9.2   Provision of energy

The electrical energy drawn from the grid is subject to strong daily and annual swings. In Germany, on the one hand, high values of electrical withdrawals from the public grid occur in the morning and evening. On the other hand, the availability of wind energy is also subject to daily and annual changes that, however, have meteorological causes and correlate only partly with the power requirements. Therefore, the best possible load and wind availability prognoses are of great importance for adherence to so-called 'grid timetables'. Thus, in the user supply, especially those that rely on large wind energy portions are subject to stochastic load and feed processes.

In the conventional electrical energy supply, a difference is made by the feeding power station and also in user values between

- base load
- medium load and
- peak load.

The base load is taken to mean the load that is not undershot over a longer period. The utilization duration (quotient of the whole work in the time period of a year to the bottleneck power) of the power stations that are used in this region is approximately 5000 to 7500 full-load hours per year (thus 8760 hours) Full-load hours are a fictitious measure for the loading of a power station. To cover the base, load use is normally made of power stations (brown coal, nuclear power plants) with high investment and low operating costs. Eight nuclear reactors were shut down in 2011 as a direct result of the renewed German exit from atomic power. The last nuclear reactor is scheduled to be taken off the grid in 2022. The supply gap that will result from this is to be closed not only by energy-efficient measures but also in the base load region by the change to renewable energies.

Medium loads are limited to about half of the year. The loading of the medium-load power station is approximately 4000 hours per year. Daily specific starting and stopping of the plants as well as part-load operation according to user demands are necessary here. At the weekend, the plants can possibly even shut down their operation.

Peak loads are limited to around 10% of the year. Peak-load power stations are operated at a loading of less than 1000 hours per year. Short access times with high feed safety are required in order to cover occurring load peaks in the grid reliably. Pumped storage hydropower stations, gas turbines and increasingly combi-power station are used. Because of the short operating times, the fuel-dependent costs can be high, but the investment costs should be kept as low as possible.

WTs in Germany reach usage duration values of approximately

- 1500 to 2300 full-load hours on land and
- 3000 to 5000 full-load hours in coastal regions

With offshore use in the North Sea, one can assume

- 4000 to 5000 full-load hours.

At windy sites such as Ireland, Spain and Morocco, these values are even reached on land, peak loads of

- 5000 to 5300 full-load hours

are also possible at coastal sites of the Red Sea (Zaferana) in Egypt and on land. Photovoltaic plants in Germany, in comparison achieve approximately 800 to 1000 full-load hours per year.

Offshore wind farms and international top locations on land thus reach continuous utilization values similar to medium, and partly, even base load power stations. However, they are more weather-dependent and do not always have increasing availability for increased user demands. Therefore, a secure electricity supply from wind energy systems has its limits. Storage systems can be of help here. However, these are associated with substantial additional costs that per unit of energy (kilowatt hour or megawatt hour) are mostly higher than the 'Generation'.

Storage systems that have been used up to now in the watt and kilowatt ranges (in exceptional cases also in the megawatt class) are

- electro-chemical storage (so-called batteries) or

in the megawatt range for minute or daily balance

- Pump storage [4.136].

(for corresponding geographical and geological conditions). Flywheel storage, supercaps, etc. are also possible in the short-term range of smaller capacities. For long-term storage, use can be made of fossil energy carriers (natural gas, diesel, etc.) also with regenerable energies in future (biogas, hydrogen etc.). In a grouping of wind energy and storage systems, it is therefore possible to ensure that the base, medium and peak loads are covered and the wind energy can be provided to the grids in a need-oriented manner all according to the given requirements.

Besides the offshore regions, the most windy parts of Germany can be found in coastal and low-land areas and only partly in the uplands. Therefore, pump storage (also for reasons of difficult approvals) can hardly be viewed as complementary energy storage for wind energy or can be mostly used in mountainous upland regions. However, large salt deposits are often found in these windy areas that can be used as storage caves. More than 250 caves are being operated and more than 50 are at the planning stage or under construction [4.137]. They are used for storing fluid and gaseous fuels (oil, natural gas), for harvesting rock salt and in a few cases for storing remainders and waste materials.

Alternative processes to pumped storage hydropower stations are compressed air energy storage (CAES). The JAVA has been used successfully for thirty and fifteen years, respectively, in Huntorf (Germany) and McIntosh (USA) [4.138, 4.139].

The compressed air energy storage power station (E.ON) was started up at the end of the 1970s. A 290-MW turbine is designed for 2 hours and a 60-MW compressor for 8 hours of operation. The compressed air is stored in two salt caves with each approximately 150 000 m³ volume at a depth of around 700 m at pressures between 50 and 70 bar.

The plant in McIntosh (Alabama, USA) was taken into operation at the start of the 1990s. Its capacity is 110 MW over 26 hours. A single cave with 538 000 m^3 provides the air storage. Further plants with up to 10 million cubic metres and 520 000 MWh storage capacity are at the planning stage or under construction.

In reference [4.137], there is a detailed description of an offshore combined power station of wind energy, lean-gas flow and compressed air storage in caves. Here, three interconnected salt deposits (Borkum, Lisa and Lollo) are shown as suitable compressed air caves. In addition, new potentials can be opened up with the offshore flow of lean gas. In this, previously uneconomical gas fields with low gas quality can be used for producing energy and with their linkage at sea can increase the stability of offshore grids. The *true* actual cost of current is given as 9 to 11 cents per kilowatt hour – depending on the rate of interest. As the external costs for offshore combined power stations are substantially lower than those of otherwise-used base power stations, from a politically economic view, this *Clausthaler concept* is seen as a good competitive alternative – equal to brown coal power stations. Due to the rising costs of conventional energy generation, the result is additional improvement in competitiveness.

4.9.3   Control and reserve power

With the integration of large wind energy in grids, additional control and reserve power are necessary. These are dependent on the accuracy of the wind power prognoses and on the deviation between prognosis and actual feed [4.140]. An average of 11 280 MW wind power was installed in 2011 in the grid of the 50-Hz transmission GmbH. The prognosis error in 74.8% of all cases was less than 5% and in 99.5% below 25%. The deviation occurred especially due to time displacement in large power gradients between prognosis and feed and was mainly 25% in excess. The deviation at present is an average of 5%. However, especially due to time displacement between prognosis and feed, it can exceed 20%.

It is necessary to keep minute and hour reserves as control and reserve power in order to equalize changes of the wind energy feed. Positive control and reserve power are required for too little and negative power up to too large wind current power. This must be kept operationally ready as positive or negative power station capacity.

Within the framework of their technical availability and in accordance with plant-specific start or pre-start times, fossil and nuclear fed thermal power stations can, on the one hand, feed the grid with the full power or, on the other hand, also the output power and can increase or decrease their rated range as desired. These features are the basis of our present very safe electricity supply. But power increases or decreases can only occur relatively slowly, i.e. 5 to 10% of the rated power per minute, which, in the case of a grid failure and other borderline situations can lead to safety problems.

In contrast, WTs are subject to the vagaries of the wind, i.e. they only permit a feed depending on the meteorology and within the power capacity of the wind. However, – just as hydropower stations – they make a start-up to full-load operation in less than a minute possible and even permit stop or switch-off processes in a space of seconds. It is these quick shut-down processes that are of great importance from a safety point of view when, for instance, wind farms of 100 MW to gigawatt class are switched off in their capacity in seconds and thus are meant to protect the grid from excess frequency [4.141].

In the year 2003, it was necessary (with 14 000 MW installed WT power) to plan for an average 1200 MW positive and 750 MW negative and a maximum of 2000 MW positive and 1900 MW negative control and reserve power.

In 2015, an average of 3200 MW positive and 2800 MW negative as well as a maximum of 7000 MW and 5500 MW negative control and reserve power is additionally expected from the wind energy feed.

According to [4.107], the control and reserve power can be covered from the mentioned wind farms so that no additional power stations need be erected and operated.

A call for control reserves leads to an increased demand for balance energy. This is associated with additional costs and O_2 emissions that can be allocated to the wind energy [4.142–4.145].

Because of the pre-control for power and renewable energies, WTs that permit very fast control access have up to now not been used as negative control reserve.

The very quick controllability of the power of WTs is given in that controlled starts and stops take place in less than a minute; changes in power of, for instance, 50% of the rated power in 1 to 2 seconds are possible and switch-off procedures can be carried out within a few seconds. Thus, WTs are best suited for provision of negative control reserves and offer the quickest possible protection against excess frequencies.

Excellent voltage and reactive power control possibilities of WTs with voltage corrections in the percentage range can be achieved at both the feed and linkage points [4.93, 4.94]. Short-term voltage collapses – so-called voltage dips – can be effectively reduced with the aid of fast control intervention of the grid-side WT converter [4.96].

The frequency and voltage support for grid collapses as they are required in the Grid Code and other regulations, can up to now mainly only be ensured by modern WTs. The momentary reserve, thus the immediate power increase with the use of spinning reserve of WT is, much greater, due to the large turbine dimensions, is about four times greater than that of conventional power stations [4.83].

Thus, WTs, in contrast to photovoltaic, are in a position to make an important contribution to grid stability.

4.9.4 Power reserve provision with wind farms (Dissertation A. J. Gesino)

The possibility for wind energy of starting to provide ancillary services which currently are being provided by conventional generation, would lead to higher penetration levels of wind power into the grids, avoiding technical current existing barriers for its large scale integration.

In any electrical system, active power has to be generated at the same time as it is consumed, keeping a constant equilibrium between consumed and generated power in order to keep the system stability. Frequency falls when demand is greater than generation and rises when generation is greater than demand. In order to manage frequency effectively, system operators utilize a range of balancing services that operate over different time horizons.

System frequency deviations are caused by unexpected unbalances between generation and demand, and are of particular concern in systems where the ratio of potential variation caused by fluctuating wind in relation to the total amount of generated energy is high. These deviations could activate a significant share of primary power reserves. A further increase of these phenomena, for example due to high wind power penetration into the grids, could create frequency deviations large enough to activate the complete available primary power reserves reducing the security margins for frequency control and putting into question the adequacy of primary reserves to limit frequency variations, and secondary reserves to restore frequency variations. Up to now an adequate frequency control depends on conventional generation resources made available by generation companies to Transmission System Operators (TSOs) for this specific porpoise [4.147].

To cover sudden unplanned disturbances that may occur in the system, TSOs procure sufficient dynamically available operating power reserve in order to secure the system with the largest acceptable losses of generation and demand. Still based on conventional generation, needed volumes of primary, secondary and tertiary power reserve have been typified and calculated by TSOs since many years ago in a reliable way. The network power frequency characteristic describes the real dependency between system frequency and power imbalance with a linear approximation. In order to ensure that the principle of joint action is observed, the network power frequency characteristic of the various control areas is taken to remain as constant as possible.

Wind energy differs from conventional generation since the production is not driven by the demand for electricity, but by meteorological conditions. This increases complexity to match consumption and production. To continuously balance supply and demand, ancillary services are defined. Whenever there is an imbalance in a power system, the frequency will start deviating from its nominal value. This activates both primary and secondary control. These control mechanisms stabilise frequency variations and restore the frequency to its nominal value. Currently, distributed generators do not participate in primary control, nor in secondary control. Balance is restored only by conventional generators. However, as the share of wind power increases, its participation in power reserve provision will become necessary [4.146].

As a result of high penetration levels of wind power into the European grids, and also expecting that these levels will be even higher in the coming years due to the European Union targets regarding renewable energies, wind power is being requested by TSOs to meet the same grid stability and grid codes requirements as conventional power plants, including balancing procedures and frequency regulation facilities.

4.9.4.1 Power reserve provision with wind farms

Considering the relation between system frequency and power equilibrium, the technical characteristics of primary and secondary power reserve and the fact that traditional schemes regarding ancillary services are not directly transferable to wind power, a new concept was developed during this research addressing how wind power could provide power reserve, both positive and negative, in a stable and reliable way.

Since the last years wind energy is not only constantly increasing its install capacity, it is also starting to replace conventional generation in some countries. This leads to situations were wind power is reaching penetration levels of 50, 60 or even 70% during a certain periods of the day.

One of the main challenges is to accept the integration of higher volumes of wind power into the European grids in coordination with other energy sources and still keep the grid stability. Because of the diversification of the European energy matrix and the introduction of several renewable energy sources, grid security has become one of the main issues to be considered. Therefore, as long as wind power is expected to play a major role into the European energy matrix, it must have the functionality required by power plants: reliable, predictable, dispatchable and controllable.

In other words: *wind power shall be manageable as a conventional power plant system*. In recent years wind energy has been considered a non dispatchable source and therefore mostly connected to the grid without any option for active communication and interaction with the grid operator or the electricity markets. However new developments and real field tests, described during this research, have shown that wind energy is manageable as a conventional power source.

4.9.4.2 Objectives and basic facts of the developed methodology

The aim was to develop a methodology for wind power to be able to provide positive and negative power reserve.

Through an analysis of the ancillary services which currently are being requested at wind farm level in the most advanced wind power countries in the world, and considering the structure of wind farm clusters developed at Fraunhofer IWES as the natural evolution for wind power management, a concept for primary and secondary power reserve provision with wind power was developed and tested in both, simulation and real test fields scenarios.

Over-frequency and under-frequency events were considered as well as frequency response procedures with wind farms were developed taking into consideration the needed quick response. It was also analyzed which control strategy could best provide the needed power reserve according to the frequency event which could occur.

How to dynamically fulfil the TSOs needs concerning the volumes of primary and secondary power reserve, keeping the current security standards, adding the needed flexibility and minimizing costs and power looses, were also discussed in [4.154].

Once an upper-frequency event occurs, negative power reserve should be activated according to the frequency variation. The related control strategies for this event were also analyzed in this research. How to contribute to balance a system with wind power after an upper-frequency event, minimizing losses and fulfilling the needed time response requirements, were addressed through the development of control strategies. The influence of available wind power forecasts and their uncertainties [4.152] [4.153] [4.149] was also considered.

Finally, through real test fields it was analyzed the current existing capability of wind farms to contribute with power reserve provision as happens nowadays with conventional generation [4.154].

Objectives of the Developed Methodology

The integration of large shares of fluctuating generation into the existing energy transmission and distribution systems requires a new strategy for the operational management of wind farms which should be equivalent (but not equal) to the strategies followed by the conventional generation power management. Some basic facts should be considered before addressing the objectives of the developed methodology and algorithm.

The considered facts are

1. Considering the technologies available nowadays, wind power has the capability to be forecasted with a 98% level of accuracy in a time frame of 8 hours ahead [4.148] [4.149].

2. The "lower interval" from each forecast has to be considered as the reference forecasted power.
3. Wind farms should be prequalified in order to be able to provide ancillary services as happens nowadays with conventional generators.
4. Due to the needed time response, an automatic algorithm is needed to monitor and control the different steps until the power reserve is provided by the selected wind farms. This algorithm may consider the interaction between the TSO, the wind farm operator and the wind farms SCADA system.

Finally, the objectives of the developed methodology are described as following:

1. Increase the power quality at low costs of those electrical systems which may have a strong penetration of wind power.
2. Allow a control of the grid in a more flexible and intelligent way, keeping the current security levels of the system requested by the TSOs.
3. Provide a secure and stable environment for all market participants to allow complex interactions between independent actors.
4. Promote the aggregation of wind farms into wind farm clusters [4.150] [4.151] in order to support the coordination between TSOs, dispatch centres, wind power producers and energy markets.
5. Promote the "multi layer control structure" as the next step for large scale integration of wind power in order to allow wind farm clusters to be monitored and controlled in real time according to the TSO needs.
6. Provide a communication structure to coordinate the bidirectional data flow and to participate in the dynamics of the power market.
7. Encourage the needed changes in the current procedures in order to allow wind power to be integrated in a flexible and secure way into the ancillary services market.

4.9.4.3 Methodology for providing power reserve with wind power

The developed methodology makes use of the advantages of the centralized and decentralized control systems, providing a two layers architecture (TSOs and dispatch centres) as well as giving the dispatch centres their own autonomy based on their own monitored and calculated data. It is also decreases the complexity of the control and decision making process with regard to the power reserve provision considering that a large volume of wind farms is expected to be able to participate in the power reserve provision markets.

This process is based on an 8 hours procedure where economical variables are considered as well as the stability of the offered power reserve, at wind farm level, is being monitored and evaluated according to the TSOs requirements. A schematic description is depicted in Figure 4.92.

Call for tenders: the needed power reserve volumes are published by the TSO one month in advance as it is described in Figure 4.93.

Tendering process: based on the requested power reserve volumes by the TSO, a "tendering process" would be opened every hour, 8 hours before the "Power Reserve Activation Time" (PRAT), as it is depicted in Figure 4.94. During these 8 hours tenders are posted, economically evaluated and finally their availability and technical stability is validated.

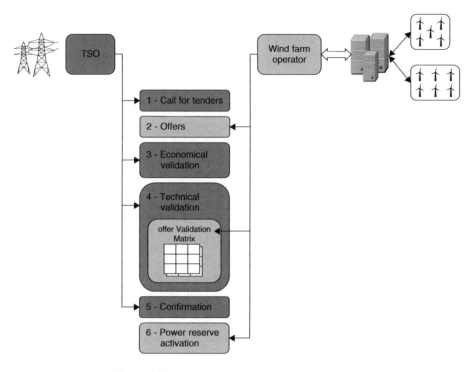

Figure 4.92   Schema of the proposed methodology

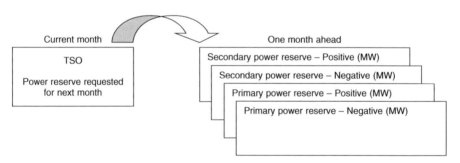

Figure 4.93   Month ahead power reserve request

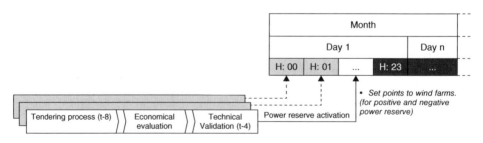

Figure 4.94   Reserve power tendering process

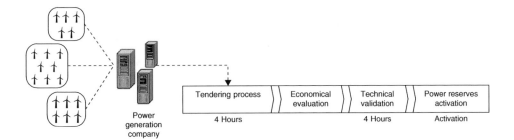

Figure 4.95 Tender process

As it is described in Figure 4.95 the "tendering process" takes place during the first four hours after the call for tenders is opened. During this period power reserve offers should be posted by each power producer based on prequalified wind farms.

Economical evaluation: once the tendering process is closed, an economical evaluation of each offer is performed. This is a market oriented process where the economical variables of each offer are being considered allowing the TSOs to optimize and reduce costs concerning power reserve provision. Those successfully evaluated economical offers are going to be technically verified during the last four hours before the power reserve activation takes place.

Validation process: the objective is to evaluate the relation between "offered power" and "available power" during the last four hours before the power reserve activation takes place. This "offer stability" validation process considers an offer to be unstable when the offered power volume is bigger than the one reported by the Lower Interval (LI) from the wind power forecast for a given wind farm.

The "Validation Process" is performed through the "Offer Validation Matrix", described in Table 4.11. Every hour the required updated information should be loaded by the wind farm operator or the dispatch centre. In order to evaluate the offer stability, two factors were developed during this PhD ("Hour Stability Factor" and "Offer Stability Factor") considering as input data for them the information available in the "Offer Validation Matrix". Based on these two factors all offers are monitored in an hourly basis as well as in a global concern regarding the whole offer and the power availability at the activation time.

Additionally to the already described schema, a weight factor was attached to each one of the four hours before the PRAT (see Table 4.12). The aim of this factor is to give a degree of

Table 4.11 Offer validation matrix with a time frame horizon of four hours

Offer Validation Matrix						
Company	Wind farm	t-4	t-3	t-2	t-1	osf
Name	Offer type	f(P)	f(P)	f(P)	f(P)	
	Install	LI (f(P))	LI (f(P))	LI (f(P))	LI (f(P))	
	capacity	–	–	–	–	
	Offered	hsf(t-4)	hsf(t-3)	hsf(t-2)	hsf(t-1)	
	power					
	Price (€)					

Table 4.12   Weigh factor

Time frame	t-4	t-3	t-2	t-1
Weigh factor (w)	0,25	0,50	0,75	1

relevance to each hour considering that potential fluctuations may occur during these hours and depending on when they occur, they could be more or less relevant with regard to the stability of the offer. This weight factor combined with the "hour stability factor" is being referred in Equation (4.58) by the "offer stability factor" calculation, which evaluates each offer as a whole based on the hourly evaluation described in Equation (4.57).

The two developed factors are explained as following:

Hour Stability Factor (hsf): evaluates the offer stability by each of the four hours previous to the power reserve activation. This factor indicates by each hour how stable the offer was keeping as a reference the offer time frame. It is based on the forecasted active power and on its Lower Interval (LI). As it is described in Equation (4.57) the "hsf" could have two possible values: 0 or 1, meaning with 0 that for a given hour the offer was not stable enough and with 1 that the offer was sufficient stable.

The needed conditions for the "hsf" to report an offer to be stable are included into Equation (4.57) and described as following:

1. For a given hour the Lower Interval from a wind power forecast should be lower than the forecasted power (as it always should be under normal conditions).
2. The Lower Interval should be above 20% of the offered power reserve

$$\overbrace{f(p - LI(f(P))) > 0}^{1} \wedge \overbrace{LI(f(P)) - 20\%(offer) >}^{2}{}_{offer} \quad hsf = 1 \qquad (4.57)$$
$$else \quad hsf = 0$$

Offer Stability Factor (osf): evaluates the complete offer based on the "hsf" results of each hour previous to the power reserve activation time and the hours weigh factor. As it is described in Equation (4.58), each offer would be evaluated at wind farm level considering its stability during the past four hours.

$$osf = hsf(t - 4) * w(t - 4) + hsf(t - 3) * w(t - 3) +$$
$$hsf(t - 2) * w(t - 2) + hsf(t - 1) * w(t - 1) \qquad (4.58)$$

With regard to the result of Equation (4.58), an offer is considered stable enough if ist final value is bigger than "2", representing that at least during the last 3 hours before the activation time of the offered power reserve from a given wind farm, the offer was always available.

4.9.4.4   Results

Wind power is, admittedly, different from other power technologies and integrating large amounts of it in the existing power systems is a challenge that requires innovative approaches to keep the sustainability of the power system operation. In the coming years the contribution of wind power to the system security will become mandatory.

Nowadays the Transmission System Operators (TSOs) are managing and operating the electric power transmission grids based on ancillary services procurement from the free market. The trend goes towards more decentralized structures and an increase in complexity due to a higher number of market participants. The provision of this kind of services also by wind energy requires innovative technical solutions.

The planned construction of numerous wind farms of the magnitude of conventional power plants calls for a new strategy in their operational management. In terms of the monitoring, connection and control conditions, wind farms should be equivalent to conventional power stations. Particularly related to the ancillary services provision, it is proved that wind farms could take over the tasks currently carried out by conventional generators.

As it has being described during in [4.154], wind power can not only be decreased but also be increased within a time frame of seconds. Widespread use of this method of operation can result in a significant frequency-supportive power reserve provided by wind farms.

Many different visions have been proposed for future power systems. Each of these visions depends on equipment, regulations, legal structures, environmental factors and many more, all with specific control needs.

The consensus no longer exists with regard to the question how a future power grid should be controlled, or what should be intermediate steps towards that direction. A new research area is currently forming on the border between electrical power engineering, industrial automation, control engineering, energy economics, communications technology and intelligent systems. The first large research projects in this area have already been launched. Some countries like Denmark or Spain, have already implemented the proposed "two layers" concept (TSO and dispatch centres), showing that wind power is controllable and predictable enough to allow larger penetration levels, according the world wide trends.

The necessity of the wide use of renewable generation is nowadays self evident. However, it is not the need of ancillary services provision based on renewable energies. Large wind farm clusters will be centrally controlled in order to coordinate and adjust the operation of many individual wind farms distributed in the field.

In order to validate the developed methodology, two simulations of power reserve provision with wind power were carried out based on real wind farm clusters [4.154]. More than 650 MW of install capacity were considered for the simulations development and the implementation of the developed methodology.

The behavior of the validation structures during the different steps of the process was also analyzed as well as the effectiveness of the complete model with regard to its control capability and the possibility of wind power to provide power reserve in a stable and secure way.

Finally, it was studied the concrete capability of wind power to provide primary and secondary power reserve during one year. The analysis was performed based on the German wind power production time series from 2009. The results of the validation process can be described as following:

- The innovative developed methodology allows increasing the power quality at low costs from those electrical systems which may have a strong penetration of wind power.
- The structure of wind farm clusters is recommended in order to reduce limitations due to potential unavailability of single units by grouping wind farms into clusters.

4.9.5   *Intercontinental grid connections*

With the use of intercontinental grid connections in Europe and also with linkage of Asia and Africa, it is possible to combine climate zones and renewable sources of energy (wind, water, solar, etc.). In this way, the so-called grid capacity or the grid short circuit power becomes very large and the effects on the grid very small. The constancy of frequency and voltage of the grid is thus given. Very different feed and user situations can be combined and excellent balances can be achieved by linking arctic, sub-tropical and tropical regions into a 'supergrid'.

With the combination of wind and hydropower, i.e. with meteorological and multiple anti-cyclic behavior, it is possible to achieve long-term auxiliary functions and thus substantial balancing effects. In addition, water storage can be used more effectively, short-term power equalization achieved and frequency regulation improved. In addition to this linkage, an intensive use of cheap wind energy from windy regions such as Egypt and Moroccan be achieved intercontinentally [4.155].

There are disadvantages in that the transmission of electricity over long distances (depending on the path and voltage level) will cause additional transmission costs and transmission losses. Furthermore, grid stability problems must be expected with very great distances and strong grid loading and these are feed- and load-dependent and can lead to frequency deviations between regions as well as reactive power swings or oscillations.

An intercontinental grid linkage will mean a grid extension that is necessary in any case. For stable grid operation, it would be possible – depending on grid extension and voltage levels – to form two or three grid zones in Europe and two or three in neighbouring West Asia and North Africa that are linked together by means of HVDC-coupling points. In addition, a ring linkage between North Africa (Morocco, Algeria, Tunisia, Libya and Egypt), the Near East (Israel, Jordan, Lebanon, Syria and Turkey) and the European ENTSO-E grid could act in an additional stabilizing manner. If required, cross-connections between North Africa and Southern Europe (e.g. Spain – Algeria, Algeria-Sardinia, Tunis – Sicily or Libya-Sicily via the Italian peninsula) could result in additional stabilization of the grid.

References

[4.1] Apel-Hillmann, F.M., *Windkraftanlagen im wissenschaftlichen Meß- und Evaluierungsprogramm (WMEP) – Datenerfassung und Netzanschluß*, Thesis, Kassel University, 1994.

[4.2] Meyer, M.: Leistungselektronik – Einführung, Grundlagen, Überblick. Springer, Berlin, Heidelberg, New York, London,Paris, 1990.

[4.3] Heier, S., Kleinkauf, W. and Sachau, J., *Power Conditioning – The Link between Solar Conversion and Consumer. Advances in Solar Energy*, Vol. 9: *An Annual Review of Research and Development*, American Solar Energy Society, Boulder, Colorado, 1994, pp. 161–244.

[4.4] Heumann, K., *Grundlagen der Leistungselektronik*, Teubner, Stuttgart, 1991.

[4.5] Bystron, K., *Leistungselektronik*, Hanser, Munich, Vienna, 1979.

[4.6] Wasserrab, T., *Schaltungslehre der Stromrichtertechnik*, Springer, Berlin, 1962.

[4.7] Meyer, M., *Leistungselektronik. Einführung, Grundlagen, Überblick*, Springer, Berlin, Heidelberg, New York, London, Paris, Tokyo, Hong Kong, Barcelona, 1990.

[4.8] Leonhard, W., *Regelung in der elektrischen Antriebstechnik*, Teubner, Stuttgart, 1974.

[4.9] Hienz, G., Neuer Direktumrichter für Industrieantriebe, *Antriebstechnik*, 1989, 28(3).

[4.10] Arlt, B., Der MOS Controlled Thyristor MCT, *Elektronik Industrie*, 1992, 11/92, 56–60.

[4.11] Bober, G. and Heumann, K., Vergleich der Eigenschaften von MCT und IGBT unter 'harten' Schaltbedingungen, Vol. 115, Books 13–14, ETZ, 1994.

[4.12] Heumann, K., Die Intelligenz hält Einzug – Umrichter in der Antriebstechnik, in *Elektronik*, Book 1, 1995, pp. 83–96.

[4.13] Salzar, L. and Jos, G., PSPICE Simulation of Three-Phase Inverters by Means of Switching Functions, *IEEE Transactions on Power Electronics*, 1994, 9(1), 35–42.

[4.14] Apel-Hillmann, F.M., *Konzeption eines Stromrichters zur Netzanbindung von permanterregten Synchrongeneratoren*, Thesis, Kassel University, 1995.

[4.15] Deisenroth, H. and Trabert, C., Vermeidung von Überspannungen bei Pulsumrichterantrieben, Vol. 114, Book 17, ETZ, 1993, pp. 1060–6.

[4.16] Schmid, W., Pulsumrichterantriebe mit langen Motorleitungen, *Antriebs- und Getriebetechnik*, Book 4, 1992, pp. 58–64.

[4.17] Aninger, H. and Nagel, G., *Vom transienten Betriebsverhalten herrührende Schwingungen bei einem über Gleichrichter belasteten Synchrongenerator*, Part 1: Theoretische Untersuchungen, Siemens Forschungs- und Entwicklungsberichte, Vol. 9, No. 1, Springer-Verlag, 1980, pp. 1–7.

[4.18] Ernst-Cathor, J., *Drehzahlvariable Windenergieanlage mit Gleichstromzwischenkreis-umrichter und Optimumsuchenden Regler*, Dissertation, TU Braunschweig.

[4.19] Schmeer, H.R., EMV 1994, in 4 Internationale Fachmesse und Kongress für EMV, February 1994.

[4.20] Bösterling, W., Jörke, R. and Tscharn, M., IGBT-Module in Stromrichtern: steuern, regeln, schützen, Vol. 110, Book 10, ETZ, 1989.

[4.21] Vereinigung Deutscher Elektrizitätswerke (VDEW), *Technische Anschlußbedingungen für den Anschluß an das Nie derspannungsnetz*, TAB 2000, VWEW, Frankfurt am Main (Hrsg), Edn 2000.

[4.22] Fördergesellschaft Windenergie e.V., *Technische Richtlinie für Windenergieanlagen*, revision 13, 1.1.2000, Part 3: *Bestimmung der Elektrischen Eigenschaften*, Fördergesellschaft Windenergie e.V. (FGW), Hamburg, 2000.

[4.23] Vereinigung Deutscher Elektrizitätswerke (VDEW), *Eigenerzeugungsanlagen am Mittelspannungsnetz. Richtlinie für Anschluß und Parallelbetrieb von Energieerzeugungsanlagen am Mittelspannungsnetz*, VWEW, Frankfurt am Main (Hrsg), 2nd Edn,1998.

[4.24] Curtice, D.H. and Patton, J.B., Analysis of Utility Protection Problems Associated with Small Wind Turbine Interconnections, *IEEE Transactions on Power Apparatus and Systems*, 1982, PAS-101(10), 3957–66.

[4.25] Dugan, R.C. and Rizy, D.T., Electric Distribution Protection Problems Associated with the Interconnection of Small, Dispersed Generation Devices, *IEEE Transactions on Power Apparatus and Systems*, 1984, PAS-103(6), 1121–7.

[4.26] Flosdorff, R. and Hilgarth, G., *Elektrische Energieverteilung*, Teubner, Stuttgart, 1994.

[4.27] Happoldt, H. and Oeding, D., *Elektrische Kraftwerke und Netze*, Springer, Berlin, 1978.

[4.28] Brown, M.T. and Settebrini, R.C., Dispersed Generation Interconnections via Distribution Class Two-Cycle Circuit Breakers, *IEEE Transactions on Power Delivery*, 1990, 5(1), 481–5.

[4.29] Webs, A., *Einfluß von Asynchronmotoren auf die Kurzschlußstromstärken in Drehstromanlagen*, VDE Technical Reports Vol. 27, VDE, Offenbach, 1972.

[4.30] Wachenfeld, V., *Netzrückwirkungen durch Windkraftanlagen*, Thesis, Kassel University, 1994.

[4.31] Krämer, T., *Untersuchung von Netzstrukturen zur Bewertung von Netzrückwirkungen durch Windkraftanlagen*, Thesis, Kassel University, 1994.

[4.32] Westinghouse Electric Corporation, *Electrical Transmission and Distribution*, Reference Book, East Pittsburgh, Pennsylvania, 1964.

[4.33] Krause, P.C. and Man, D.T., Transient Behavior of Class of Wind Turbine Generators during Electrical Disturbance, *IEEE Transactions on Power Apparatus and Systems*, 1981, PAS-100(5), 2204–10.

[4.34] Durstewitz, M., Heier, S., Hoppe-Kilpper, M. and Kleinkauf, W., *Meßtechnische Untersuchungen am Windpark Westküste. Untersuchung der elektrischen Komponenten und ihrer Integration in schwache Netze. BMFT-Abschlußbericht*, April 1992.

[4.35] Enßlin, C., Durstewitz, M., Heier, S. and Hoppe-Kilpper, M., Wind Farms in the German '250 MW Wind'-Programme, in European Wind Energy Association Special Topic Conference '92 on *The Potential of Wind Farms*, Herning, Denmark, 1992, pp. B4.1–B4.7.

[4.36] Enßlin, C., Hoppe-Kilpper, M., Kleinkauf, W., Koch, H. and Schott, T., The WMEP in the German '250 MW-Wind' Programme – Evaluations from the large scale WMEP measurement network, in ISES Solar World Congress, Budapest, August 1993.

[4.37] Heier, S., *Wind Power Generation. Decentralized Energy. Options and Technology*, Omega Scientific Publishers, New Delhi, 1993, pp. 65–76, ISBN 81-85399-26-3.

[4.38] Tande, J.O.G. and Landberg, L., A 10 Sec. Forecast of Wind Output with Neural Networks, in European Wind Energy Conference, 1993.

[4.39] Menze, M., *Leistungsprognose von Windkraftanlagen mit Neuronalen Netzen*, Thesis, Kassel University, 1996.

[4.40] Rohrig, K., Online Monitoring of 1700 MW Wind Capacity in a Utility Supply Area, in European Wind Energy Conference and Exhibition, Nice, France, March 1999.

[4.41] Beyer, H.G., Heinemann, D., Mellinghoff, H., Mönnich, K. and Waldl, H.-P., Forecast of Regional Power Output of Wind Turbines, in 1999 European Wind Energy Conference and Exhibition, Nice, France, March 1999.

[4.42] Rohrig K., Online Monitoring and Short Term Prediction of 2400 MW Wind Capacity in a Utility Supply Area, in *Wind Forecast Techniques*, 33 Meeting of Experts, Technical Report from the International Energy Agency, R&D Wind, Ed. S.-E. Thor, FFA, Sweden, July 2000, pp. 117–119.

[4.43] Rohrig K., Ernst, B., Ensslin, C. and Hoppe-Kilpper, M., Online Supervision and Prediction of 2500 MW Wind Power, in *Wind Power for the 21st Century*, Special Topic Conference and Exhibition, Convention Centre, Kassel, Germany, 25–27 September 2000.

[4.44] Landberg, L., Joensen, A., Giebel, G., Madsen, H. and Nielsen, T.S., Zephyr: The Short Term Prediction Models, in *Wind Power for the 21st Century*, Special Topic Conference and Exhibition, Convention Centre, Kassel, Germany, 25–27 September 2000.

[4.45] Moehrlen, C, Jorgensen, J.U., Sattler, K. and McKeogh, E.J., On the Accuracy of Land Cover Data in NWP Forecasts for High Resolution Wind Energy Prediction, in 2001 European Wind Energy Conference and Exhibition, Bella Center, Copenhagen, Denmark, 2–6 July 2001.

[4.46] Moehrlen, C., On the Benefits of and Approaches to Wind Energy Forecasting, invited speaker at the Irish Wind Energy Association Annual Conference on *Towards 500 MW*, Ennis, 25 May 2001.

[4.47] Focken, U., Lange, M. and Waldl, H.-P., Previento – A Wind Power Prediction System with an Innovative Upscaling Algorithm, in 2001 European Wind Energy Conference and Exhibition, Bella Center, Copenhagen, Denmark, 2–6 July 2001.

[4.48] Rohrig, K., Ernst, B., Schorn, P. and Regber, H., Managing 3000 MW Wind Power in a Transmission System Operation Center, in *2001 European Wind Energy Conference and Exhibition*, Bella Center, Copenhagen, Denmark, 2–6 July 2001.

[4.49] Heier, S., *Grid Connected Wind Energy Converters. Commercialization of Solar and Wind Energy Technologies*, Amman, Jordan, April 1992, pp. 519–32.

[4.50] Heier, S., *Technical Aspects of Electrical Supply Systems for Villages. Decentralized Energy. Options and Technology*, New Delhi, 1993, pp. 107–27, ISBN 81-85399-26-3.

[4.51] Caselitz, P., Hackenberg, G., Kleinkauf, W. and Sachau, J., *Windenergieanlagen in elektrischen Energieversorgungssystemen kleiner Leistung*, BMFT-Final Report, Kassel University, 1985.

[4.52] Caselitz, P., Giebhardt, J. and Mevenkamp, M., On-line Fault Detection and Prediction in Wind Energy Converters, in European Wind Energy Association Conference – Macedonia EWEA '94, Thessaloniki, Greece, October 1994.

[4.53] Heier, S., *Elektrotechnische Konzeptionen von Windkraftanlagen im Vergleich. Windenergie Bremen '90*, DGS-Sonnenenergie Verlags-GmbH, Munich, 1990, pp. 91–106.

[4.54] Heier, S., Grid Influences by Wind Energy Converters, in International Energy Agency (IEA) Expert Meeting, Göteborg, October 1991, pp. 37–50.

[4.55] Heier, S., Windkraftanlagen im Netzbetrieb, in German Wind Energy Conference DEWEK 92, Wilhelmshaven, 28–29 October 1992, pp.141–5.

[4.56] Heier, S., Netzintegration von Windkraftanlagen, in Fördergesellschaft Windenergie (FGW)-Workshop on *Netzanbindung von Windkraftanlagen*, Hanover, 23 March 1993.

[4.57] Heier, S., Technical Aspects of Wind Energy Converters and Grid Connections, in British Wind Energy Association (BWEA) Workshop on *Wind Energy Penetration into Weak Electricity Networks*, 10–12 June 1993, Rutherford Appleton Laboratory, Chilton, Didcot, pp. 38–55, ISBN 1-870064-17-8.

[4.58] Heier, S., Netzeinwirkungen durch Windkraftanlagen und Maßnahmen zur Verminderung, in Husumer Wind Energy Conference, 22–26 September 1993, Husum, pp. 157–68.

[4.59] Heier, S., Grid Influences by Wind Energy Converters and Reduction Measures, in American Wind Energy Association 24th Annual Conference, Minneapolis, Minnesota, 1994.

[4.60] Dangrieß, G., Heier, S., König, V., Kuntsch, J. and Müller, J., Konzeptionen zur Auslastung der Netzkapazität, in German Wind Energy Conference DEWEK '94, Wilhelmshaven, 22–23 June 1994, pp. 163–70.

[4.61] Dangrieß, G., Heier, S., König, V., Kuntsch, J. and Müller, J., *Konzeption zur Ausnutzung der Netzkapazität*, Neue Energie Heft 11/94, Steinbacher Druck GmbH, Osnabrück, 1994, pp. 44–45.

[4.62] Krengel, U., *Untersuchung der Rückwirkungen eines 1 kW-Photovoltaik-Wechselrichters auf das Versorgungsnetz*, Thesis, Kassel University, 1990.

[4.63] E.ON-Netz GmbH: Netzanschlussregeln für Hoch- und Höchstspannung. Elektronisches Dokument (PDF), 1. April 2006. http://www.eon-netz.com/pages/ehn_de/Veroeffentlichungen /Netzanschluss/Netzanschlussregeln/ENENARHS2006de.pdf.

[4.64] Nunez, M.: Untersuchung der Rückwirkungen von LVRT-Versuchen auf das Verbundnetz, 2011.

[4.65] Chun, S. and Damm, F., *Netzrückwirkung im Windpark Westküste – Oberschwingungen*, Thesis, Kassel University, 1990/1991.

[4.66] Götze, F., Schäfer, H. and Schulz, D., *Impedanz-Simulation beliebiger Netzkonfigurationen*, Project II, Kassel University, 1991.

[4.67] Heier, S. and Kleinkauf, W., Grid Connection of Wind Energy Converters, in European Community Wind Energy Conference, Lübeck–Travemünde, March 1993, pp. 790–793, ISBN 0-9521452-0-0.

[4.68] Cramer, G., Durstewitz, M., Heier, S. and Reinmöller-Kringel, M., *1,2 MW-Stromrichter am schwachen Netz. Filterauslegung zur Reduzierung von Stromoberschwingungen*, SMA info 10, April 1993, pp. 10–11.

[4.69] Oort, H.A. van, Numerical Model for Calculating the Power Output Fluctuations from Wind Farms, *Journal of Wind Engineering and Industrial Aerodynamics*, 1988, 27.

[4.70] Cretcher, C.K. and Simburger, E.J., Load Following Impacts of a Large Wind Farm on an Interconnected Electric Utility System, *IEEE Transactions on Power Apparatus and Systems*, 1983, PAS-102 (3), 687–92.

[4.71] Büchner, J., Beyer, H.-G., Eichelbrönner, M., Haubrich, H.-J., Steinberger-Willms, R., Stubbe, G. and Waldl, H.-P., *Modellierung der Netzbeeinflussung durch Windparks. Energiewirtschaftliche Tagesfragen 43. Jg.*, 1993, Book 5, pp. 332–5.

[4.72] Enßlin, C., Durstewitz, M., Heier, S. and Hoppe-Kilpper, M., Wind Farms in the German '250 MW-Wind'-Programme, in European Wind Energy Association Special Topic Conference '92 on *The Potential of Wind Farms*, Herning, Denmark, 1992, pp. B4.1–B4.7.

[4.73] Arnold, G. and Heier, S., Netzregelung mit regenerativen Energieversorgungssystemen, in Kasseler Symposium Energie-Systemtechnik '99, ISET, Kassel, 1999, pp. 166–77.

[4.74] Heier, S., Arnold, G., Durstewitz, M., Perez-Spiess, F., Meyer, R. and Juarez-Navarro, A., Grid Control with Renewable Energy Sources, in European Wind Energy Association (EWEA) Special Topic Conference on *Wind Power for the 21st Century – The Challenge of High Wind Power Penetration for the New Energy Markets*, International Conference, Kassel, Germany, 25–27 September 2000.

[4.75] Arnold, G. and Heier, S., Grid Control with Wind Energy Converters, in Second International Workshop on *Transmission Networks for Offshore Wind Farms*, Royal Institute of Technology, Electric Power Systems, Stockholm, Sweden, 30–31 March 2001.

[4.76] Arnold, G., Heier, S., Perez-Spiess, F. and Lopez-Manzanares, L., Grid Control with Renewable Energy Sources – Results to the Field Tests, in European Wind Energy Conference, Copenhagen, Denmark, 2–6 July 2001.

[4.77] Apel-Hillmann, F. M.: Windkraftanlagen im wissenschaftlichen Mess- und Evaluierungsprogramm (WMEP) – Datenerfassung und Netzanschluss. Studienarbeit, Universität Gh Kassel, 1994.

[4.78] Thurner, L., *Entwicklung eines Regelungsalgorithmus zur Begrenzung der Spannung an Windenergieanlagen im Parknetz*, Master's thesis, University of Kassel, 2013.

[4.79] Al-Awaad, Ahmad-Rami Khalil: Beitrag von Windenergieanlagen zu den Systemdienstleistungen in Hoch- und Höchstspannungsnetzen. Dissertation, 2009.

[4.80] Verband der Netzbetreiber VDN e.V. beim VDEW: *Netzcodes*, Elektronische Dokumente (PDF), 2004; www.vdn-berlin.de/netzcodesl.asp.

[4.81] Arnold, G., *System zur stützung von Elektrizitätsnetzen mit windkraftan lagen und anderen Erneuerbaren Energien*, Dissertation, Universität Kassel, 2004.

[4.82] E.ON-Netz GmbH, *Netzanschlussregeln für Höch- und Höchstspannung*, Elektronisches Dokumente (PDF), 1 August 2004; www.con-netz.com/Ressources/downloads/ENE_NAR _HS_01082003.pdf.

[4.83] Arnold, G.: System zur Stützung von Elektrizitätsnetzen mit Windkraftanlagen und anderen Erneuerbaren Energien. Dissertation, Universität Kassel, 2004.

[4.84] Heier, S.: Wind plant development and state of the art of grid connected systems. In: The World Renewable Energy Conference IX and Exhibition, Florence, Italy, August 19–25, 2006.

[4.85] Heier, S.: Integration großer Windleistungen ins Netz. In: 7th Austrian Wind Energy Symposium St. Pölten, Austria, 20.–21. Oktober 2005.

[4.86] Heier, S.: Overview on the development of wind technology. In: Power Generation from Renewable Energy: Practical Approaches, BITEC,Bangua, Bangkok, July 8, 2005.

[4.87] Heier, S.: Integration of Wind Power Plants in Electrical Grids. In: Internationales wissenschaflich-technisches Seminar, Polytechnische Universität Tomsk, 10.–11. April 2006. Seiten 43–44, ISBN 5-98298-075-7.

[4.88] Heier, S.: Interconnection issues for wind turbines. In: ECPE Seminar Power Electronics e. V. Renewable Energies, Kassel, Germany, February, 9–10, 2006.

[4.89] Heier, S.: State of the Art and Outlook on Wind Energy utilisation. In: Internationales wissenschaflich-technisches Seminar, Polytechnische Universität Tomsk, 10.–11. April 2006. Seiten 41–42, ISBN 5-98298-075-7.

[4.90] Crotogino, F. und B. Prevedel: Komprimierte Luft speichert Windstrom. Druckluftspeicher-Gasturbinen-Kraftwerk zum Ausgleich fluktuierender Windenergie-Produktion. Erneuerbare Energien, 13(11):41–43 November 2003.

[4.91] Biermann, K., B. Ernst, F. Fischer, S. Hartge, S. Heier, L. Hofmann, M. Hoppe-Kilpper, K. Rohrig, B. Valov und W. Winter: Tools and concepts to integrate german offshore potential into electrical energy supply. In: The international technical Conference DEWEK 2004, Wilhelmshaven, October 20–21, 2004.

[4.92] Biermann, K., B. Ernst, F. Fischer, S. Hartge, S. Heier, L. Hofmann, M. Hoppe-Kilpper, K. Rohrig, Y. Saßnick und W. Winter: New concepts to integrate german offshore wind potential into electrical energy supply. In: European Wind energy Conference, London, November 2004.

[4.93] Arnold, G., S. Heier und B. Valov: Spannungsänderungen und Stabilisierungsmöglichkeiten in Versorgungsnetzen mit erneuerbaren Energieanlagen. In: 48th Int'l. Scientific Colloquium Technical University of Ilmenau, Ilmenau, 22.–25. September 2003.

[4.94] Arnold, G., S. Heier und B. Valov: Spannungsregelung in dezentralen Multisupply-Strukturen. In: VDE-Kongress, Berlin, 18.-20. Oktober 2004. Seiten 599–603. Band 1.

[4.95] Fischer, F., S. Hartge, S. Heier, R. Jursa, K. Rohrig, F. Schlögl und M. Wolff: Advanced control strategies to integrate german offshore wind potential into electrical energy supply. In: Fifth International Workshop on Large-Scale Integration of Wind Power and Transmission Networks for Offshore Wind Farms, Glasgow, April 7–8, 2005.

[4.96] Arnold, G., S. Heier und R. Saiju: Voltage dips compensation by wind farms equipped with power converters as decoupling element. In: 11th European Conference on Power Electronics and Applications, Dresden, Germany, September 11–14, 2005.

[4.97] Hartge, S., S. Heier, F. Fischer, B. Lange, K. Rohrig, F. Schlögl, B. Valov und M. Wolff: Extra large scale virtual power plants (xlsvpp) - new concepts to integrate german wind potential into electrical energy supply. In: European Wind Energy Conference & Exhibition, Athens, Greece, February 27 – March 2, 2006.

[4.98] Heier, S. und B. Valov: The complete use of receiver capacity of a grid through the control of the power generation units. In: 51. Internationales Wissenschaftliches Kolloquium, Technische Universität Ilmenau, 11.-15. September 2006. Proceedings TU Ilmenau, ISBN 3-938843-16.0.

[4.99] Heier et al.: Windenergieforschung am Offshore-Testfeld – Netzinegration von Offshore-Windparks (RAVE). Interner Bericht, 2008–2011. Forschungsvorhaben gefördert durch das BMU.

[4.100] Fischer, F., G. Füller, S. Heier, L. Hofmann, B. Lange, R. Mackensen, K. Rohrig, B. Valov und M. Wolff: Advanced operating control for wind farm clusters. In: 6th Int'l Workshop on Large-Scale Integration of Wind Power and Transmission Networks for Offshore Wind Farms Technical University Delft, Oktober 26–28, 2006. Seiten 188–192, Proceedings TU Delft.

[4.101] Fischer, F., G. Füller, S. Heier, L. Hofmann, B. Lange, R. Mackensen, K. Rohrig, B. Valov und M. Wolff: Extended operation control to integrate german (offshore) wind farms. In: The International Technical Conference DEWEK 2006, Bremen, Germany, November 22–23, 2006.

[4.102] Heier, S. und B. Valov: Integration großer Offshore-Windparks in elektrische Versorgungssysteme Interner Bericht, 2004–2009. Gefördert durch das Bundesministerium für Wirtschaft und Technologie, FKZ-Nr. 0329925C.

[4.103] Heier, S. und B. Valov: Software for analysis of integration possibility of renewable energy units into electrical networks. In: The 5th International Conference Electric Power Quality and Supply Reliability, Viimsi, Estonia, August 23–25, 2006. Seiten 173–177, Proceedings Tallin University of Technology, ISBN 9985-59-647-1.

[4.104] Bock, C.: Netzanschlussanalyse von Offshore-Windparks. Diplom-Arbeit, Universität Kassel, 2007.

[4.105] Bock, C., S. Heier, B. Lange, K. Rohrig und B. Valov: 25GW Offshore-Windkraftleistung benötigt ein starkes Energieübertragungssystem auf der Nordsee. Wind Kraft Journal & Natürliche Energien, 28(1):14–20 2008.

[4.106] Konermann, V.: Wie umweltverträglich ist die Netzanbindung von Offshore-Windparks? WWF Deutschland, Frankfurt am Main, 1. Auflage, Januar 2006.

[4.107] dena: Energiewirtschaftliche Planung für die Netzintegration von Windenergie in Deutschland an Land und Offshore bis zum Jahr 2020. Studie, Deutsche Energie-Agentur GmbH (dena), Berlin, Mai 2005.

[4.108] Bundesrepublik Deutschland: Gesetz zur Beschleunigung von Planungsverfahren für Infrastrukturvorhaben. Bundesgesetzblatt Jahrgang 2006, Dezember 2006. Teil 1, Nr. 59, S. 2833–2853 Fassung vom 9.12.2006.

[4.109] BSH: Nordsee: Offshore-Windparks. Elektronisches Dokument, 2009. http://www.bsh.de/de
/Meeresnutzung/Wirtschaft/CONTIS-Informationssystem/ContisKarten/NordseeOffshoreWind
parksPilotgebiete.pdf

[4.110] Wensky, D.: Netzanschluss von Offshore-Windparks. In: Fachtagung Windenergie und Netzintegration, Hannover, 17.–18.02.2005 2005.

[4.111] dena: Energiewirtschaftliche Planung für die Netzintegration von Windenergie in Deutschland
an Land und Offshore bis zum Jahr 2020. Studie Deutsche Energie-Agentur GmbH (dena)
Berlin, Mai 2005.

[4.112] Brakelmann, H.: Netzverstärkungs-Trassen zur Übertragung von Windenergie: Freileitung
oder Kabel? Studie im Auftrag des Bundesverband WindEnergie e.V., 2004. Universität
Duisburg-Essen.

[4.113] Schreiber, M., M. Gellermann und G. Gerdes: Vermeidung negativer ökologischer Auswirkungen bei der Netzanbindung. Elektronisches Dokument, 2005. www.bmu.de/files/pdfs/allgemein
/application/pdf/schreiber_20.pdf, Projektbericht, Bundesumweltministerium.

[4.114] Sobottka, M.: Kabeltrassen für Offshore-Windenergieparks in Schutzgebieten des nieder-
sächsischen Küstenmeeres. Elektronisches Dokument, Vortrag am 20. Juni, Fachgespräch
Verlegung von Seekabeln zum Netzanschluss von Offshore Windparks in Schutzgebieten im
Meer, Juni 2006. http://www.offshore-wind.de/page/fileadmin/offshore/documents/StAOWind
_Workshops/Kabel_in_Schutzgebieten/Kabel_in_Schutzgebieten_Vortrag_Sobottka.pdf,
Nationalparkverwaltung Niedersächsisches Wattenmeer.

[4.115] Bauer, M.: Kabeltrassen für Offshore-Windparks im schleswig-holsteinischen Küsten-
meer Elektronisches Dokument, Vortrag am 20. Juni, Fachgespräch Verlegung von
Seekabeln zum Netzanschluss von Offshore Windparks in Schutzgebieten im Meer,
Juni   2006.   http://www.offshore-wind.de/page/fileadmin/offshore/documents/StAOWind
_Workshops/Kabel_in_Schutzgebieten/Kabel_in_Schutzgebieten_Vortrag_Bauer.pdf, Nation-
alparkverwaltung Schleswig-Holstein.

[4.116] Niedersächsische Energie Agentur GmbH: Untersuchung der Wirtschaftlichen und
energiewirtschaftlichen Effekte von Bau und Betrieb von Offshore-Windparks in der
Nordsee auf das Land Niedersachsen. Elektronisches Dokument (PDF), 14. Juni 2001. In
Zusammenarbeit mit Deutschem Windenergie-Institut GmbH, Niedersächsischem Institut für
Wirtschaftsforschung e.V., Projekt Nr. 2930.

[4.117] Brakelmann, H. und M. Jensen: Neues sechsphasiges Übertragungssystem für VPE-isolierte
HVAC-See- und Landkabel hoher Übertragungsleistung. ew, 105(4):34–43 2006.

[4.118] Brakelmann, H., C. Burges, M. Jensen und Th. Schütte: Bipolar transmission systems with
XLPE HVAC submarine cables. In: 6. Int. Workshop for Offshore Windfarms, Delft, October
2006. Seiten 165–169.

[4.119] Tang, Xiaoyan: Drei- und Einphasen-Transformatoren im Offshore-Einsatz. Diplomarbeit I,
Universität Kassel, Kassel, 2007.

[4.120] Lönker, O.: Die Steckdose auf dem Meer. *Neue Energie*, 16(12):22–25 2006, ISSN 0949–8656.

[4.121] AG, ABB: Xlpe cable systems - user's guideXLPE Cable Systems - User's Guide, 2009.

[4.122] Balzer, G.: Energieversorgung - teil 1. vorlesungsskript, 2008.

[4.123] Kadry, S., TITLE = Thyristorgeschaltete Kondensatorbank school = Patent: DE19528766C1
year = 1997 month = Januar day = 16 Siemens AG und.

[4.124] Siemens AG: The Efficient Way - SVC PLUS, 2009.

[4.125] Simond, J.J. und A. Sapin: FACTS, welche Vorteile für die elektrischen Netze?

[4.126] Merkele, M.: Dynamische Modellierung von Verbrauchergruppen und statischer Blindleis-
tungskompensatoren zur Untersuchung der Spannungsstabilität in Netzen, 2002.

[4.127] Schwab, A.J.: Elektroenergiesysteme: Erzeugung, Transport, Übertragung und Verteilung elektrischer Energie, 2006.

[4.128] Lönker, O.: Ein Stromnetz für Europa. Neue Energie, 16(12):16–21 2006, ISSN 0949–8656.

[4.129] Manthey, A.: Windkraft-Energietransport mit Hochspannungsgleichstromübertragung Diplomarbeit, Universität Kassel, Kassel, 2008.

[4.130] Diedrichs, V.: Möglichkeiten der Erhöhung der Anschlussleistung durch Lastflussmanagement. In: Husum Wind 99: Fachmesse und Fachkongress zur Windenergie, Husum, 22.–26. Sept. 1999.

[4.131] Diedrichs, V.: Energieversorgung mit dezentralen Kleinkraftwerken in leistungsbegrenzten Versorgungsnetzen. Informationen aus dem Forschungsschwerpunkt, Okt. 1999. Fachhochschule Oldenburg, Standort Wilhelmshaven.

[4.132] Kleinkauf, W., O. Haas, S. Heier und Ph. Strauß: Zukunftsaspekte erneuerbarer Energien und die Rolle der Photovoltaik. In: VDI-Gesellschaft Energietechnik (Herausgeber): Fortschrittliche Energiewandlung und -anwendung – Schwerpunkt: Dezentrale Energiesysteme, Düsseldorf, März 2001. VDI-Berichte; 1594, Seiten 3–16, VDI-Verlag.

[4.133] Heier, S., W. Kleinkauf und J. Sachau: Wind energy converters at weak grid. In: Commission of the European Communities (Herausgeber): European Community Wind Energy conference Herning (Denmark), Luxembourg, June 1988. Seiten 429–433.

[4.134] Cramer, G.: Modulare autonome elektrische Energieversorgungssysteme werden zunehmend interessanter. SMA info, 111(4):1–6 1990.

[4.135] Siemens AG (Herausgeber): Formel- und Tabellenbuch für Starkstrom-Ingenieure. Girardet, Essen, Auflage, 1965.

[4.136] Bogenrieder, W.: Moderne Pumpspeicherwerke im Gigawattbereich – Darstellung am Beispiel des PW Goldisthal. In: Institut für Solare Energieversorgungstechnik (ISET) e.V. (Herausgeber): 7. Kasseler Symposium Energie-Systemtechnik – Energiespeicher und Transport, Kassel, 14.–15. Nov. 2002. Seiten 39–59, ISET.

[4.137] Dietz, P.: Netzintegration von Offshore-Großwindanlagen – Grundlast von der Nordsee Papierflieger, Clausthal-Zellerfeld, 2008, ISBN 978-3-89720-978-7.

[4.138] Crotogino, F.: Druckluftspeicher-Gasturbinen-Kraftwerke/Geplanter Einsatz beim Ausgleich fluktuierender Windenergie-Produktion und aktuellem Strombedarf. In: Institut für Solare Energieversorgungstechnik (ISET) e.V. (Herausgeber): 7. Kasseler Symposium Energie-Systemtechnik – Energiespeicher und Transport, Kassel, 14.–15. Nov. 2002. Seiten 26–38, ISET.

[4.139] Crotogino, F. und B. Prevedel: Komprimierte Luft speichert Windstrom. Druckluftspeicher-Gasturbinen-Kraftwerk zum Ausgleich fluktuierender Windenergie-Produktion. Erneuerbare Energien, 13(11):41–43 November 2003.

[4.140] Enßlin, C.: The influence of Modeling Accuracy on the Determination of Wind Power Capacity Effects and Balancing Needs. Kassel University Press, Kassel, 2007, ISBN 978-89958-248-2. zugl. Dissertation Univ. Kassel 2006.

[4.141] Heier, S.: Large scale wind energy application. In: Energia Eolica – una Oportunidad que no se Puede Desperdiciar, Buenos Aires, Argentinien, 23 de Julio 2008. Workshop.

[4.142] Auer, H., C. Huber, G. Resch et al.: Action plan for an enhanced least-cost integration of res-e into the european grid. Wp10-report, Project GreenNet, February 2005. Available on www.greennet.at.

[4.143] Dany, G. und H.J. Haubrich: Anforderungen an die Kraftwerksreserve bei hoher Windenergieeinspeisung Energiewirtschaftliche Tagesfragen, Seiten 890–894, Dezember 2000.

[4.144] Milborrow, D.: The real cost of integration wind. Windpower Monthly, Seiten 35–39, February 2004.

[4.145] Pantaleo, A., A. Pellerano und M. Trovato: Technical issues on wind energy integration in power systems: Projections in italy. In: (EWEA), European Wind Energy Association (Herausgeber): Proceedings of the European Wind Energy Conference 2003, Madrid, Spain, Brussels (Belgium), June 16–19 2003. EWEA.

[4.146] Kapetanovic, T., B.M. Buchholz, B. Buchholz, V. Buehner, "Provision of ancillary services by dispersed generation and demand side response – needs, barriers and solutions",

CIGRE session 2008, Paris, France, 2008. Internet reference: http://www.springerlink.com/content/dp3h792nt3666436/fulltext.pdf (Last accessed: 10.03.2010).

[4.147] UCTE AD-HOC Group, "Frequency quality investigation", Final report, 30th August 2008.

[4.148] Dobschinski, J., A. Wessel, B. Lange, K. Rohrig, L.v. Bremen, Y.M. Saint-Drenan, "Estimation of wind power prediction intervals using stochastic methods and artificial intelligence model ensembles", Dewek 2008.

[4.149] Jan Dobschinski, Ennio De Pascalis, Arne Wessel, Lüder von Bremen, Bernhard Lange, Kurt Rohrig, Yves-Marie Saint Drenan, "The potential of advance shortestterm forecasts and dynamic prediction intervals for reducing the wind power induced reserve requirements", EWEC 2010. Internet reference: http://www.ewec2010proceedings.info/papers/1251755017.pdf (Last accessed:08.07.2010).

[4.150] Dobschinsk J., Gesino A., Lange B., Mackensen R., Mata J. L., Braun M., Quintero C., Pestana R., Rohrig, K., Wessel A., Wolff M., "Wind Power Plant Capabilities – Operate Wind Farms like Conventional Power Plants", European Wind Energy Conference 2009, Marseille, France, 16–19 March 2009. Internet reference: http://www.ewec2009proceedings.info/allfiles2/272_EWEC2009presentation.pdf (Last accessed: 06.01.2009).

[4.151] Gesino A., Lange B., Mackensen R., Quintero C., Rohrig K., Wolff M., "Wind farm Cluster Management System", XV Energie Symposium, Fachhochschule Stralsund, Stralsund November 6–11th 2008.

[4.152] Rohrig K., "Online-monitoring and prediction of wind power in German transmission system operation centres", World Wind Energy Conference, Cape Town, South Africa, 2003.

[4.153] Reinhard Mackensen, Bernhard Lange, Florian Schlögl, "Integrating Wind Energy into public power supply systems - German state of the art".

[4.154] Alejandro J. Gesino, "Power reserve provision with wind farms", Internet reference: http://www.uni-kassel.de/upress/online/frei/978-3-86219-022-5.volltext.frei.pdf (last accessed: 17.09.2012).

[4.155] Czisch, Gregor: Szenarien zur zukünftigen Stromversorgung. Dissertation, Universität Kassel, 2005.

[4.156] Heier, S., Grid Integration of Wind Energy Converters and Field Applications, in Wind Energy Symposium, Alacati-Izmir, Turkey, 5–7 April 2001, pp. 151–164, ISBM 975-395-425-5.

5

Control and Supervision of Wind Turbines

Intervention into turbine driving power is possible at all times in thermal power stations [Chapter 3, Reference 14 and Chapter 4, References 26, 27]]. In this connection we must differentiate between reactions to necessary long-term changes by energy feed and short-term output smoothing carried out to a limited degree by the steam circuit or a corresponding energy transport route. In diesel units, gas turbines and other systems (see Figure 5.1(a)), the supply of fuel can be adjusted to suit the grid state and the plant control system (or drooping characteristic), thus matching energy supply in the long and short term to changing consumer conditions within the given output framework.

Wind turbines, on the other hand, are dependent upon airflows, which are themselves subject to weather conditions and local effects. This results in corresponding variations in the primary energy supply over which the turbines have no influence. Output can only be changed by reducing power generation. The grid is therefore not only influenced by fluctuations on the power consumption side, but also – in the case of uncoordinated feed from wind turbines – by the effects of the weather on the energy supply (see Figure 5.1(b)).

On the power-supply side, short-term wind-speed changes such as gusts have a particularly pronounced effect on the behavior of wind turbines. They can lead to high component loading and fluctuations in electrical output variables (voltage, frequency, power). These transient processes thus influence the control characteristics of the system. Moreover, the demands on component groups and their reaction characteristics can be determined to ensure the functionality and integration of a turbine.

Unlike the control unit, the management system must both provide up-to-date desired values and react to medium- and long-term variations in the ranges from minutes to years. High availability can be achieved by load adaptation or energy storage (not considered further here). Moreover, energy management can be used to create power reserves.

Consumption values and wind conditions seem, at first glance, to be completely random variables. However, certain relationships exist, based on consumer habits in connection with environmental influences and physical conditions at the corresponding wind speeds. Structuring these precisely, converting them into layouts and using them in supply/load management, e.g. as shown in Figure 5.2, must take on particular importance if the value of wind power is to be increased.

Grid Integration of Wind Energy: Onshore and Offshore Conversion Systems, Third Edition. Siegfried Heier.
© 2014 John Wiley & Sons, Ltd. Published 2014 by John Wiley & Sons, Ltd.

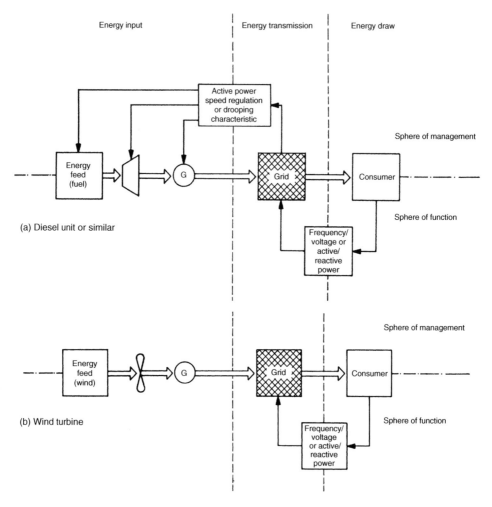

Figure 5.1 Energy flow and differences between the function and control of electricity supply systems based upon (a) diesel generators and (b) wind turbines in uncontrolled supply

To the wind turbine control system, only short-term wind-speed changes are of importance, whereas the management system must also take into account variations in the medium term. The management system, together with the control system, must take account of both consumer characteristics and conditions, and those of the turbine and its components. We shall briefly describe these demands below.

5.1 System Requirements and Operating Modes

In the control and management of a wind turbine, both internal conditions (characteristics of component groups and their interaction) and external variables (consumer desires, conditions for grid-parallel operation) must be taken into account.

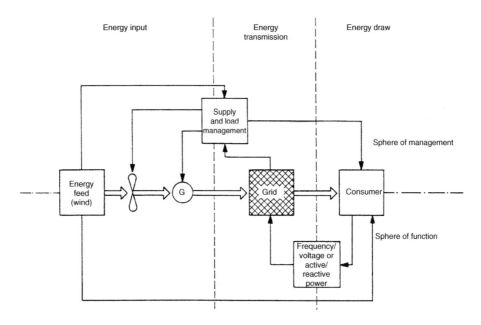

Figure 5.2   Energy flow in an electrical supply system with a controlled wind energy supply

The management system takes decisions on logical connections. It monitors whether process plans are followed and limit values adhered to. The control system, on the other hand, must maintain the values specified for the turbine by the management system. Accordingly, it is necessary to ensure that management system decisions are not transferred directly to final control elements (e.g. the blade pitch adjustment device). Where this is compatible with the required reaction speed, the values determined by the management system should be implemented by the control system, which takes the dynamics of components and the turbine as a whole into account when intervention is called for. Exceptions should only be permitted for safety reasons (rapid shut-down processes in the event of faults, etc.).

Along with the normal characteristics for energy conversion plants, additional requirements for the operation of wind turbines are:

- automatic start-up and shut-down depending upon wind and turbine conditions;
- safety monitoring of turbine components by a management unit, with remote interrogation and fault indication available to the operator or maintenance service;
- the option of controlling turbine speed and electrical output;
- separate protection, independently of the control system, to prevent wind turbine output from increasing too fast in high winds;
- the characteristics of all electrical turbine components must be adapted to suit the electricity purchaser with regard to grid reactions, etc.;
- in future, the prerequisites for fault prediction should be met and the resulting advantages used.

Requirements differ for isolated and grid operation. In isolated operation the purely turbine-specific conditions must be taken into account. Moreover, the demands of the consumer must be considered, although these can only be precisely defined for individual cases. The relevant electrotechnical regulations must definitely be fulfilled, in particular those concerning earthing and protection against overvoltage, etc. (VDE 0100 and IEC 555). In addition to the above, for grid operation the local conditions for the parallel operation of electricity generating plants with the grid [Chapter 3, Reference 1 and Chapter 4, References 21–23] must be fulfilled.

The control system of a wind energy converter represents the link between the management unit and the actual turbine and its components. It must be oriented in particular towards the dynamic characteristics and load possibilities of the turbine, so that it can fulfil its adjustment task. Turbine-specific subsystem behavior patterns, in particular, must be taken into account (e.g. hub and generator inertia, blade pitch adjustment moments, etc.).

The numerous different wind turbine configurations and applications require different types of control system [5.1–5.4]. These can be divided into isolated, grid and combined operation types to permit the allocation of different types of control system and thus allow application-specific differences to be discussed.

This is necessary in order to define the demands that will be placed on the system components and the operating options for the planned wind turbine. In addition, numerous criteria can be used, such as total turbine efficiency, costs, effects on the grid, operating reliability, use of proven mass-produced components, etc. [5.5].

The interaction between input systems and users strongly influences the behavior of the entire configuration in weak grids, and in particular in isolated operation. This will be briefly described below.

5.2   Isolated Operation of Wind Turbines

In so-called isolated operation the wind turbine is not connected to an electrical supply grid, but feeds the connected consumers directly. The voltage-regulated synchronous generator is particularly suited for mechanical–electrical energy conversion in standalone operation, i.e. in a supply system with only one supply unit. To achieve the desired constant voltage with an asynchronous generator would require the availability of a regulated source of reactive power for excitation. The demands of the consumer regarding the maximum fluctuation of voltage and frequency at the generator and the maximum speeds – dictated by turbine components – limit the variation options and determine the design of the control system.

If sufficient wind is available, wind turbines in standalone operation can supply electrical consumers directly. The speed of the wind turbine is regulated by influencing the turbine output, thus maintaining the frequency of the generator. A prerequisite for this is that the required voltage can be maintained in the mechanical–electrical conversion system or at the consumer site. This is achieved by the adjustment of the required reactive power by varying the generator excitation, by the use of capacitors or by the use of static or rotating phase shifters (see Figure 5.3). The frequency value and the associated voltage are interdependent for synchronous generators. In a simplification, two separate control circuits can be formed, possibly using decoupling networks [5.6, 5.7].

According to the assignment of the spheres of influence within the generator of

- active power and frequency f_1 as well as
- reactive power and voltage U_1 (in connection with the electrical output power P_{el})

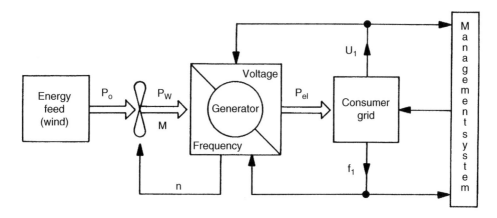

Figure 5.3 Spheres of influence in the separate operation of wind turbines

as shown above, both grid state variables (f, U) can be viewed, within certain limits, as largely decoupled effects. The power P_0 in the wind is passed on from the wind turbine as component P_w or as torque M according to the turbine state. The interaction between the drive and load torque, caused by the load, gives rise to the generator frequency or speed. If the consumer load exceeds the output capacity of the wind turbine, operation can only be maintained if

- the load is reduced and adjusted to the given output value of the prevailing wind conditions or
- an additional supply is connected.

With the help of the turbine management system

- users can be disconnected according to their supply priority,
- a battery with a bidirectional converter can be connected for short- or medium-term equalization or
- a diesel generator can be operated in the case of a long-term deficit.

The management systems, which prototypes are in operation and will become important, will be briefly mentioned here.

5.2.1 Turbines without a blade pitch adjustment mechanism

Due to the significant cost of a blade pitch adjustment, small turbines are usually designed with fixed blades. If the electrical output variables are not subject to any special demands, the simplest layout can be used for isolated operation, as shown in Figure 5.4(a). Self-excited synchronous generators are normally used for mechanical–electrical energy conversion. In this configuration the voltage and frequency of the generator vary with wind speed and load. Simple electricity consumers can be supplied. The generator must be designed such that the power in the wind can be utilized up to the shut-down speed v_{shut}. The turbine can be braked

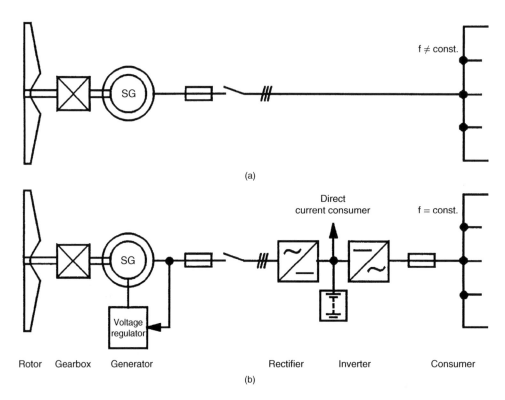

Figure 5.4   Control principles for wind turbines without a blade pitch adjustment mechanism for (a) a turbine without control intervention and (b) a turbine with variable speed and constant output frequency

to low speeds via a mechanical cut-off device (e.g. an air brake mechanism on the blades) at speeds above v_{shut}.

Figure 5.4(b) shows a system in which the generator output current is rectified and direct current consumers are directly supplied. Sensitive alternating current consumers can be supplied with three-phase current of constant frequency and voltage via a self-commutated inverter.

The main disadvantage of all fixed-pitch turbines is that the maximum usable wind speed depends upon the size of the generator and the number of consumers connected at the time. As significantly lower wind speeds usually prevail, such turbines almost always operate in the lower part-load range due to their overdimensioned generators.

5.2.2   Plants with a blade pitch adjustment mechanism

Depending upon the mechanical design, wind turbines with blade pitch adjustment can be operated up to very high wind speeds regardless of the power consumption at the time. Their start-up characteristics can also be influenced by the blade pitch.

At a sufficiently high level of wind, a blade pitch adjuster ensures that the turbine speed is kept roughly constant by altering the blade pitch angle. For reasons of stability and to reduce component loading, it is often advisable (in the case of larger turbines) to include a circuit

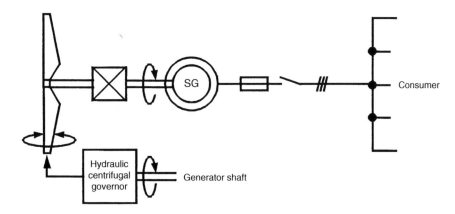

Figure 5.5 Control principle for wind turbines with blade pitch adjustment and a centrifugal governor for the direct connection of alternating current consumers

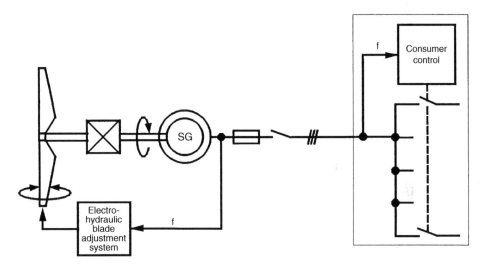

Figure 5.6 Control principle for wind turbines with blade pitch adjustment and electrohydraulic speed regulation for direct consumer connection via load management

for blade pitch adjustment and/or a circuit to control the rate and acceleration of blade pitch adjustment in the speed control circuit (see Figure 5.12).

A simple design for speed regulation by varying the blade pitch can be achieved using a hydraulic or mechanical centrifugal governor (Figure 5.5). Using such a device, the generator speed and thus the frequency can be controlled within a range of approximately ±10%. This is adequate for the supply of numerous robust electrical consumers (e.g. simple motors, heat-exchanger units, etc.). Sensitive loads such as electronic devices, on the other hand, can only tolerate very small frequency fluctuations.

A clear improvement in control behavior and frequency stability can be achieved by using an electrical or electrohydraulic blade pitch adjustment device. This can achieve high blade pitch adjustment speeds and (using electronic regulators) an exact matching of the control dynamics to the behavior of the control system (Figure 5.6). Frequency fluctuations can, for example, be limited to a maximum of $\pm 1\%$.

5.2.3   Plants with load management

In order to account for the normal requirements of electrically sensitive consumers, generator voltage and frequency, and thus speed, should be kept almost constant, even in the part-load range of the wind turbine. To achieve this, the load must always be less than the power available from the wind. Thus the supplied consumers must be connected and disconnected according to frequency (Figure 5.6). However, power consumption need not be continuously variable, but can be altered in stages, i.e. by the connection and disconnection of individual consumers or groups of consumers. Excessively frequent switching processes, and the associated load shocks, are to be avoided.

5.2.4   Turbine control by means of a bypass

A reliable operation of 'stall-controlled' wind turbines is only possible if the turbine speed is maintained by the generator. This requires – as for grid operation – that the energy draw by the consumer is always ensured. If the consumer capacity is insufficient to maintain the drive train at its design speed and to operate the turbine in stall mode, additional consumers (so-called dump loads) must be connected via a bypass (Figure 5.7). This facilitates good frequency control at a reasonable cost for power electronics with the aid of three-phase a.c. power controllers. The generator is preferably an electronically regulated or permanently excited synchronous machine.

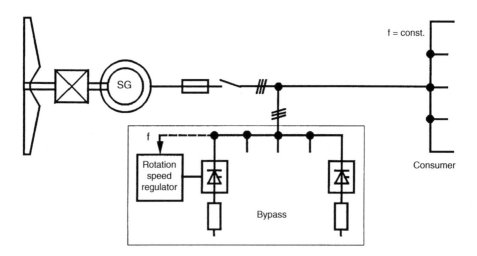

Figure 5.7   Control principle for stall-controlled wind turbines with bypass control

Control principles that provide further options for the integration of wind power into the grid can be derived from the designs for isolated operation.

5.3   Grid Operation of Wind Turbines

When operating wind turbines in the rigid combined grid, it is usually assumed that the output provided by the turbine, up to the generator capacity, can always be directly transferred to the grid. A high degree of energy utilization can be achieved in this manner. In order to protect the drive train, generator and grid connection from overload, it is sufficient to limit the output to the design-dependent nominal or maximum value. As shown earlier, this can be achieved in the short-term for control purposes by

- active intervention at the turbine by changing the rotor blade position relative to the wind or flow direction or
- passive design measures by operating in stall.

For long-term regulation by the management system, it is possible to turn the rotational plane of the turbine into the wind (storm protection).

'Stall-controlled' wind turbines with asynchronous generators connected directly to the grid always exhibit inherent behavior under different meteorological conditions (wind speed, wind direction, air density, etc.) and grid states (frequency, voltage), due to the constructional layout of the turbine and generator parameters. Moreover, they are subject to turbine-specific conditions caused by long-term changes such as dirty rotor blades, temporary influences such as blade icing or periodic fluctuations such as tower shadowing.

Site-specific harmonics or resonant frequencies in the grid can, if necessary, be reduced by a suitably designed generator winding. Operation-related intervention with regard to slight grid effects (e.g. power fluctuations) are, however, not possible.

Stall-controlled wind turbines of up to approximately 200 kW are encountered in large numbers in Californian and Danish wind farms, and stall-controlled turbines in the megawatt class are also used. The options for influencing this type of turbine by the control and management systems are largely limited to connection and disconnection processes, which can take place via a power converter (soft start) or at defined points in time. These processes can also be influenced by remote monitoring or control.

As described in Section 4.3.2, if this type of turbine is installed and operated in large numbers, greater smoothing effects are achieved in the case of power fluctuations as the number of turbines, and thus the area, increases. The operation of stall-controlled turbines with low individual output power connected to a common supply point is therefore entirely feasible.

If, however, grids are to be heavily loaded by (large) individual turbines or small groups (which for cost reasons is usually desirable in windy areas that previously had a low supply of power), the grid reactions of individual plants or of favourable combinations of systems that complement or compensate for each other (see Figure 4.45) must be assigned great importance.

In contrast to Figure 5.8, in Figure 5.9 the wind power or the drive torque can be brought into line with operating requirements by the management and control systems by adjusting the generator frequency f_G, i.e. the turbine speed n. The significance of remote monitoring systems

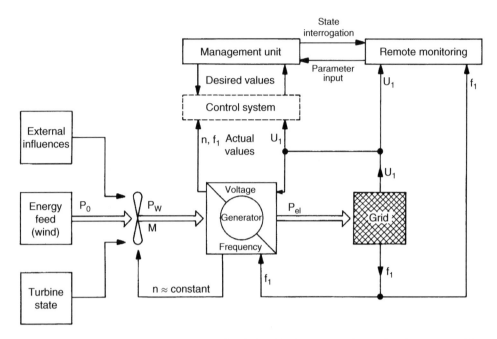

Figure 5.8 Control and management range of a wind turbine kept at a grid-specified fixed speed without blade pitch adjustment

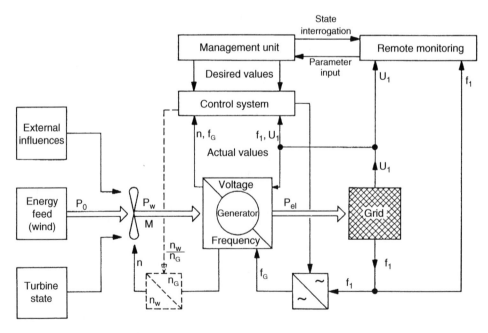

Figure 5.9 Control and management range for a variable-speed wind turbine without blade pitch adjustment

could therefore increase greatly in the future for the presetting of maximum output values, etc. Such adjustments are most commonly achieved by frequency conversion by means of power electronic converter systems, with or without a constant current d.c. link. Alternatively, as shown by the dashed section in Figure 5.9, the supplied power can be altered by the turbine speed by varying the gear transmission [5.8]. In this manner, synchronous generators with a rigid connection to the grid can be used to support the grid, maintaining voltage and frequency and thereby achieving low grid reactions.

Blade pitch adjustment, as shown in Figure 5.10, permits the energy flow to be adjusted by means of the blade pitch angle β for the regulation or limitation of power output. State interrogation by the management system allows switching and regulation parameters to be stipulated by the remote monitoring system and suitably modified target values to be provided by the management system. External influences that affect the flow of power must be monitored at all times by the control system if enough wind energy is available. By contrast, deviations from the original turbine state, e.g. ice or dirty blades, can only be corrected by monitoring and correcting the turbine parameters.

Speed regulation by blade pitch adjustment, which is common in isolated operation, can also be used in grid operation to run up the wind turbine for synchronizing the generator. Unlike small turbines, which are only sometimes fitted with blade pitch control, wind turbines above approximately 500 kW nominal output are increasingly operated with this mechanism. The rotor blades are either adjusted along their entire length or only at the tip – the particularly active section. The wide range of aerodynamic effects on turbine loading and torque generation will not be considered in more detail here.

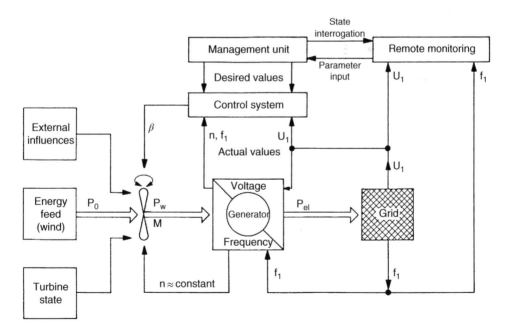

Figure 5.10 Control and management range of a wind turbine with a fixed-speed grid connection with blade pitch adjustment

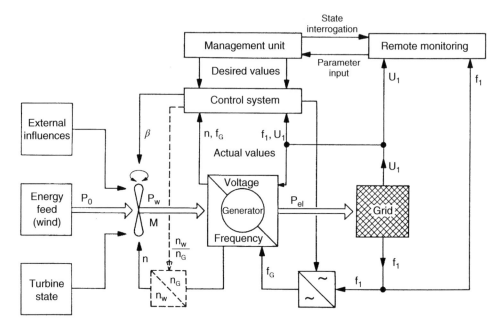

Figure 5.11 Control and management range of a variable-speed wind turbine with blade pitch adjustment

With regard to reducing grid reactions, the effects caused by output fluctuations can be reduced with particular effectiveness if, as is the case for this type of turbine, the energy uptake can be actively altered. According to Figure 4.29, the greatest output fluctuations can be expected from individual turbines in the part-load region. In these operating states, power limitation by blade pitch adjustment is rarely or never used. Figure 4.29, also shows (in the right-hand section) that in the operating range above the nominal wind speed, output variations can be significantly reduced. Blade pitch adjustment is fully utilized here.

Power variations, and the associated grid reactions in low-power or heavily loaded grids, can therefore be reduced if, for example, data supplied to the management unit by the remote monitoring system allow modified target values to be achieved by the control system. This process is, however, always associated with a reduced level of wind energy utilization. Moreover, the blade pitch adjustment mechanism of the type currently prevalent can be highly loaded by frequent interventions, and its service life thus significantly reduced. Wind turbines of the type shown in Figure 5.11, which are particularly common in medium- and high-output units, provide the option of influencing the wind turbine output and the torque at the drive train, either by adjusting the blade pitch or by influencing generator torque and thus turbine speed. Such units therefore possess two independent intervention systems. This significantly increases operating reliability. Through splitting the influence options as targeted, mechanical regulation, procedures can be significantly reduced when the rotating mass is used as a fly wheel during speed variation. Furthermore, it reduces operational demands on the blade pitch adjustment system on the one hand and on the turbine, drive train and generator on the other hand. Wind turbines that can be operated in this manner can therefore reduce grid reactions caused by output fluctuations and minimize component loading. As a result, structural conditions can be set up much more favourably than is the case for other systems. The results of attempts to quantify this precisely are available. However, options for the definition of the construction and design of relevant elements call for extensive long-term investigations.

Particular attention must be paid to the safety of entire systems and subcomponents in connection with designs with built-in redundancy. In this respect, regulation systems of the type shown in Figure 5.11 with different intervention options offer great advantages. Knowledge of the faults and failures occurring in wind turbines (currently mainly caused by the control and electrical systems due to the as-yet low running times) can give rise to measures for improving reliability and increasing the service life of turbines. Particular attention should be paid to critical components with regard to service life in large-scale use. Owing to the low-cost contribution of, for example, electrical components, fundamental improvements to the entire system can be achieved with great effect at a relatively low financial cost. Repair and maintenance costs can thus be saved and plant availability increased.

Viewed as a whole, wind turbines can be set up using management and control systems such that they are suited for supplying electrical energy to isolated networks and particularly to the supply structures in the grid. This requires appropriate control concepts. These will be briefly described below.

5.4   Control Concepts

Wind turbines must provide reliable operation in all operating states. As well as appropriate dimensioning of the turbine, drive train and tower, the control and management systems, which are described in the next section, are of particular importance. For this reason, it is necessary to develop a local, turbine-specific profile of requirements, to expand this into a strategy that takes into account the desires of operators, manufacturers, power supply companies, etc., and to translate this into a control plan. This can be used for the dimensioning of the regulator, the simulation of the turbine and the design of the control mechanisms from a hardware point of view. Now that the requirements and possible strategies have been highlighted, the most important control concepts will be described below.

5.4.1   Control in isolated operation

Figure 5.12 shows the control diagram for a wind turbine fitted with a synchronous generator in isolated operation. A power regulator influences the power to be drawn from the wind by varying the blade pitch, allowing output to be held almost constant if the wind speed is sufficiently high. By limiting regulator output, this power control circuit can control a speed regulator. Controlled running-up and shut-down processes are thus achieved in the same manner as speed constancy within a tolerance range predetermined by the regulator and adjustment system.

For reasons of stability with regard to turbine characteristics and for the reduction of component loading, it is advisable, particularly for larger turbines, to add a blade pitch and/or blade pitch adjustment speed regulation circuit (internal cascade) and a blade pitch adjustment acceleration regulation circuit to the speed or output control circuit. The output voltage can

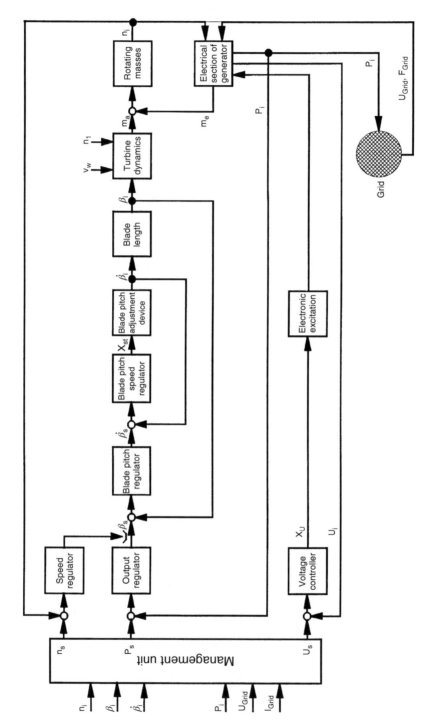

Figure 5.12   Structure for the control of a wind turbine with a synchronous generator in isolated operation

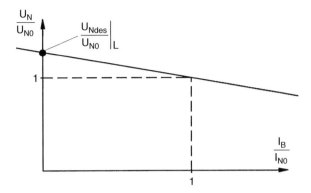

Figure 5.13 Voltage regulation by the reactive power drooping characteristic

be adjusted within narrow limits by using a separate voltage regulator if the generator excitation can be externally controlled. This regulation circuit is often integrated into the generator system. Voltage can thus be regulated to constant values or (as is usually the case) altered in accordance with the characteristic in Figure 5.13 with the aid of a reactive power drooping characteristic and matched to the reactive power conditions. A low inductive load leads to the predetermined desired value for the open-circuit grid voltage $U_{Ndes,O}$. Highly inductive consumers, on the other hand, bring about voltage reductions.

An adjustment of voltage can take place such that the actual value of the grid voltage U_{Nact} arises synthetically from the grid voltage U_N by reactive power correction according to the quantity $\beta I_B/I_{N0}$ (Figure 5.14(a)) [Chapter 3, Reference 13]. Alternatively, voltage regulation with a reactive power-dependent desired value is possible (Figure 5.14(b)). The grid voltage desired value U_{Ndes} is made up of the no-load desired value of grid voltage $U_{Ndes,L}$.

Moreover, it is possible in a similar way to correct the turbine rotational speed or the grid frequency with the aid of an active power drooping characteristic according to the load state, as shown in the characteristic in Figure 5.15, in such a way that low loads lead to slightly increased rotation speeds and frequency values (e.g. 52 Hz at no-load). These fall to 48 Hz at high loads, for example, when operating at nominal ratings. Thus the actual values as well as the desired values (Figure 5.16) can be corrected, depending upon the load.

The characteristic of a supply system is altered by the active or reactive power drooping characteristic, such that the user is supplied at a high frequency and high voltage at low loads. At high loads these grid variables are reduced correspondingly. The control of the supply system is therefore supported by the stabilizing behavior of the drooping characteristic.

The structure shown in Figure 5.12 can also be used for the regulation of the blade pitch, output and speed of a wind turbine with an asynchronous generator in isolated operation. The voltage regulation shown in Figure 5.17(b) can take place within a large fluctuation range using capacitors that can be connected in stages, as shown in Figure 5.17(a). The voltage change can be reduced with an increasing number of capacitor stages. In practice $n = 2$ to 5 stages are used. Larger variation ranges, e.g. with 12 stages, are the exception. Voltage and frequency [5.9, 5.10] can also be kept within narrow limits using rapid switching systems (Figure 5.17(c) and (d)).

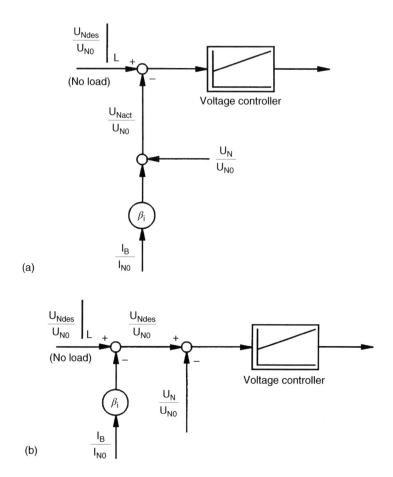

(a)

(b)

Figure 5.14 Voltage regulation with (a) reactive power-dependent actual value and (b) reactive power-dependent desired value

Just as in Figure 5.17(a), the capacitors provide the excitation reactive power for the asynchronous machine. The quasi-continuous bridging of the additional resistor R_z, e.g. by using IGBT switches, also permits the control of the stator frequency within narrow limits by slip variation.

For small wind turbines, sensing the blade pitch is relatively costly. To avoid this, the inclusion of pitch position and pitch speed regulation circuits can be dispensed with (Figure 5.18). The speed or power control circuit thus acts directly upon the blade pitch adjustment mechanism and regulates the blade pitch angle according to the prevailing speed and output values, without knowing the actual value of the angle.

However, a simple limiting effect can still be imposed upon the blade pitch adjustment speed in this case. This can be achieved by specifying the maximum value for the speed of the electric positioning motors or the flow in the hydraulic positioning cylinders. This system simplification, however, does not permit the highly dynamic characteristics of the design shown in Figure 5.12.

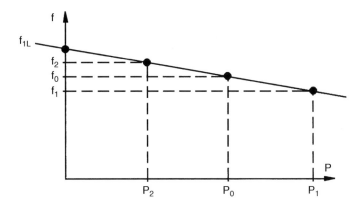

Figure 5.15   Load-dependent speed regulation of a wind turbine by the active power drooping characteristic (characteristic)

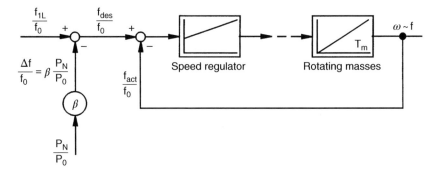

Figure 5.16   Load-dependent speed regulation of a wind turbine by the active power drooping characteristic (circuit with a desired value correction)

5.4.2   Regulation of variable-speed turbines

Control structures for variable-speed turbines can easily be derived from the concepts developed for the control of wind turbines in separate operation. Instead of a fixed speed, or a speed predetermined by the power drooping characteristic, as shown in Figure 5.12, the speed producing the optimum output is determined (see Figures 2.58 and 3.35) and the turbine is run as close to this speed as possible. This type of system must therefore be designed with two different types of speed control circuit (Figure 5.19).

The regulation circuit in the upper section of the diagram has the task of limiting the input power and speed of the turbine at full-load operation to the nominal value, e.g. by adjusting the blade pitch. The second speed control circuit in the lower section of the diagram (e.g. as described in Section 4.1.5.6), on the other hand, must control the turbine speed by controlling the generator electric torque such that the turbine output takes on optimal values or achieves a reliable operating state and behavior that protects components from excess loading. Using electric actuators, the output power can be adjusted ten times faster than the turbine output.

Variable-speed turbines thus provide the option of reducing the load on drive-train components by rapid intervention. This can be achieved by targeted regulation or by limiting the generator torque and using the transient effects of all rotating masses in the drive train according to Equations (2.92) and (2.93).

The control structure as shown in Figure 5.19 is therefore suitable for both grid and isolated operation. The prerequisite for isolated operation is, however, that a self-commutated and self-clocked inverter, e.g. as shown in Figures 4.21 to 4.23, is used that is capable of forming a grid and covering the active and reactive power.

Different options exist for the desired output optimization. The energy conversion systems in Figure 3.4 (b) and (c) can be used and the systems in Figure 3.4 (i), (j), (k) and (l) are particularly suitable. Two well-tested methods are, firstly, the management of rotational speed according to a family of characteristics that depend upon the actual turbine values as shown in Figure 2.58 (c) or (d) and, secondly, approaching the optimum output by means of a search process based upon incremental changes to the rotational speed. The two methods can also be used in combination. Moreover, there are also optimization options that include wind speed and output values based upon the characteristic fields found.

These processes for the improvement of energy yields are not applicable in the case of systems with a rigid connection to the grid, as shown in Figure 3.4(a) and (g). Plants operated with variable slip, as shown in Figure 3.4(d) to (f), which will be considered below, only permit such optimization methods if the mechanical–electrical converter systems allow appropriately large turbine speed ranges, e.g. from 40% of the nominal speed.

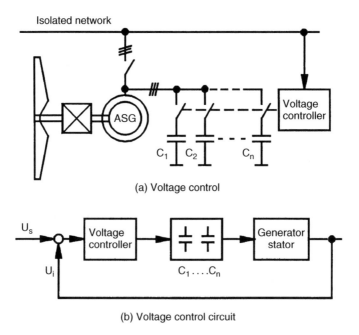

(a) Voltage control

(b) Voltage control circuit

Figure 5.17   Voltage and frequency control of asynchronous generators

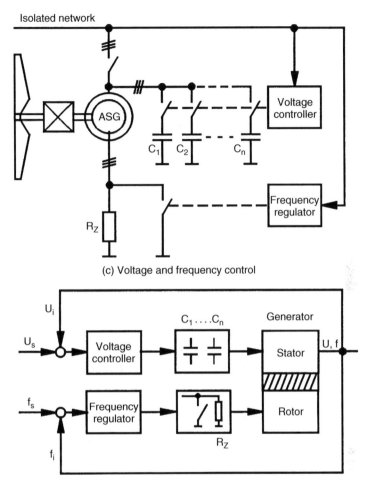

(c) Voltage and frequency control

(d) Voltage and frequency control circuits

Figure 5.17 (*continued*)

5.4.3 Regulation of variable-slip asynchronous generators

Asynchronous machines permit operation in the oversynchronous and undersynchronous range, i.e. above and below the synchronous speed, according to Equation (3.22). The speed of asynchronous machines can be varied if their generators have multiphase coils ($m = 2, 3, \ldots$) and current transfer via slip rings. Speed can also be varied inductively, and thus without brushes, via additional coils, by

- up to around 10% with dynamic slip control (Figure 3.4(d));
- approximately 30% in oversynchronous operation using power converter cascades (Figure 3.4(e));

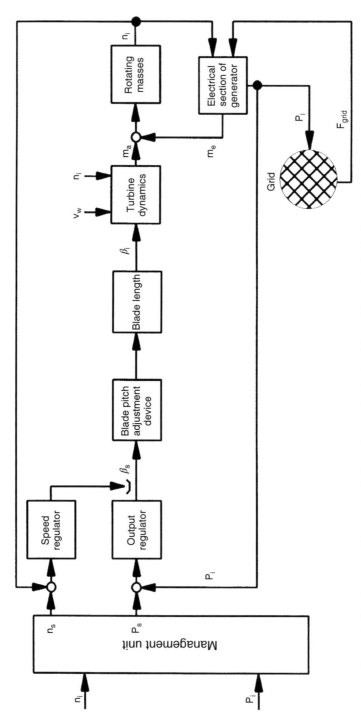

Figure 5.18   Structure for the regulation of a small wind turbine without blade pitch sensing in a separate operation

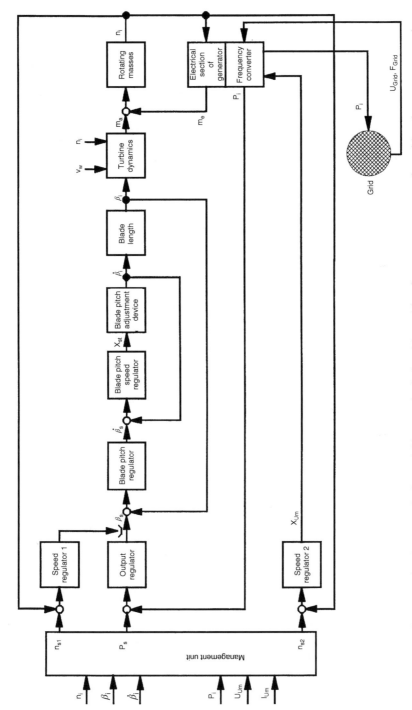

Figure 5.19 Structure for the control of a variable-speed wind turbine with a frequency converter for grid and separate operation

- around 40% in under- and oversynchronous operation in double-fed systems (Figure 3.4(f)), i.e. a total of 0.6 to 1.4 times the synchronous speed.

So-called slip-ring rotor asynchronous machines give the option of output branching. In this system, the majority of the converted energy, e.g. 95%, is passed directly from the stator to the grid. Rotor output, which is usually much lower and is proportional to slip, is drawn off into additional resistors in the form of loss by dynamic slip control. In the power converter cascade and double-fed asynchronous machines, on the other hand, this slip output is conditioned for transfer to the grid by frequency converters in oversynchronous operation. In double-fed machines, which will be considered below, the rotor power is taken from the grid via frequency converters in undersynchronous operation.

5.4.3.1 Double-fed asynchronous generators

Asynchronous machines with slip-ring rotors are generally designed with three-phase a.c. windings in the stator and rotor. Thus the windings in the stator and rotor can be supplied from three-phase systems of different frequency and voltage. Moreover, the magnetization (as in synchronous machines) can be wholly or partly supplied through either of the two electric circuits.

The stator is normally connected directly with the supply grid. The rotor is then supplied via a frequency converter. Thus the angular velocity of the stator rotary field

$$\frac{\omega_1}{p_1} = \omega_{\text{mech}} \pm \frac{\omega_2}{p_2} \tag{5.1}$$

is equal to the sum of angular velocity of mechanical rotation ω_{mech} and the rotor current frequency ω_2. Depending upon direction of the supply frequency, the machine can thus be operated in either the undersynchronous or oversynchronous mode. The pole numbers of the stator and rotor (p_1 and p_2 respectively) at the angular velocity of the rotary fields must be allowed for here.

The rotor active power

$$P_R = sP_\delta \tag{5.2}$$

is obtained from the product of the slip s and air-gap power P_δ. Thus the power that the rotor circuit transfers to or from the grid via a frequency converter is also proportional to slip. The slip-dependent speed-regulating range $\Delta n = s/n_0$ of the generator, e.g. $0.6 \leq n/n_0 \leq 1.4$, therefore determines maximum output (and thus to a large extent the cost) of the required frequency converter.

The magnetizing current of asynchronous machines is made up of the sum of stator and rotor currents as shown in Figure 3.16. Unlike, for example, short-circuit rotors, double-fed rotors allow the rotor current to be altered. Different phase positions can thus be achieved for the stator current. Similarly to synchronous machines, neutral and overexcited generator states as well as the normal underexcited operation can be achieved in this type of double-fed arrangement. This allows the machine speed and thus also the turbine speed to be varied. Figure 5.20 shows the structure for the regulation of a double-fed asynchronous generator and illustrates the complexity of such systems.

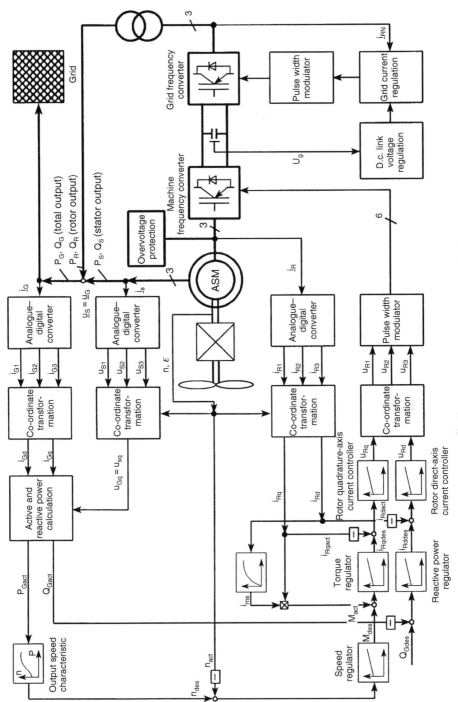

Figure 5.20 Block diagram for the control of double-fed asynchronous generators

The transformation of the active and reactive components of the total output of large machines into the two-axis field coordinate system with d as the direct-axis and q as the quadrature-axis component (Figure 5.20, top centre) leads, if we disregard stator resistance, to a complete disconnection between the active power, or the largely corresponding machine torque variable, and reactive power [Chapter 2, References 21, 22]. Thus the two state variables can be regulated separately without mutual interaction.

In order to make use of the advantages of variable-speed conversion systems, the total output of the machine P_G should be influenced by the regulation system, and the generator and turbine speed should be dictated in a controlled manner. To control the rotational speed (upper-left part of the block diagram) the speed–output characteristic (e.g. as shown in Figure 2.64(c)) can be used. Speed is influenced by torque, as is normal for rotating systems. This can be adjusted by means of the quadrature-axis component of the rotor current i_{Rq} by suitable control of the machine frequency converter via the rotor voltage according to the field [5.11, 5.12]. The reactive power can be predetermined in a similar manner, e.g. as a fixed desired value, or can be guided according to the requirements of the power generation company by means of the direct-axis component of rotor current i_{Rd} with the aid of the machine frequency converter (Figure 5.20, bottom left). Instead of reactive power Q_{Tdes}, the power factor can be predetermined and thus adjusted.

In both control circuits (i_{Rd} and i_{Rq}) the rotor current (as in the stator circuit) is detected in at least two or in all three phases by rotating pointer i_R, converted into digital values and transformed into the field coordinates i_{Rd} and i_{Rq}. Thus the double-fed asynchronous generator can be adjusted or regulated via the pulse-controlled inverter on the machine side for the desired values of speed, torque and active power, as well as reactive power or the power factor.

In order to avoid disruptive connections between the grid-side pulse-controlled inverter and the overall regulation system via the machine-side pulse-controlled inverter, the grid frequency converters can be operated by the d.c. link voltage (see Figure 5.20, bottom right). This can be kept roughly constant by a preset hysteresis band [Chapter 2, Reference 22]. In this manner, the grid-side frequency converter adapts to the value and phase of the required active power and works within the preset tolerance band in neutral operation. The grid-side reactive power in the rotor can therefore be kept close to zero and the total reactive power can thus be adjusted by means of the reactive power regulation system at the machine frequency converter.

High-speed current changes can occur in the stator circuit, particularly if the generator is suddenly disconnected from the grid for some reason, e.g. by the tripping of protective switches or by rapid auto-reclosure. These current changes induce high-voltage peaks in the rotor winding. To prevent this, overvoltage protection must be fitted in the rotor (see Figure 5.20, centre).

The double-fed asynchronous machine is particularly suitable for use as a generator in wind turbines. Due to the connection of the stator to the grid, these machines combine the beneficial electrical characteristics of synchronous machines with mechanical advantages associated with speed variability at very good efficiency. They thus allow the mechanical loading of the drive train to be significantly lessened and the fluctuations of electrical output power to be greatly reduced. By the use of IGBT pulse-controlled inverters in the rotor circuit, grid reactions caused by harmonics can be reduced to a noncritical level. Other options for capacitive and inductive operation can be used for grid support. Thus wind turbines with double-fed asynchronous generators can, particularly in weak grids, give access to significant connection capacities.

The double-fed asynchronous machine was first used with a direct frequency converter made up of thyristors, in the GROWIAN wind turbine, which was installed in Kaiser-Wilhelm-Koog on the North Sea coast at the beginning of the 1980s. The American MOD 5 B turbine in Hawaii was also fitted with an appropriate converter device.

During the operation of the 3 MW turbines it was possible to demonstrate the excellent regulating characteristics of these systems by the extremely well-smoothed output power [5.13]. The complete utilization of the so-called ceiling voltage proved its worth here, since it determines the rotation speed range of the generator and the reactive power requirement of the frequency converter [5.1]. By the capacitive operation of the stator, the phase control reactive power of the direct frequency converter can be generated to a large degree by the generator itself. However, the harmonic content of the generator current increases with rising capacitive stator reactive power, with the fifth, seventh and eleventh harmonics of grid frequency, in particular, dominating at high rotor frequencies and at high firing angle settings of the direct frequency converter.

Static reactive power compensation of the fundamental reactive power by filter circuits tuned to dominant harmonics provides a favourable compromise. However, this requires precise knowledge of grid characteristics in order to avoid possible grid resonances [Chapter 3, Reference 54].

Both oversynchronous and undersynchronous operations are possible if the rotor of a slip-ring rotor machine, for example, supplies into a four-quadrant indirect converter. The load-side power converter can be designed with self-commutated thyristor valves or with a self-commutated or machine-clocked pulse-controlled a.c. converter. The regulation of the machine can be based on the block diagram in Figure 5.21. Thyristor valves, however, create high harmonic currents in the rotor of such systems, which, in combination with the high

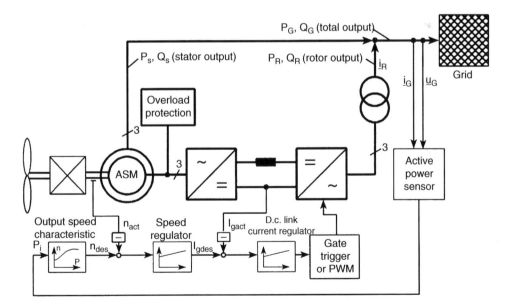

Figure 5.21 Block diagram for the control of an asynchronous machine with a slip-ring rotor and oversynchronous power converter cascade

cost of the power converter (compared with oversynchronous cascades), makes this concept appear largely unsuitable for use in wind turbines [5.1].

Double-fed asynchronous machines with IGBT frequency converter systems allow particularly favourable operating characteristics to be achieved with regard to grid reactions. At the beginning of this section the advantages were described of how this type of converter device can be fully exploited in this type of system. Such systems are becoming increasingly important in turbines of the 0.5 to 5 MW class (e.g. DeWind 4 shown in Figure 1.28 to N80/90 shown in Figure 1.24(d) or V80/90 shown in Figure 1.25(f)) due to the possibilities that currently exist in power electronics and computing. The wear associated with slip-ring transmission is prevented if field-oriented currents are transmitted to the rotor via additional windings. Such machines are already available on the market. The additional delays that occur, however, play a decisive role in the field control.

If undersynchronous generator operation and reactive power regulation of the machine is avoided, the system based upon an oversynchronous power converter cascade, which operates in a narrow speed range, can be used for wind turbines. This system is described below.

5.4.3.2 Asynchronous generators with oversynchronous power converter cascades

Compared with double-fed asynchronous generators, much simpler conversion systems are achieved if slip-ring rotor machines are only operated in the oversynchronous speed range. In this system, the slip power of the rotor can be drawn from the rotor and supplied to the three-phase grid via an uncontrolled or controlled rectifier and via a line-commutated or self-commutated inverter (Scherbius principle). This type of converter system is characterized by its robust and low-maintenance construction and, moreover, represents a very reasonably priced variant.

Figure 5.21 shows the relatively simple structure for the control of the converter system. The machine-side rectifier converts the variable-frequency alternating variables of the rotor (voltage $\underline{u}_R$ and current $\underline{i}_R$) into direct variables (u_g, i_g). The direct current of the d.c. link is inverted to grid frequency by an inverter and matched to the grid voltage by a transformer. This system therefore brings about a complete decoupling of rotor speed and grid frequency. Rectifiers and inverters can take the form of diode rectifiers or thyristor-controlled or pulse-controlled a.c. converters.

Active power speed regulation is similar to that in a double-fed machine. However, the speed variations are limited to a much narrower range. Reactive power regulation is no simple matter. Overload protection, e.g. in the form of a switchable resistor, protects the d.c. link from overload and impedes uncontrolled operation of the generator.

The simple design of the system generally ensures unproblematic commissioning and operation. Moreover, if the d.c. link is favourably designed excellent dynamic converter characteristics can be achieved.

This type of conversion system was installed in the Spanish–German AWEC 60 on the north-west tip of Spain in Cabo Vilano at the end of the 1980s. The mechanical–electrical converter was able to demonstrate fully its good regulation potential and excellent dynamic characteristics in the 1.2 MW wind turbine with a 60 m rotor diameter, even under extreme operating conditions. This turbine was of almost identical design to the two German WKA 60 I and II wind turbines.

However, as is the case for cage rotor machines (see Figures 3.9 and 3.10), the reactive power for the excitation of the generator must be provided by the grid or by an additional compensation unit (see Figures 3.11 and 3.12). In an oversynchronous power converter cascade the commutation reactive power of the rotor-side rectifier, the phase-control reactive power of the inverter and the distortion reactive power of the entire turbine must also be provided if, for example, thyristor valves are used.

The phase-control reactive power of the inverter can be reduced to low values by specifying a narrow oversynchronous speed range. To achieve this, the inverter must be adjusted to the inverter stability limit when the generator is running at its maximum speed. Thus a reserve voltage must be dispensed with in order to fully utilize the d.c. link.

The slip-dependent stator current harmonics caused by the rotor currents represent a further disadvantage. These give rise to considerable grid reactions and mechanical loading of the machine, particularly due to oscillating torque. Moreover, the harmonics give rise to additional losses in the generator and reduce the efficiency of the conversion system [5.1]. It must also be borne in mind that the rising and falling edges of the rotor current are smoothed by the commutation of the rotor-side rectifier (see the sketch in Figure 5.22). This causes a reduction of harmonics in the generator current. However, the commutation causes a phase shift between the fundamental component of voltage and current in the rotor. This results in a weakening of the electrical moment and an increase in stator reactive power around the commutation component of the rectifier. The utilization of the slip-ring rotor machine is thus reduced.

Significant disadvantages of this system can be prevented by the use of pulse-controlled a.c. converters [5.14, 5.15]. These give rise to greater frequency converter losses. Because of the small contribution of the rotor conversion system to total output (e.g. 0 to 15%) these are hardly noticeable, since the greater part of the increase in losses can be largely balanced out by more favourable operating conditions (near-sinusoidal currents, reduction in oscillating torque, etc.).

The system is further simplified by dispensing with the return of a slip power predetermined by the slip regulation system from the rotor to the grid. The desired slip can be adjusted to the values necessary for dynamic adjustment (as low as possible). These will be considered below.

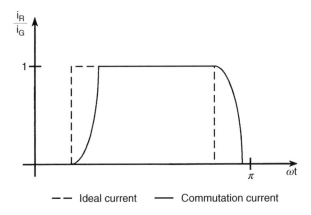

Figure 5.22   Sketch of the rotor current path

5.4.3.3   Asynchronous generators with dynamic slip control

Asynchronous generators with wire or bar windings in the rotor usually permit intervention via the slip rings. Moreover, the energy induced in the rotor can be converted in the rotor system or inductively transferred to the stator. As well as the methods that have already been described for converting slip energy and feeding it into the grid, there is also the option of completely dispensing with the use of this energy component in order to keep the cost of the conversion system low. Wind turbines are, however, subject to random and periodic output fluctuations due to wind-speed fluctuations, tower-shadowing effects, natural resonances of components, etc. In a fixed-speed connection of the generator to the grid, these are passed on to the grid at almost their full level via the drive train and the generator coil (see Figure 3.15(a)). Large slip values significantly reduce the load on the drive train and greatly reduce output fluctuations (see Figure 3.15(b)). However, if the slip energy is not recovered high losses in the rotor circuit are the result according to Equation (5.2). To achieve a high total efficiency, the rotor slip should be kept as low as possible in this relatively simple system. A favourable conversion system should therefore combine the two seemingly contradictory options of

- low slip energy losses and
- high dynamic speed and slip flexibility (see Figure 3.16).

   Slip-ring rotor asynchronous machines and systems based upon the same operating principle permit three operating methods, as shown in Figure 5.23:

- In the case of a short-circuited rotor winding, the generator operates at the lowest slip and thus achieves the best machine efficiency. However, this results in high drive-train loading and large output fluctuations.
- By connecting resistors in the three phases of the rotor circuit, high slip values and thus high elasticity of speed can be achieved if the resistors are correctly dimensioned, with increased slip bringing about better dynamic characteristics but poorer efficiency.
- With a continuous adjustment of slip to the prevailing wind and grid conditions, a rapid change of rotor circuit resistance aided by alternating switching processes between short-circuited rotor winding and full resistance in the rotor circuit can bring about output-smoothed or efficient operating ranges [5.16]. Thus, in the part-load range, low slip values can be set and adjusted slightly (see Figure 3.16) in order to achieve a high level of efficiency.

   Torque fluctuations arising from the drive train can be largely smoothed out in this manner [Chapter 3, Reference 3]. If the wind supply is above the level required to produce nominal output, large losses due to increased slip can be accepted without sacrificing energy yield. Thus the advantages of torque- and output-smoothing characteristics can be combined with adequate speed-regulation reserves [5.17, 5.18].

   Figure 5.23 illustrates the excellent characteristics of a system with slip control in comparison with the unregulated layout both with short-circuited rotor and with additional resistors connected in the rotor circuit. This shows that a slip-controlled system can smooth output fluctuations – caused, for example, by tower shadowing or similar effects – excellently. At the same time, average slip values, and thus losses, can be kept low.

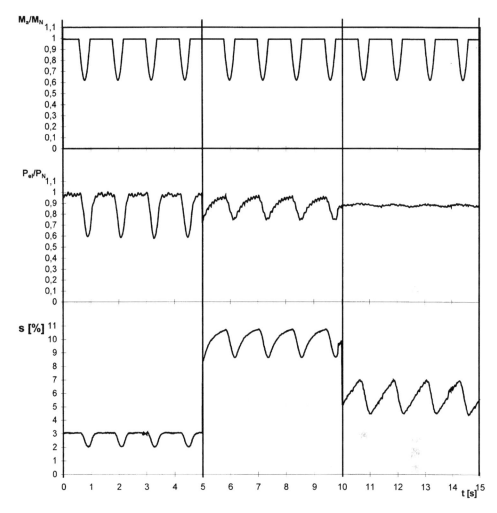

Figure 5.23 System behavior of asynchronous generators for (a) the short-circuited rotor, (b) the rotor with fixed additional resistors and (c) slip control

Unlike the systems described previously, power electronic components do not affect the components that are electrically connected to the grid. They only act in weakened form via the air gap and the coupling inductances in the machine (see Figure 3.7(a)). This significantly reduces grid harmonics [5.16, 5.17]. The measurements in Figure 5.24 provide clear proof of this. The system configuration later in Figure 5.26 was used in these investigations. A comparison of the frequency spectra of the rotor and stator currents and voltages shows the 'filtering characteristics' of the generator.

Figure 5.25 shows the conventional concept for wind turbines with slip control, based upon the industry standard robust asynchronous generator with a slip-ring rotor. Using this system it is possible to design the rotor circuit with a rectifier bridge and single-phase pulsing (e.g. using

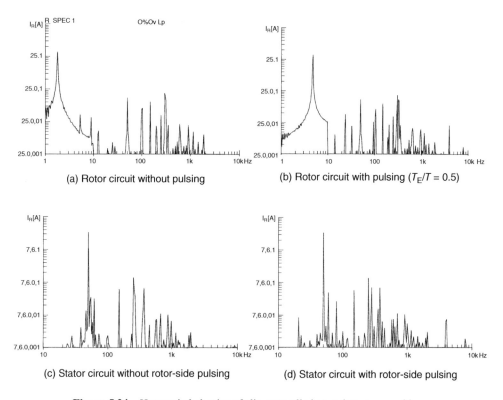

(a) Rotor circuit without pulsing

(b) Rotor circuit with pulsing ($T_E/T = 0.5$)

(c) Stator circuit without rotor-side pulsing

(d) Stator circuit with rotor-side pulsing

Figure 5.24   Harmonic behavior of slip-controlled asynchronous machines

IGBT) at the additional resistor (Figure 5.25(a)) or to design the rotor system with additional resistors and pulsing in all three phases (Figure 5.25(b)).

The additional rotor resistors and the slip regulation and its drive are generally fitted as a separate unit outside the generator. This makes it easy to perform maintenance and any necessary repairs to components. Moreover, the subsystem can be designed as a modular unit.

The investigations in references [Chapter 3, Reference 3] and [5.16] have shown that a simple single-phase design, as shown in Figure 5.25(a), tends to cause mechanical vibrations in the drive train. These result from the rectification of rotor currents and, due to the square-wave current in the case of a B6 rectifier bridge, for example, take on values six times the slip frequency. The rotor current harmonics this causes are also transferred to the stator and give rise to increased losses in the generator. Moreover, remedial measures are necessary to remove the vibrations. These disadvantages can be avoided by a symmetrical system with a three-phase rotor circuit design, as shown in Figure 5.25(b). However, due to the three-phase synchronized antiparallel design of the frequency converter that this calls for, the cost is increased to such a degree that, at only a little extra cost (rectifier, transformer), power recovery – and thus the utilization of the slip energy – is also possible.

By a combination of the two variants, the design shown in Figure 5.26 with a three-phase resistor and single-phase pulsing can be used to build a relatively simple system which can achieve good operating results at a reasonable cost. Unlike the single-phase system in

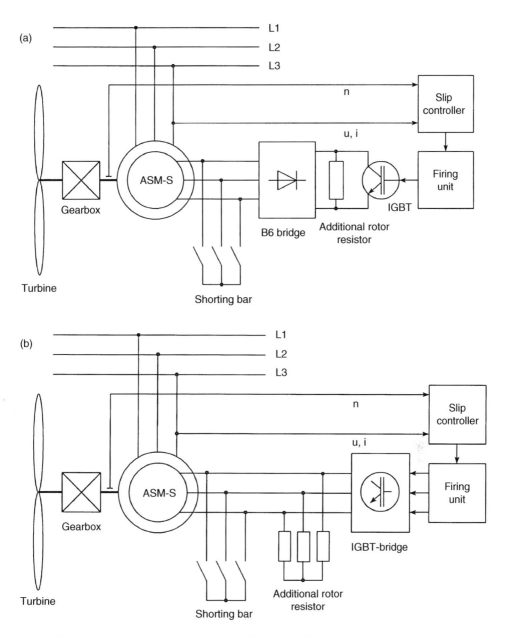

Figure 5.25 Dynamic slip regulation of wind turbines with slip-ring rotor asynchronous machines for (a) a rotor circuit with a rectifier bridge and single-phase pulsing at the additional resistor and (b) a rotor circuit with three-phase pulsing of the additional resistor

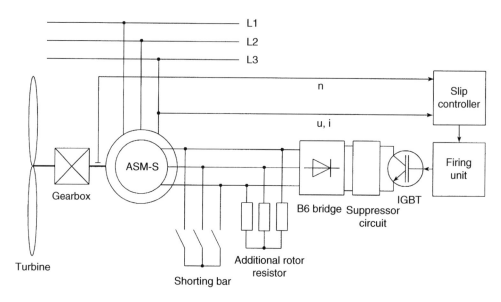

Figure 5.26 Dynamic slip regulation of wind turbines by three-phase additional resistors with direct current pulsing

Figure 5.25(a), this configuration does not cause mechanical vibrations in the drive train. However, due to the lack of a resistor in parallel with the 'IGBT switch' an additional protective circuit is required to protect against the overvoltage that is created in the direct current circuit due to current pulsing.

An annual running time of approximately 8000 hours must be assumed for generators in wind turbines. Components, particularly the slip-ring transmission system, are subject to wear. Thus, when these are used, regular maintenance and easy replacement of wearing elements must be ensured.

Brushless designs – normal in synchronous machines (see Figure 3.6) – are also desirable in asynchronous machines. It is necessary to make a fundamental differentiation here between brushless inductive transmission of the rotor-slip energy by secondary windings (Figure 5.27(a)) and the complete redesign of the power section with additional resistors, power electronics and regulation in the rotating part of the generator (Figure 5.27(b)), as in the OptiSlip® regulation system in the Vestas V 44 (600 kW) to V 66 (1.65 MW) and the V80/90 (2 or 3 MW), which is the US export variant.

The generators specifically designed for the OptiSlip® system are fitted with a wound rotor and an integrated current regulation system in the rotor (RCC = rotor current controller). This is installed at the rear of the generator on the end of the shaft, and consists of additional resistors, power electronics, current sensors and a microprocessor controller. The communications signals between the management system (VMP = Vestas multi-processor) and the current-regulation device are carried via a maintenance-free fibre-optic cable.

Figure 5.28 shows the block diagram for the dynamic slip regulation system of an asynchronous machine. During the running-up and shut-down processes the generator is disconnected from the grid. The speed of the turbine can be controlled and, if necessary, limited by the speed regulator using blade pitch adjustment.

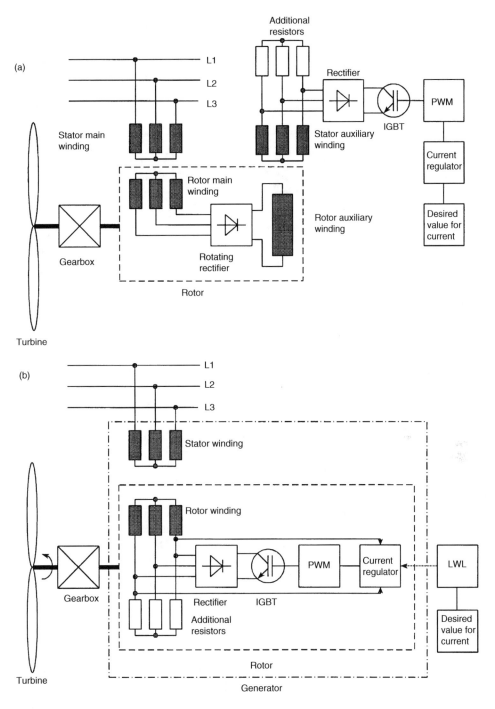

Figure 5.27 Dynamic slip regulation of wind turbines using asynchronous machines without slip-ring systems with (a) slip energy transmission by secondary windings and (b) power section and regulation in the rotor (Vestas)

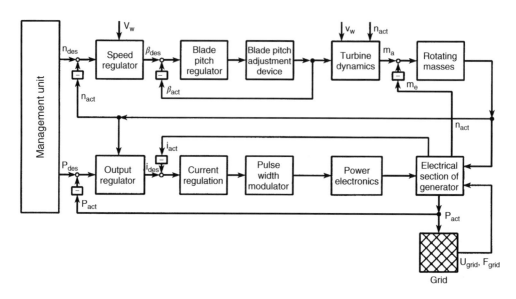

Figure 5.28   Block diagram for dynamic slip control of an asynchronous machine in the grid

Once the generator has been connected to the grid, the wind turbine can be operated at part load (i.e. below nominal output) at a constant blade pitch or, as in the Vestas design, brought into the optimal range according to the prevailing wind speed. Thus, with the aid of this so-called OptiSlip® function, the greatest possible energy yield can be achieved at part load.

Generator output is adapted to the momentary operating state (part or full load) with the aid of slip by means of current regulation, pulse-width modulation and the power controller in the rotor. Thus, low slip values (e.g. 2%) can be set in the part-load range and high slip values (e.g. 5%) can be set at nominal operation, so that the generator speed can be varied to smooth output power and drive-train torque.

If the above regulation options and their benefits with regard to reducing the loads on components, yield optimization, etc., are dispensed with, a very simple and robust system can be designed using wind turbines in rigid connection to the grid. This system is most widespread for turbines up to 600 kW nominal output. In wind turbines in the MW class, on the other hand, there is a clear trend in favour of flexible-speed units.

5.4.4   Regulation of turbines with a rigid connection to the grid

For plants with a largely fixed-speed connection to the grid neither power optimization nor the option of power smoothing are available. Turbine speed is predetermined by the grid via the generator (Figure 5.29). The speed regulation circuit therefore only has the task of running the turbine up and limiting its speed. All further functions of the circuit for the regulation of power, blade pitch and blade pitch adjustment speed are similar to those in wind turbines in isolated operation, as shown in Figure 5.12. For small turbines, the system structure can be further simplified by not including a blade pitch sensing system, as shown in Figure 5.18, if

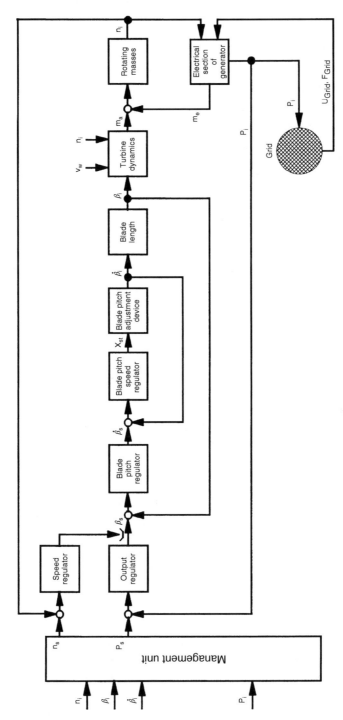

Figure 5.29   Structure for the control of a wind turbine with a fixed-speed grid connection (grid operation)

the regulation dynamics are suitably limited. This must be particularly taken into consideration in the dimensioning of the system components and the regulator.

5.4.5 Wind turbine control using hydrodynamic variable-speed superimposing gears

As already described in Section 3.4 and Figure 3.19, rigidly rotating grid-coupled synchronous generators can be operated in a vibration-free and stable manner by means of hydrodynamic superimposing gears.

Here, the superimposing gears take over the decoupling effect between the turbine and the grid by means of the rigidly operated synchronous generator positioned between the turbine with a fixed gear step.

In reference [5.19], the structure of the drive train as well as the control and operation of the plant for wind turbines of the 2 to 8 MW class are described from the point of view of plant and control. The emphasis is on the capacity control. On the basis of the structure and the draft of the control, the practical testing based on a 2 MW plant is presented. Structure and quality requirements are carried out for the pitch regulation, and possibilities of optimizing with a view to load reduction measures are discussed and the simulation tests are shown.

5.5 Controller Design

The reliable operation of wind turbines is possible only if physical and technical limit values are not exceeded. Turbine control and adjustment mechanisms must therefore be capable of maintaining specified speed and output limits at all times, both in normal operation and in extreme situations. Therefore particular attention must be paid to design of these systems.

The options for the dimensioning of actuators have already been described in Section 2.3.2.5. The following considerations therefore relate to the representation of control circuits, the determination of their parameters and the design of the required controller. The reader is referred to the description in Section 2.1 with regard to the forces acting upon the rotor blade and the driving torque and power at the turbine. Reference will also be made to blade pitch adjustment, described in Sections 2.3.2.1 to 2.3.2.3, drive-train influences, described in Section 2.4, and the behavior of the generator and supply systems, as described in Chapters 3 and 4. The regulation structures in Sections 5.4.1 to 5.4.4 form the basis for further calculations.

It is evident from Figure 2.4 and Equations (2.24) and (2.25) that the tangentially acting driving force and the axially acting thrust value can be approximated by

$$\begin{bmatrix} dF_t \\ dF_{ax} \end{bmatrix} = z\frac{\rho}{2}v_r^2 t_B \, dr \begin{bmatrix} \sin\delta & -\cos\delta \\ \cos\delta & \sin\delta \end{bmatrix} \begin{bmatrix} c_a(\alpha) \\ c_w(\alpha) \end{bmatrix} \tag{5.3}$$

for a turbine circumferential element. The following equation describes the relationship between the blade pitch angle β, the resulting direction of flow δ and the angle of flow over the profile α:

$$\beta = \frac{\pi}{2} - \vartheta = \frac{\pi}{2} - \delta + \alpha \approx \frac{\pi}{2} - \arctan\left(\frac{v_2}{v_u}\right) + \alpha. \tag{5.4}$$

Therefore, with the aid of a system for altering the blade pitch angle, the tangential component of the force acting on the rotor blades, and therefore the torque, can be influenced and the output and rotational speed of the turbine can be controlled. Furthermore, by deliberately influencing the thrust acting in the axial direction, the rotor blade and tower bending can also be limited.

By the superposition of the processes, it is also possible to reduce or damp the tower vibrations. However, this requires that system-specific changes and the resulting influences on the rotor blades during rotation (e.g. due to airstream height gradients, flow disturbance at the tower and partial gust effects on the surface of the turbine) be taken into account. Local conditions and the time-variant position of the turbine system thus have a decisive effect upon measures to limit or damp bending and vibrations in the rotor blades and the tower. For such configurations, forces and moments on one blade can be represented according to

$$\begin{bmatrix} F_{t1} \\ F_{ax1} \end{bmatrix} = \int_{R_i}^{R_a} \frac{\rho}{2} v_r^2(r, \varepsilon, t) t_B(r) \begin{bmatrix} \sin \delta & -\cos \delta \\ \cos \delta & \sin \delta \end{bmatrix} \begin{bmatrix} c_a(\alpha, \varepsilon) \\ c_w(\alpha, \varepsilon) \end{bmatrix} dr \qquad (5.5)$$

and

$$\begin{bmatrix} M_{t1} \\ M_{ax1} \end{bmatrix} = \int r \begin{bmatrix} dF_t \\ dF_{ax} \end{bmatrix} = \int_{R_i}^{R_a} r \frac{\rho}{2} v_r^2(r, \varepsilon, t) t_B(r) \begin{bmatrix} \sin \delta & -\cos \delta \\ \cos \delta & \sin \delta \end{bmatrix} \begin{bmatrix} c_a(\alpha, \varepsilon) \\ c_w(\alpha, \varepsilon) \end{bmatrix} dr. \qquad (5.6)$$

The vast majority of all installed wind turbines are designed with three-blade rotors. For this design, as shown in Figure 5.30, the total values at the turbine, which are dependent upon the rotor position, are found to be

$$\begin{bmatrix} F_t \\ F_{ax} \end{bmatrix} = \begin{bmatrix} F_{t1}(\varepsilon, \alpha, t) + F_{t2}(\varepsilon + 2\pi/3, \alpha, t) + F_{t3}(\varepsilon + 4\pi/3, \alpha, t) \\ F_{ax1}(\varepsilon, \alpha, t) + F_{ax2}(\varepsilon + 2\pi/3, \alpha, t) + F_{ax3}(\varepsilon + 4\pi/3, \alpha, t) \end{bmatrix} \qquad (5.7)$$

and

$$\begin{bmatrix} M_t \\ M_{ax} \end{bmatrix} = \begin{bmatrix} M_{t1}(\varepsilon, \alpha, t) + M_{t2}(\varepsilon + 2\pi/3, \alpha, t) + M_{t3}(\varepsilon + 4\pi/3, \alpha, t) \\ M_{ax1}(\varepsilon, \alpha, t) + M_{ax2}(\varepsilon + 2\pi/3, \alpha, t) + M_{ax3}(\varepsilon + 4\pi/3, \alpha, t) \end{bmatrix} \qquad (5.8)$$

and these values have a decisive effect upon measures and control strategies to limit bending or minimize vibrations. The short-term interventions necessary to achieve these goals can usually only be carried out by means of the blade pitch adjustment system. Although changes to flow

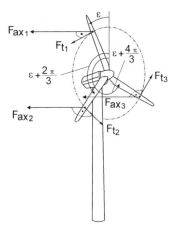

Figure 5.30 Dynamic effects and angular relationships at the turbine

can be initiated very quickly (at speeds below the millisecond range) on the generator side by rotational speed changes, such interventions are only conditionally suitable due to the high rotor system time constants that they require.

From the point of view of control engineering, wind turbines with blade pitch adjustment systems thus provide the option of actively altering the flow of energy in response to both changes in wind speed and changes in conditions in the turbine and supply system. Furthermore, the mathematical treatment of the control circuits in such designs covers a very comprehensive field of required dimensioning. The following discussions therefore relate mainly to turbines with blade pitch adjustment systems.

Very different procedures for the determination of the controller parameters are possible, depending upon the control philosophy. We can assume that a controller presetting is used that ensures stable system behavior in relation to the design state or the operating ranges under consideration. This initial dimensioning can be carried out using the normal measuring procedures such as optimum amount, symmetrical optimum, etc., in accordance with references [5.20–5.22] and [5.23], and the stability of the control circuit can be investigated using the familiar methods of frequency response analysis, root locus curve procedure, Hurwitz criterion of stability, etc. However, this requires that considerable simplifications (see Section 2.3) and linearizations are made in the control systems, and partial restrictions of the operating range must be accepted. This default setting of the controller, which will be described in more detail in the following part, can be checked by means of simulation calculations and system trials, and can be empirically fine-tuned with regard to the strategies and goals of the control system. The following section considers the control circuits and preliminary dimensioning of controllers, based on the adjustment processes at the rotor blades.

5.5.1   Adjustment processes and torsional moments at the rotor blades

The necessary definitions and the effect of moments during the adjustment of turbine rotor blades have already been described in Section 2.3.2. The following representations are based upon this section.

If the inertia caused by accelerating air masses, which plays only a very minor role in comparison with the rotor blade component, is disregarded then the following relationship is found for nonteetering hub turbines, based on Equation (2.68):

$$J_{Bl}\frac{d^2\beta}{dt} + \left(\frac{dJ_{Bl}}{dt} + k_{DB} + k_{RL}\right)\frac{d\beta}{dt} + \left(\frac{dk_{DB}}{dt} + \frac{dk_{RL}}{dt}\right)\beta$$

$$+ M_{Pr} + M_{lift} + M_T + M_{bend} = M_{st}. \tag{5.9}$$

For deflection-resistant blades, dJ_{Bl}/dt and M_{bend} can also be disregarded. Furthermore, for predimensioning, the time derivatives of the damping and frictional components (dk_{DB}/dt and dk_{RL}/dt) can also be disregarded. This gives the greatly simplified differential equation

$$J_{Bl}\frac{d^2\beta}{dt^2} + (k_{DB} + k_{RL})\frac{d\beta}{dt} + M_{Pr} + M_{lift} + M_T = M_{St}. \tag{5.10}$$

This includes moments arising as a result of inertia, damping, friction, propeller effects, lift and torsion due to aerodynamic resetting. These will be described in more detail in the following section.

5.5.1.1 Propeller moments

For turbine systems with non-bending rotor blades that rotate with their blade axis in the plane of rotation with no cone angle, the propeller moment as shown in Figures 2.37 and 5.31 can be determined for each blade by the equation

$$M_{\text{Pr}} = -\int_{R_i}^{R_a} \omega_R^2 a_p^2 \sin(90° - \beta)\cos(90° - \beta)\, dm. \tag{5.11}$$

The geometrical positions of the axis of rotation and the centre of gravity on the blade profile are of particular significance here. As shown in Section 2.1.4, calculations can again be based upon 20 subelements.

Figure 5.32 illustrates the different magnitudes and directions of propeller moments for a turbine of 40 m rotor diameter as a function of the blade pitch angle and speed for three different blade axes of rotation. Very different maximum moment values are found for the above turbine at different blade rotational axis positions, i.e.

$$\text{at } t_B/8, \quad M_{\text{Pr max } t/8} = -1700 \text{ N m},$$

$$\text{at } t_B/4, \quad M_{\text{Pr max } t/4} = -400 \text{ N m},$$

and near the blade centre of gravity

$$\text{at } 3t_B/8, \quad M_{\text{Pr max } 3t/8} = -1.2 \text{ N m}.$$

In this connection, negative torques characterise a twisting of the blades in the direction of increasing blade pitch angle β according to the definition in Figure 2.34.

5.5.1.2 Torsional moments due to lifting forces

According to Figure 2.36, the torsional moment caused by lift, M_{lift}, is proportional to the lifting force and the distance a_a between the blade axis of rotation and the point of action of the lifting force or the active component $a_a \cos \alpha$. Equation (2.22) therefore yields

$$M_{\text{lift}} = \int_{R_i}^{R_a} a_a \cos\alpha \frac{\rho}{2} t_B(r) c_a(\alpha) v_r^2 \, dr. \tag{5.12}$$

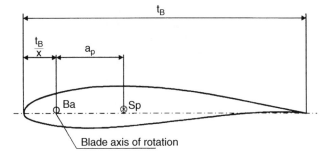

Figure 5.31 Geometrical position of the axis of rotation and the centre of mass on the blade profile (Ba = blade axis, Sp = centre of gravity)

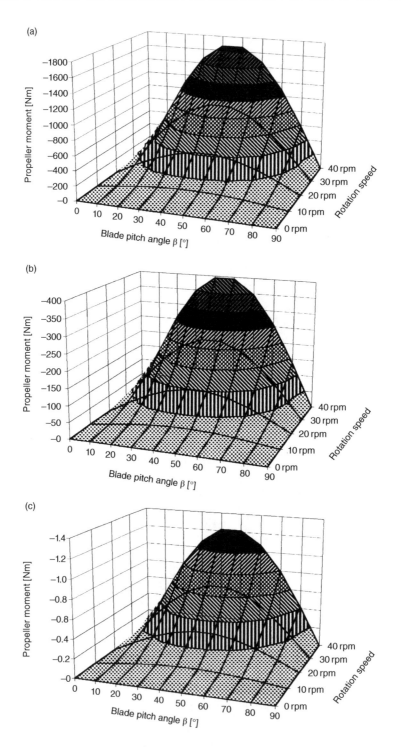

Figure 5.32 Propeller moments as a function of the blade pitch angle and the turbine rotation speed, with the blade axis of rotation at (a) $t_B/8$, (b) $t_B/4$ and (c) $3t_B/8$

According to Section 2.1.4, the calculation can be based upon 20 blade elements in accordance with Equation (2.60). Figure 5.33 shows the torsional moments caused by lift on the profile of a turbine with a 40 m rotor diameter, as a function of wind speed and rotational speed, with the blade pitch angle as a parameter. The different effects caused by the positioning of the axis of rotation in front of or behind the point of action of the lifting force (at $t_B/8$ or $3t_B/8$) are again evident here. Furthermore, the above equation illustrates that, with an axis of rotation at $t_B/4$, no moments caused by lift are present. This state only exists, however, with completely deflection-resistant blades that have a centre-of-pressure resistant profile.

5.5.1.3 Torsional resetting moments

Similarly to Equation (5.12), using Equation (2.61), the torsional moment caused by the resetting effect of the profile in the airstream based on $t_B/4$ can be determined by

$$M_T = \int_{R_i}^{R_a} c_t(\alpha) v_r^2 \frac{\rho}{2} t_B^2 \, dr \qquad (5.13)$$

Here, c_t represents the angle-of-flow-dependent coefficient of the torsional moment for a profile. Figure 5.34 shows the result of a calculation with 20 blade elements for a 40 m turbine. For blade axes of rotation deviating from $t_B/4$, appropriate recalculations of the torsional moments are necessary.

5.5.1.4 Total moments

Figure 5.35 illustrates the strong dependence of the blade pitch adjusting moment on the position of the blade axis of rotation for three characteristic operating states. These were selected close to the starting point of 6 m/s wind speed and a rotation speed of 14 rpm, in the nominal range (12 m/s, 28 rpm) and shortly before the shut-down of the turbine (24 m/s, 28 rpm). It is clear from this that the torque can take on high positive or negative values, or can be largely cancelled out at an axis position of $t_B/4$.

As well as the torsional moments mentioned here, frictional moments caused by the blade bearings, aerodynamic damping components and any moments caused by pull-back springs, which are mainly dependent upon the blade pitch angle, should be taken into account as described in Section 2.3.

5.5.2 Standardizing and linearizing the variables

The variables of torsional moment, wind speed, blade pitch angle, rotation speed, output, etc., have different types of dimensions. It is therefore advisable to standardize all variables, e.g. to nominal or maximum values, thereby obtaining dimensionless units in similar ranges, such as between zero and one.

Due to nonlinear relationships in the transmission, a quantitative analysis for the determination of control parameters is no simple matter. However, close to its operating point continuous families of characteristics can be approximated by the linear standardized elements, i.e. the

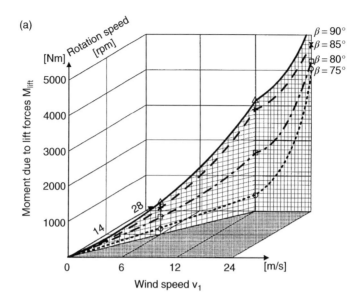

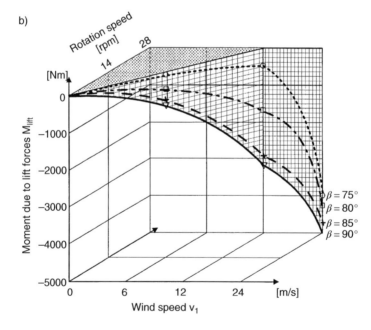

Figure 5.33   Torsional moments caused by lift with the blade axis of rotation at (a) $t_B/8$ and (b) $3t_B/8$

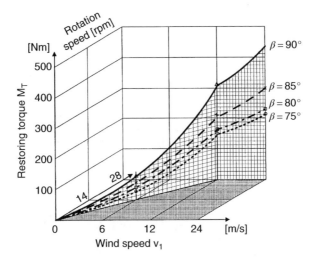

Figure 5.34 Torsional resetting moment as a function of wind speed and rotational speed, with the blade pitch angle as a parameter

start of a Taylor expansion in standard form, according to the equation

$$\Delta\left(\frac{M}{M_{\tau N}}\right) \approx (v_1 - v_{1N})\frac{\partial f}{\partial\left(\frac{v_1}{v_{1N}}\right)}\Bigg|_{v_{1N},\beta_N,n_N} + (\beta - \beta_N)\frac{\partial f}{\partial\left(\frac{\beta}{\beta_N}\right)}\Bigg|_{\beta_N,v_{1N},\,n_N}$$

$$+ (n - n_N)\frac{\partial f}{\partial\left(\frac{n}{n_N}\right)}\Bigg|_{n_N,v_{1N},\beta_N} \tag{5.14}$$

Nonlinear characteristics around the operating or nominal operating point are hereby linearized [5.20]. We can thus obtain the linearized form of the standardized torsional moments due to:

- lifting force

$$\frac{M_{\text{lift}}}{M_{\tau N}} \approx \Delta\frac{M_{\text{lift}}}{M_{\tau N}} = k_{11}\frac{\beta}{\beta_N} + k_{12} + k_{13}\frac{v_1}{v_{1N}} + k_{14}\frac{n}{n_N}, \tag{5.15}$$

- resetting effects

$$\frac{M_T}{M_{\tau N}} \approx \Delta\frac{M_T}{M_{\tau N}} = k_{21}\frac{\beta}{\beta_N} + k_{22} + k_{23}\frac{v_1}{v_{1N}} + k_{24}\frac{n}{n_N}, \tag{5.16}$$

- propeller moments

$$\frac{M_{\text{Pr}}}{M_{\tau N}} \approx \Delta\frac{M_{\text{Pr}}}{M_{\tau N}} = k_{31}\frac{\beta}{\beta_N} + k_{32} + k_{33}\frac{v_1}{v_{1N}} + k_{34}\frac{n}{n_N} \tag{5.17}$$

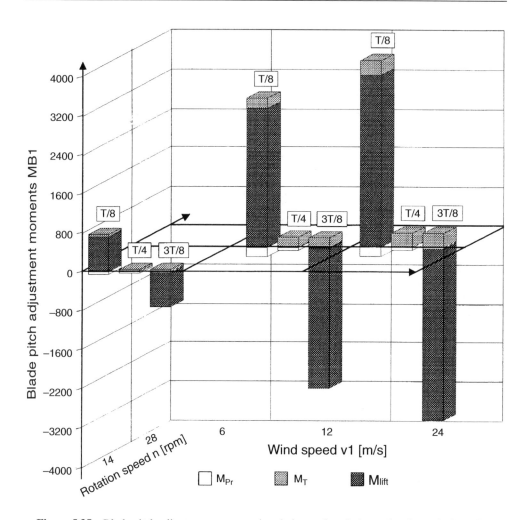

Figure 5.35   Blade pitch adjustment moments in relation to the wind speed and speed of rotation

- and resetting moments by springs (if these are used)

$$\frac{M_F}{M_{\tau N}} \approx \Delta \frac{M_F}{M_{\tau N}} = k_{41}\frac{\beta}{\beta_N} + k_{42} + k_{43}\frac{v_1}{v_{1N}} + k_{44}\frac{n}{n_N}. \tag{5.18}$$

Furthermore, standardizing and linearizing the motion equation (5.10) with the moments acting on the rotor blades yields the following equation:

$$k_b\frac{d^2\left(\frac{\beta}{\beta_N}\right)}{dt^2} + k_d\frac{d\left(\frac{\beta}{\beta_N}\right)}{dt} + k_\beta\frac{\beta}{\beta_N} + k_1 + k_v\frac{v_1}{v_{1N}} + k_n\frac{n}{n_N} = \frac{M_{St}}{M_{\tau N}} \tag{5.19}$$

where the acceleration-dependent component is

$$k_b = \frac{J_{Bl}(v_1)\beta_N}{M_{\tau N}},$$

the damping value, which depends upon the speed of blade pitch angle adjustment, is

$$k_{\mathrm{d}} = \frac{(k_{\mathrm{DB}} + k_{\mathrm{RL}})\beta_{\mathrm{N}}}{M_{\tau\mathrm{N}}},$$

the angle-dependent variable is

$$k_\beta = \sum k_{\mathrm{v1}},$$

the constant component is

$$k_1 = \sum k_{\mathrm{v2}},$$

the wind-speed equivalent is

$$k_v = \sum k_{\mathrm{v3}},$$

and the value proportional to rotation speed is

$$k_{\mathrm{n}} = \sum k_{\mathrm{v4}}.$$

The dynamic behavior of the transmission elements close to the selected steady-state operating point is therefore described by a linear differential equation with constant coefficients. This can be solved in closed form, and general statements can be made, for example, concerning stability [5.20].

Linearization is, however, not meaningful if marked nonlinearities exist. For noncontinuous characteristics, encountered, for example, in switching amplifiers, linearization is completely inappropriate and must be avoided.

The torque matrix of the control system for rotor blade pitch adjustment is found from Equations (5.15) to (5.19):

$$\frac{M_{\mathrm{St}}}{M_{\tau\mathrm{N}}} = \frac{\sum M}{M_{\tau\mathrm{N}}} = \frac{1}{M_{\tau\mathrm{N}}} \begin{bmatrix} M_{\mathrm{b}} \\ M_{\mathrm{d}} \\ M_{\mathrm{lift}} \\ M_{\mathrm{T}} \\ M_{\mathrm{Pr}} \\ M_{\mathrm{F}} \end{bmatrix}$$

$$= \begin{bmatrix} k_{\mathrm{b}} & 0 & 0 & 0 & 0 & 0 \\ 0 & k_{\mathrm{d}} & 0 & 0 & 0 & 0 \\ 0 & 0 & k_{11} & k_{12} & k_{13} & k_{14} \\ 0 & 0 & k_{21} & k_{22} & k_{23} & k_{24} \\ 0 & 0 & k_{31} & k_{32} & k_{33} & k_{34} \\ 0 & 0 & k_{41} & k_{42} & k_{43} & k_{44} \end{bmatrix} \begin{bmatrix} \ddot{\beta}/\beta_{\mathrm{N}} \\ \dot{\beta}/\beta_{\mathrm{N}} \\ \beta/\beta_{\mathrm{N}} \\ 1 \\ v_1/v_{1\mathrm{N}} \\ n/n_{\mathrm{N}} \end{bmatrix} \qquad (5.20)$$

Depending upon the rotor configuration, the coefficients of the matrix take on characteristic values, which can be zero. For example, the lift-dependent components disappear in the case of a blade rotational axis position of $T/4$. Then, for nonbending blades,

$$k_{11} = k_{12} = k_{13} = k_{14} = 0.$$

Moments brought about by springs are usually independent of the prevailing wind speeds and rotational speed values. Therefore, in addition,

$$k_{43} = k_{44} = 0.$$

Moreover, in systems without return springs

$$k_{41} = k_{42} = 0.$$

If the blade axis of rotation is selected at the centre of mass of the section then some of the coefficients vanish, i.e.

$$k_{31} = k_{32} = k_{33} = k_{34} = 0.$$

Furthermore, torsional moments caused by blade bending can be considered in the components through lift M_{lift} with coefficients k_{11} to k_{14} and propeller effects M_{Pr} with coefficients k_{31} to k_{34} as through an increase in the moment of inertia in k_a.

5.5.3 Control circuits and simplified dimensioning

The linear transmission element for rotor blade pitch adjustment can be described using the second-order differential equation according to Equation (5.19) or in the form

$$\frac{k_b}{k_\beta} \frac{d^2 \left(\frac{\beta}{\beta_N} \right)}{dt^2} + \frac{k_d}{k_\beta} \frac{d \left(\frac{\beta}{\beta_N} \right)}{dt} + \frac{\beta}{\beta_N} = \frac{1}{k_\beta} \left(\frac{M_{\text{St}}}{M_{\tau N}} - \frac{M_\tau}{M_{\tau N}} \right) \tag{5.21}$$

where $M_\tau/M_{\tau N} = k_1 + k_v v_1/v_{1N} + k_n n/n_N$. Using the Laplace transform, this relationship can also be represented as a complex frequency response, by the transmission function of the so-called control system, where $s = \delta + j\omega$, or, for the special case $s = p = j\omega$,

$$F_S(p) = \frac{\frac{\beta}{\beta_N}}{\frac{M_{\text{St}} - M_\tau}{M_{\tau N}}} = \frac{\frac{1}{k_\beta}}{T_1 T_2 p^2 + (T_1 + T_2) p + 1} = \frac{V_\beta}{(T_1 p + 1)(T_2 p + 1)} \tag{5.22}$$

i.e. it is possible to substitute

$$\frac{k_b}{k_\beta} = T_1 T_2, \qquad \frac{k_d}{k_\beta} = T_1 + T_2 \qquad \text{and} \qquad \frac{1}{k_\beta} = V_\beta$$

Figure 5.36 shows the block diagram for the blade pitch control system of a wind turbine. Here the Laplace transforms characterize the following quantities:

β_s/β_N the command variable,
$\beta/\beta_N = \beta_i/\beta_N$ the regulating variable or its actual value,
$M_\tau/M_{\tau N}$ the disturbance variable of the control circuit.

When designing the system, the question of whether the actuator is centrally controlled and acts upon all blades simultaneously or each blade is separately adjusted must be taken into account.

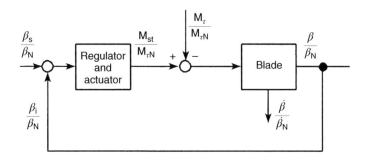

Figure 5.36 Control system with the control circuit for blade pitch adjustment

The transfer function of the open control circuit F_K is found to be the product of the transfer function of the controller F_R and the controlled section F_S:

$$F_K(p) = F_R(p)F_S(p). \tag{5.23}$$

By superimposing the two stimuli of command and disturbance variables, the regulation variable is found to be

$$\frac{\beta}{\beta_N}(p) = \frac{F_K}{1 + F_K}\frac{\beta_S}{\beta_N}(p) + \frac{F_S}{1 + F_K}\frac{M_\tau}{M_{\tau N}}(p). \tag{5.24}$$

This represents the sum of two products: the product of the so-called command transfer function $F_g = F_K/(1 + F_K)$ and the command variable and the product of the disturbance transfer function $F_{gz} = F_g/F_R$ and the disturbance variable of the control circuit, or

$$\frac{\beta}{\beta_N}(p) = F_g(p)\frac{\beta_S}{\beta_N}(p) + F_{gz}(p)\frac{M_\tau}{M_{\tau N}}(p) \tag{5.25}$$

The purpose of the control system is to bring the regulated variable into line with the command variable over as wide a frequency range as possible, and to suppress the influence of the disturbance variable [5.20]. This means that the following should hold:

$$|F_g| \approx 1 \tag{5.26}$$

and

$$|F_{gz}| \approx 0. \tag{5.27}$$

Furthermore, if damping is sufficient, there must be stable control-circuit behavior and adequate control speed. Therefore, controller characteristics are carefully selected in an attempt to compensate completely or partially for large delays in control sections, which are particularly detrimental to the control-circuit dynamics.

The conditions according to Equations (5.26) and (5.27) are fulfilled if $|F_K| \gg 1$ in the fundamental frequency range. This condition can be fulfilled by high amplification in the circuit or controller or with integrating controllers, the transfer function of which has one pole at zero frequency, so that $F_R(0) \rightarrow \infty$ if there are no differentiating sections and $F_S(0) \neq 0$.

Thus the dynamic characteristics of second-order control sections as shown in Figure 5.35 can be significantly improved by 'compensating' for large delay components with the aid of the derivative action of a controller (PI (proportional-integral), PID (proportional-integral-derivative)). This means that, in the case of dimensioning according to the so-called optimum amount system, the delay with the greater time constant T_1 is eliminated. The resulting transmission link would then only exhibit a parasitic delay of around $T'_v \approx 0.1 T_1$ and the remaining smaller time constant T_2.

The 'system time constants' T_1 and T_2 can, however, vary greatly – due in particular to blade deformation and the increased moments of inertia that this causes – compared with nonde-formed blades (with T_0), e.g. $T_1 T_2 = 1 - 5(T_0)^2$. It should also be considered that measuring the blade angle is very expensive and is avoided in small systems in particular for reasons of cost. Blade pitch speeds or their equivalents such as the speed of actuators, the mass flow of hydraulic mechanisms and so on, on the other hand, are relatively simple to sense. It is there-fore completely feasible to install blade pitch speed or even blade pitch acceleration control circuits in order to utilize fully the dynamic characteristics of this control system for speed or output control.

5.5.3.1 Blade pitch speed control circuit

The output or rotation speed of a turbine can be influenced, within the range of the available flow provided by the wind, by adjusting the blade pitch. Desired power or rotational speed gradients, on the other hand, can be maintained by appropriate blade pitch speeds. Figure 5.37 shows the simple structure for controlling the blade pitch speed of a wind turbine. The control circuit consists of the controller and the 'actuator' and 'blade' system sections. Therefore the open-loop transfer function is

$$F_{K\beta} = F_{R\beta} F_{s\,St} F_{S\,Bl}. \tag{5.28}$$

For a system controlling purely output and rotational speed respectively it is sufficient to use the dynamic characteristics of the blade pitch speed control circuit without knowing the precise blade pitch and its speed and accurately adjusting these values. In the simplest case this can be achieved using a P (proportional) controller, and the control processes can be designed using a converter-fed motor. The amplification of the actuating system and the delay time for the torque generation of the actuator motor (approximately 10 to 50 ms) should be taken into account here. The integration time constant of the blade is

$$T_{Bl} = J\dot{\beta}_N / M_{rN} \tag{5.29}$$

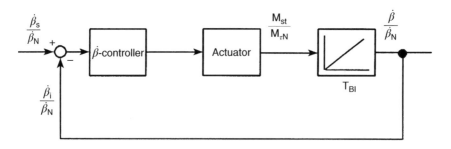

Figure 5.37 Structure of a blade pitch speed control circuit

Therefore, by the selection of the controller circuit damping (e.g. $D_\beta = 1, 1/\sqrt{2}, \dots$), the controller amplification can be determined, or, at a given amplification, the damping behavior and therefore the stability of the circuit ascertained. As described in Section 5.4, the blade pitch control circuit can also be dispensed with. In order to avoid control deviations the $\dot{\beta}$ controller should take over the integral effect. On the other hand (as is often the case) if output or speed tolerances are permissible then the immunity ranges can be adjusted by the selection of controller amplification. In order to maintain the predetermined turbine output and speed gradients can also be maintained on the drive side, for example, as well as controlling the speed the acceleration of the blade pitch should also be controlled using a cascade control circuit. However, this control system will not be described in further detail in what follows. A further option for the control of output and speed gradients is to control these variables on the frequency converter side via the drive train. The higher-level control circuit is described briefly in what follows.

5.5.3.2 Blade pitch control circuit

The so-called pitch speed integrator, which is characterized by the time constant $T_\beta = \beta_N/\dot{\beta}_N$, is connected after the closed blade pitch speed control circuit as an additional controlled system. A blade pitch controller is added in front of the control system to create a higher-level cascade as shown in Figure 5.38. This should exhibit integrating behavior and derivation action (e.g. PI, PID controller) in order to maintain the output or speed of the turbine as precisely as possible and to adjust these quickly. The transfer factor of the open circuit is

$$F_{K\beta} = F_{R\beta}F_{g\beta}F_{S\beta}. \tag{5.30}$$

A controller is connected in series with the $\dot{\beta}$ control loop described above, which can, for example, be represented as a second-order delay element. The controlled system therefore represents a delayed integrator. However, there is a delay element of the second order or higher. The integral term of the blade pitch controller, which is necessary for precise control of disturbance variables, leads, in connection with the system integrator, to a double pole at $p = 0$ in the transfer function of the circuit. The derivation action of the controller is necessary here, in order to be able to stabilize the system at all.

In order to dimension the control circuit in a simple manner, the blade pitch speed circuit with its higher-order delay can be approximated by a first-order substitution function. A common

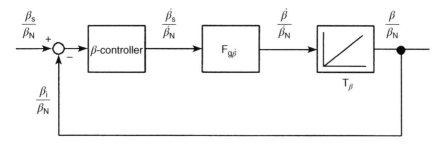

Figure 5.38 Structure of a blade pitch control circuit

method of approximating the measured or calculated step response of a higher-order controlled system, for example, by means of a first-order delay element is the creation of a substitution function. The relationship

$$f_{sub}(t) = V_{sub}(1 - e^{-t/T_{sub}})$$
(5.31)

is approximated by the step response for the same final value and equivalent control area. If the time constants of the system components differ greatly, e.g. between the actuator and the blade, with values of 20 ms compared with 500 ms, there is the option of replacing the blade pitch speed control circuit by the partial first-order function with the greatest time constant as an approximation. Stabilization of the system is only possible if the time constants in the integral section of the controller are greater than the delay time of the controlled system [5.20]. For the control circuit, a PI controller can be simply dimensioned according to the symmetrical optimum system. Strong overshooting of the step response in the control circuit, which is to be expected if the control variable is altered, can be eliminated by the connection of a desired value delay.

The circuits for the control of the turbine speed and output can be dimensioned in a similar manner. These will, however, not be described further here. However, the designs shown here – for which linearizations have been carried out and simplifying assumptions made – can, when dimensioned for safe damping behavior, lead to critical operating states in certain operating ranges due to the sluggish control system. On the other hand, critical designs would lead to the risk of instabilities. In order to safely control all the operating ranges of the turbine, the control system therefore requires refinement.

5.5.4   Improving the control characteristics

Control procedures that can react to different operating states are designed to protect components, are orientated towards effectiveness criteria and can automatically perform adjustments in response to altered system and environmental conditions as well as improve the control characteristics and therefore increase the operating reliability and service life of wind turbines.

Figure 5.39 shows the structure for the control of turbine rotational speed. This includes the speed controller, the closed blade pitch adjustment circuit, the turbine driving torque generation and the inertia of the rotating masses of the drive train. Their integration and running-up time constant T_R can be determined by Equation (2.93). With regard to controller design, particular attention must be paid to the highly nonlinear characteristics of control sections. This

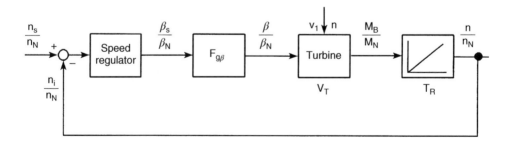

Figure 5.39   Structure of the rotation speed control circuit of the turbine

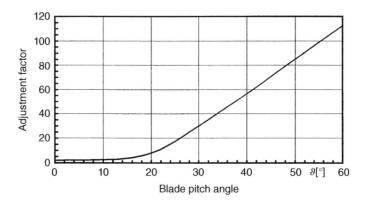

Figure 5.40 Adjustment factor for the regulation of turbine output as a function of blade pitch angle

particularly affects the generation of driving torque and driving power. The nonlinear $c_p-\lambda$ family of characteristics (see Figure 2.6 or 2.8) causes a highly variable amplification of this system. This is the result of output changes in relation to blade pitch adjustment. Furthermore, it is necessary to differentiate here between idling and nominal operation of the turbine. The range of values covers around two powers of 10 (e.g. amplification factors of approximately 1 to 100), with low values being encountered in nominal operation.

If fixed regulation parameters were to be used, dimensioning would have to be based on the highest amplification – i.e. on the highest permissible wind speed – in order to ensure stable regulation behavior of the wind turbine in all operating ranges. This would mean that only very inadequate regulation characteristics could be achieved between idling and nominal operation. In particular, in the operating range in which the turbine is operated for most of the time, unfavourable types of behavior and the resulting negative effects would be unavoidable.

Good regulation behavior can be achieved in all operating ranges if the regulation parameters are adapted to the current operating state. This can be achieved in different ways. The simplest option for achieving good results is a fixed characteristic of the adaptation factor, e.g. in relation to the blade pitch adjustment angle (Figure 5.40).

5.5.4.1 Fuzzy controllers

The application of fuzzy logic [5.24] and the use of fuzzy controllers offers an innovative method for the control and management of wind turbines [5.25]. Unlike digital systems, which can only differentiate between and evaluate 'true' and 'false' values or equivalent contrasts, this technology, being based on imprecise technology, can also take into account intermediate levels. Such systems are therefore also attributed human characteristics. So-called membership functions represent the connection between linguistic statements and numerical values, and basic predictions can be linked to complex values by logical connections. Therefore linguistic statements on the input side can lead to a 'control engineering' operating instruction on the output side. The sum of all linguistic rules therefore represents the basis of the controller. In the design of a fuzzy controller the controller structure forms the framework within which the knowledge base can be defined in substages.

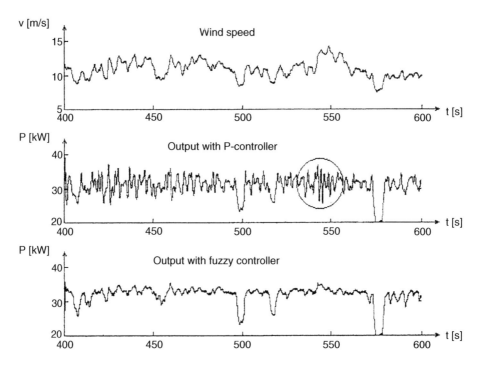

Figure 5.41   Output behavior of a wind turbine with a P controller and a fuzzy controller as a function of the associated wind speed (simulation results)

Linguistic variables and terms must be selected such that the complexity of the controller is kept low and the selected formulations must be expressively structured. Furthermore, the membership functions of the input variables must be defined and the controller base made up in matrix form. Performance characteristic controllers are the result of the design process.

The simulation results shown in Figure 5.41 for one such system [5.25] illustrate the different output characteristics of turbines with P controllers compared to those with fuzzy controllers. The comparison shows markedly smoother reactions for fuzzy controllers. Even at more rapid wind speed changes the fuzzy controller variant does not tend to oscillate. Similar results can, however, also be achieved with conventional control devices, e.g. with controllers that also have integral parameters and are designed for smooth system behavior.

5.5.4.2   Self-tuning control systems

Surface effects on the rotor blades, e.g. icing, roughening, dirt and moisture, should also be taken into account, as well as physical changes to the operating-state-dependent section amplification of wind turbines. Designing a controller set-up that takes all operating ranges and states of the turbine into account to the same high level is therefore no simple matter. Good controller behavior can, however, be achieved at all turbine states if the linearized transfer behavior of the turbine is identified at different operating points. Therefore, with the aid of the system parameters that have been acquired, control systems can be built for the controller that are dimensioned for the controller and are capable of learning.

The creation of a self-tuning turbine controller can, as described in reference [5.26], be based upon the identification of the transmission link between the wind speed and electrical power output. This process has the disadvantage that a prediction of the wind speed is required and the dynamics of the turbine must be known. In fact, it is almost impossible to fulfil both conditions in practical operation.

The process described in reference [5.11], on the other hand, requires only the measured output power and the desired value for the blade pitch adjustment speed. If computers are used for the control and supervision of turbines, these variables are already available within the system. No additional measuring devices are required for their identification. This method takes as its basis the general discrete-time process representation based upon the controlled and manipulated variables, a white noise signal, and the wanted and interference signals of the controlled variable. Identification in a closed control circuit takes place with an initially approximately adjusted controller, so that even extreme gusts cannot damage the turbine. System natural frequencies of interest are stimulated by a test signal at the controller output and identified with the aid of correlation analysis.

The estimated impulse response of the model exhibits a high level of scattering due to wind-speed fluctuations. The values are smoothed by regression. This process produces the coefficients of the discrete impulse transfer function. For this, the scanning time and the test signal must be determined in a suitable manner, since these parameters influence the effectiveness of the identification. After the identification of the impulse transfer function of the controlled system model, a model of the control circuit can be simulated on the computer in parallel to the actual control system. Thus predetermined responses by the controlled system model output variables to interference can be achieved by controller optimization.

By identifying the controlled system model transfer function at discrete operating points, optimal controller parameters can be determined and expanded to a control system with managed adaptation, i.e. to an adjustment of parameters. Between the detected values, the controller parameters can be found by interpolation at any operating point. However, particularly during turbulent wind conditions, the adaptation of controller parameters to each sampling step must be avoided. Good operating results can be achieved using quasi-continuous controller adjustment by averaging the values over a time period, e.g. in the seconds range [5.27].

5.5.4.3 System-oriented controller design

In addition to the classical methods for controller design given in references [5.1] and [5.28] to [5.33], etc., which are particularly oriented towards the function and stability of the control system, procedures can be employed that use control mechanisms to maintain maximum and minimum values of component or system parameters. Thus, for example, the use of 'linear–quadratic' (LQ) optimization [5.34] aims to reduce periodically induced blade noises, which are particularly associated with influences due to tower shadowing and the wind height profile. Furthermore, significant inherent values are taken into account in reference [5.35], while the use of Kalman filters to estimate the turbine state is considered in references [5.36] and [5.37], and references [5.38] and [5.39] take account of multivariable controllers. The use of the linear–quadratic output-feedback method (LQOFB) in reference [5.40] was aimed at reducing the torques in the rotor shaft of a fixed-speed turbine. In references [5.41] to [5.43] blade loading is reduced in variable-speed turbines by the use of linear–quadratic methods.

In a numerical design process for the optimization of quality vectors developed by Kreißelmeier and Steinhauser [5.44] the controllers are optimized on the basis of simulation results. This requires no special (e.g. linear) controller structure. Design criteria can be defined as turbine- or operation-specific. For this, control aims must be quantified in the form of quality criteria and summarized as quality vectors. These control aims may include the limiting of electrical output and rotor speed at full load, the maximization of electrical output at part load, the reduction of output fluctuations, the minimization of actuating processes and the reduction of load alternations in the drive train and rotor, at individual blades or at the tower [5.45–5.49].

Therefore particular requirements are imposed on the control system that can lead to conflicting goals. Moreover, since the limits of realization work are generally not known at the design stage, goals can only be defined within limits. Besides, the different results on the controller adjustment, which arise through many essential optimizations, demand many objective selection criteria.

A control structure must be determined before the use of the procedure. A quality criterion is linked to each required system characteristic, with lower values indicating better fulfilment of the criterion. The standard deviation, the control area and the displacement of a dominant pole of the transfer function from the desired pole position are examples of variables that can be used for evaluation.

The individual quality criteria and the resulting quality vector are determined by simulation and compared with the desired values or the corresponding desired vector. The aim of the optimization of controller parameters is to minimize all components of the quality vector. However, a performance function defined for this purpose exhibits step changes at points where the criteria change. For practical applications, therefore, if there are 10 to 20 criteria it is advisable to minimize the natural logarithm of the sum of all exponential functions of quotients of quality and preset values.

From general principles in references [5.50] and [5.51] and in representations, for example, according to references [5.52] and [5.53], so-called 'state monitors' can be introduced for the recovery of unacceptable state variables. Thus the 'estimation' of complete state vectors can take place, from which the best possible approximation can be expected.

An estimator is used in references [5.47] to [5.49] to take into account the mechanical loading variables in the control system. In reference [5.49] this derives, with the aid of a simulation model for the wind turbine, the required estimated values for the aerodynamic loading of the turbine and the effective wind speed in the rotor circuit from the electrical output, generator speed and blade pitch, these variables being available by measurement. In this manner, significant mechanical loading variables in the turbine, drive train and tower can be combined with the available measured values, to reconstruct and be included in the control process.

In reference [5.49] control is performed by two independent subcontrollers. A drive-train controller influences the generator torque and a turbine controller acts exclusively upon the blade pitch adjustment system. Both subcontrollers can be active. In this manner, the loading of the turbine and drive train can be kept within preset values independently of the operating state of the generator by limiting wind turbine output. Insignificant energy sacrifices thus lead to significantly reduced peak values of the rotor blade shock moment and tower deflection moments. Furthermore, with these methods it is possible to achieve better types of behavior during the transitions between the part-load and nominal-load range than is possible when conventional controllers are used.

A one-year test run of the control process on an experimental system belonging to the Institut für Solare Energieversorgungstechnik (ISET, Institute for Solar Energy Supply Technology) in Kassel demonstrated an enormous alleviation of the load on the rotor. A comparison of the measured load collective between conventional and modified controllers at fixed-speed operation indicated a statistical extension of the service life of rotor blades on the order of magnitude of 30%. For variable-speed turbines, the expected service life of rotor blades can be increased by around 60% and the service life of the rotor shaft can be increased by around 40% [5.54].

5.5.4.4   Neural networks

Innovative control procedures for wind turbines, e.g. with self-adjusting, component-oriented or system-oriented load-limiting controllers, usually require the processing of large amounts of measured data and consideration of as much additional information as possible. However, favourable modes of behavior for components and systems remain largely limited to the selected design goals. A cumulation of as many advantages as possible is only attainable to a limited degree with a technology-orientated procedure.

Biological systems cope with the processing and compression of very different types and almost unlimited amounts of information by evolution. The human skin can be cited as an example, with approximately 500 000 touch receptors. It can be viewed as a comprehensive interface with our environment. The numerous very different sensors initiate influence-specific body reactions.

Neuroanatomical investigations in references [5.55] and [5.56] yielded topological relationships between neighbouring receptors and cerebral cortex areas. Spatial information was therefore retained in both sensing areas. It was also found that the size of the cerebral cortex areas was related to the density of the receptors and not the spatial extent of the skin zone, indicating that the sensing signal areas are represented according to their importance.

Neural connections, which are genetically influenced only in their basic structure, are largely created by self-organization processes between the cells [5.57]. Their activity depends upon a variety of environmental stimuli.

Biologically accurate models for self-organizing processes in systems involving topology have been developed for the simulation of neural processes. A generally valid principle in accordance with references [5.58] to [5.61] is based upon a two-dimensional neuron layer in the form of a grid. Equidistant nodes (so-called 'neurons') symbolize the nerve cells. Excitations are transmitted by the comparison between defined pattern vectors and the input signal vector. The excitation response therefore decreases as the distance from the excitation centre increases. Limited learning steps ensure that the 'map' containing topology achieves quasi-stationary states, and slow changes in the input signal area can take place. Ritter and Schulten carried out investigations into the dynamics of the map generation processes and highlighted various possible applications [5.62–5.69].

The use of neural networks is particularly beneficial in systems where functional relationships cannot easily be sensed analytically or using measuring technology. Their application for storing characteristic lines for sensing the state of lead storage batteries is a classic example

[5.70]. With a network of 40×40 neurons, the state of discharge of batteries in a typical application in hybrid systems can be illustrated with sufficient precision.

Neural networks can be characterized by the following valuable aspects: they can learn relationships between inputs and outputs (e.g. the $c_p - \lambda$ family of characteristics of a wind turbine) without knowing the physics of the processes involved or the associated mathematical equations. Furthermore, they are capable of generalizing characteristics and thus reacting in an appropriate manner to unpredicted, i.e. nonlearned, events by interpolation or extrapolation. Moreover, if the task or problem formulation is altered, only the pattern files need be updated. The developed programs can, unlike in other procedures, be retained.

Investigations into the use of neural networks for the learning of power or power coefficient characteristic diagrams [5.71] have delivered satisfactory results in terms of precision, if a suitable network structure is selected. Owing to the excellent interpolation characteristics, even with a coarse interpolation-point density the characteristic diagrams can easily be determined. Furthermore, these networks have excellent generalization capabilities. For wind turbines, therefore, correspondingly good operating characteristics can be expected from a system-orientated controller. Progressive adaptation in the future for taking into account external and internal parameter changes could lead to the importance of neural networks increasing greatly.

5.5.5 *Control design for wind turbines*

Starting from the description of the problem, reference [5.72] provides an overview of applied and usable processes, e.g. use of multi-variable control and handling of periodic error signals for control design. A preview of new developments (floating wind turbine plants, rotor blades with distributed actuators, etc.) rounds off the comprehensive subject.

The pitch angle of the rotor blades is kept as near as possible constant at $0°$ for **control of the part-load region**. In the simplest case, a rotation torque or rotation power curve (e.g. Figure 3.21 or Figure 2.64) is implemented in place of the variable-speed generator system so that the optimum speed of rotation is set for all stationary operating regions [5.73]. In order to counter uncertainties and the concomitant energy losses, improvements can be made with adaptive processes [5.74]. In contrast, only limited power increases can be achieved by means of rotor acceleration-rated value prescriptions or estimates of the rotor-effective wind velocity [5.75, 5.76] with dynamic rotation tracking or so so-called 'Disturbance Tracking Control' [5.77].

In the **full-load region,** the turbine power must be limited to the nominal value by means of blade setting. Besides effective controls (PI-, PID controllers [5.73]), use is made of filters in order, for instance, to prevent periodic excitation from the rotor rotation. The design is oriented, on the one hand, to maintain a rotation tolerance band for wind velocity changes and, on the other hand, the extreme loads for pitch system rotor blades and tower must be limited.

Disturbance size estimated and disturbance size switching are meant to increase the gusting resistance of wind turbines. Estimates for rotor-effective wind velocities are described in reference [5.78] or acquired on the basis of blade-root bending torques [5.79].

In addition, in reference [5.80], the advantages of non-linear adaptive controllers compared to classic PID controllers for gusting with large changes of wind velocity are presented.

The **use of LIDAR technologies** (**li**ght **d**etection **a**nd **r**anging) offers the possibility of measuring wind fields far in front of the turbine at some hundreds of meters and to use this for disturbance switching [5.81] in order to obtain better results than with estimates [5.82]. Here, both fatigue and extreme load reductions as well as limitations in the control behavior [5.83] must be considered.

The **adaptive feed-forward control for wind turbine plants** [5.84] uses LIDAR wind measurements in a feed forward control using filters in order to reduce variations in rotation or the loading of the wind turbine plants.

The **integration of load and vibration signals** for extreme and fatigue load reduction leads to a multi-variable control that also considers these signals besides the speed of revolution. Load-reducing controls developed at the Fraunhofer IWES for active tower damping and for tilt and yaw moment compensation with the aid of individual adjustment was successfully tested in the field on a 5-MW plant [5.85].

New developments for controlling wind turbines are also suitable for **floating offshore plants.** These can also be used in countries with steep coastlines. The additional excitation of the plants due to the force of waves and the degrees of freedom of floating platforms leads to complex load situations. Compared to fixed foundations on land, for instance, the eigen frequencies of the tower (approximately 0.3 Hz) for floating plants is about a tenth lower (approximately 0.03 Hz) in order to lie outside the excitation wave spectrum. Besides the increase in the variations of the revolutions [5.86]; in reference [5.87], control systems with the generator torque as actuator size for rotation speed are discussed.

Error diagnostic and error-tolerant controls are a further emphasis for research. Already in the middle of the 1990s, use was made of signal-based error diagnostic-processes in wind energy [5.88]. Condition monitoring with CMS (condition monitoring systems) and SHM (structural health monitoring) has been used commercially for a long time [5.89, 5.90] but up to now no mature products are available especially for wind turbine plants [5.91, 5.92]. Also, error-tolerant processes for controlling wind turbine plants are not yet mature [5.92–5.94].

The aerodynamic properties of the rotor blade profile and thus the aerodynamic forces can be locally influenced by means of **local actuators**. In principle, several distributed actuators allow the forces to influence the rotor over the radius. Local actuator elements have the advantage of requiring only small changes to the blade geometry. In this way, only small masses need be moved. Thus, fast control circuits that can also influence high-frequency loads in the rotor blades are possible. However, more care must be taken due to the high adjustment dynamic of instationary effects of the aerodynamics that are stronger than for conventional control processes. Simulation calculations show approximately the halving of damage-equivalent of impact bending moments in rotor blades [5.84, 5.95]. Thus, the rotor surface and the energy yield can be substantially increased [5.96]. Overviews of the state of the development of adaptive rotor blades for helicopters and wind turbines and many actuator principles are discussed in references [5.97] and [5.98]. A promising method of local influencing of aerodynamic forces are flaps distributed along the length of the blade [5.99] that can actively influence the local drive force. Feedbacks from the flap movements and control dimensioning are discussed in [5.95, 5.100] as well as [5.101].

5.6 Management System

The management system must ensure the reliable and automatic operation of wind turbines. To achieve this, the relevant components and system variables must be monitored continuously. By maintaining permissible values and value ranges for system variables, the management system can bring about predetermined operating states and recognition in emergency or fault situation.

To achieve this, the turbine management system must influence the operating behavior of the wind turbine based on the preset control signals and desired values, and react to changes in system variables or to malfunctions. Along with reliable operation, another goal is to achieve the optimal compromise between the output and low mechanical and electrical loading of the turbine and its components [5.1, 5.29, 5.30].

Figure 5.42 shows the structure of the management system for a variable-speed wind turbine with a frequency converter supply (see Figure 5.19) and gives an overview of the most important operating states and changeovers, which will be outlined in what follows as an example based upon the main procedures.

5.6.1   Operating states

The turbine normally runs in automatic mode. However, manual and semi-automatic operating modes with the manual input of desired values are necessary during commissioning and maintenance.

Transient operating states may last for a limited period only. Their duration is therefore monitored. After the predetermined maximum periods have been exceeded a fault shut-down is initiated, since it must be assumed that there is a malfunction.

The duration of steady-state operating states is not monitored by the management system. The turbine remains in these states as long as all normal operating conditions are fulfilled.

In all operating states the conditions for normal operation must be continuously interrogated. Only one condition is required for the transition to the immobilization, shut-down, fault shut-down or emergency shut-down operating states. In contrast, all conditions must be fulfilled for the initiation of the start-up or running-up operating states.

5.6.1.1   Turbine testing (transient)

After commissioning the management system, the monitored components, influencing variables and command variables must be checked and recorded. Figure 5.43 shows the turbine testing structure and the associated Table 5.1 lists the most important messages. The outputs of all subsystems must be interrogated for standstill values and all mechanical actuators driven for test purposes. The correct reactions of configurations can be checked by sensors. If faults occur, these must be recorded. Faults lead to the suspension of further operation until the faults have been rectified and the turbine manually released.

All turbine components and their limit values must be checked in all operating states. This system check-tests whether all systems are functioning properly, whether temperatures are within the operating range and whether the message 'system OK' is universally present. After successful testing, the turbine goes over into the next operating states; otherwise the testing of the turbine operating state is repeated until all release conditions are fulfilled, such as operator commands, unlocking after emergency shut-down, grid available and OK, component functionality, temperatures and limit values.

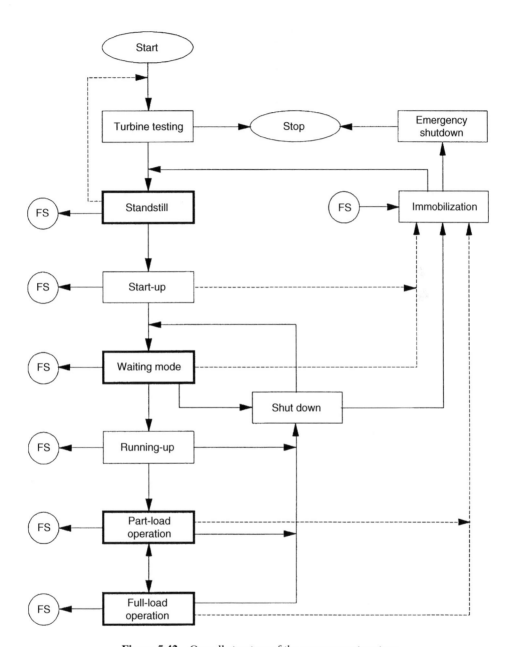

Figure 5.42 Overall structure of the management system

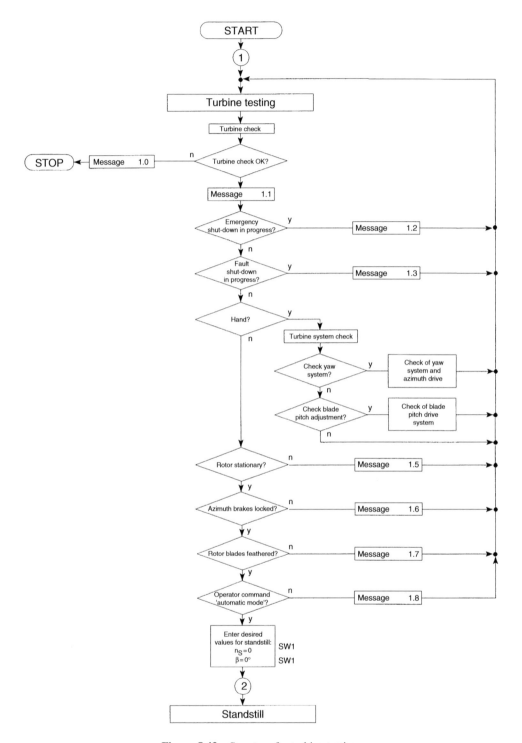

Figure 5.43   Structure for turbine testing

Table 5.1   Messages during turbine testing

Output	Meaning
1.0	STOP: turbine check negative
1.1	Turbine check positive
1.2	Emergency shut-down in progress
1.3	Fault shut-down in progress
1.4	STOP: maximum azimuth angle exceeded
1.5	STOP: rotor not stationary
1.6	STOP: azimuth brakes not locked
1.7	Rotor blades not in the feathered position
1.8	No operator command 'automatic mode'

5.6.1.2   Standstill (steady-state)

The standstill turbine state is characterized by the stationary rotor. Moreover, in this operating state the rotor brakes are activated, the rotor blades are in their feathered position and the nacelle of the wind turbine is yawed out of the wind. If cable twist in the tower must be rectified, this can be carried out at low wind speeds. The electric generator is switched off and disconnected from the supply grid. First of all, those conditions that have prevented the turbine from going over into the start-up operating state are checked. Then a system check is carried out. If all conditions are now fulfilled the start-up conditions are interrogated. If these are also fulfilled, the turbine goes over into the start-up operating state. As in the turbine testing operating state, the messages 'supply grid disconnected', 'cable twist', etc., are identified by the appropriate operating state number.

5.6.1.3   Start-up (transient)

When starting the turbine at no load, i.e. with the rotor brakes released, the turbine is driven, by the wind alone with no power being drawn via the frequency converter, from standstill to the speed that has been predetermined by the control system. At this point, the rotor blades are driven by the blade pitch regulation mechanism from the feathered position to a defined angle. In a repeating sequence, the conditions for fault shut-down and immobilization are then checked and the appropriate routines initiated if required. Speed is also checked during this sequence. As soon as the minimum waiting speed is achieved, the turbine goes over into the steady-state waiting mode. During the start-up operating state the nacelle is again yawed out of the wind.

5.6.1.4   Waiting mode (steady-state)

In the waiting mode all the components of the wind turbine are ready for operation. The rotor speed lies within a range determined by the management system and is influenced by the blade pitch control system. The generator system is not yet connected to the supply system. The fault shut-down, immobilization and running-up conditions are checked one after the other.

If the appropriate conditions are fulfilled the relevant operating states are initiated. The speed is maintained within a defined permissible range by the adjustment of the rotor blade pitch according to desired values. If the waiting mode is maintained for a long period then the operator is notified and, after a certain period, e.g. one day, a further turbine test is carried out. Moreover, in this operating state the conditions to be fulfilled are continuously checked and the nacelle is yawed in the direction of the wind.

5.6.1.5 Running-up (transient)

If the wind speed is high enough, the rotor speed of the wind turbine can be run up to a value at which it is possible to connect the generator system to the grid (see Figure 5.44). The frequency converter is first checked for its readiness for power so that the grid protection system can be connected. Then, with the aid of the blade pitch adjustment system, the rotor speed is adjusted to the speed determined by the management system. During running-up, the fault shut-down and immobilization conditions are continuously checked and the nacelle yawed according to the wind direction.

When the required desired speed is attained, the generator and frequency converter system is connected to the supply grid and electrical power can be supplied. The turbine is now in part-load operation. The necessary messages and limit values during this process are listed in Table 5.2.

5.6.1.6 Part-load operation (steady-state)

In part-load operation (see Figure 5.45) the generator system supplies electrical energy into the supply grid. The blade pitch is set or adjusted to an optimal value, so that the maximum power output or minimum component loads are possible. The management system sets a value for output power in relation to speed (see Figure 2.64(c)). In part-load operation, the generator system's frequency converter regulates the speed and power output. In Table 5.3, no changes are made to the desired values within the adjustment range.

When the control reserve value is reached, the desired value for speed is altered according to the power–speed characteristic line. The nacelle is also continuously yawed into the wind. The blade pitch control system functions as part of the safety system, braking the rotor in the event of an emergency. Given a high enough wind speed, the turbine automatically goes over into the steady-state full-load operating state. Again, all conditions for normal operation are checked in part-load operation and, if necessary, the appropriate procedures are initiated.

5.6.1.7 Full-load operation (steady-state)

If the wind speed is high enough, the turbine will go over from part-load operation to full-load operation (see Figure 5.46). In this operating state the management system sets desired values for nominal speed, its fluctuation range and the nominal output of the system. Speed and power output are regulated by blade pitch adjustment.

In full-load operation the frequency converter can maintain both the power output and the generator moment at a constant level or change them in relation to a specified function. Output fluctuations at the turbine therefore give rise to slight speed changes. The speed is maintained

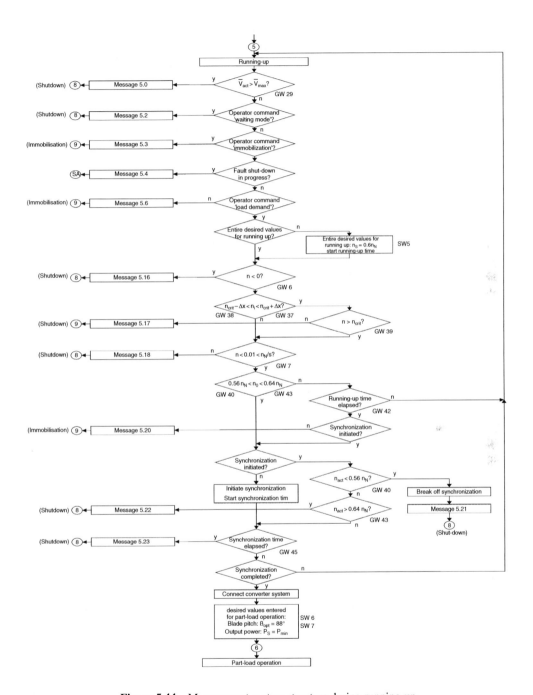

Figure 5.44   Management system structure during running-up

Table 5.2   Messages during running-up

Output	Meaning
5.0	RUNNING UP
5.1	SHUT-DOWN: average wind speed too high $v_{\text{act av}} > v_{\text{ru av max}}$
5.2	IMMOBILIZATION: supply grid disconnected
5.3	SHUT-DOWN: operator command 'waiting mode'
5.4	IMMOBILIZATION: operator command 'immobilization'
5.5	FAULT SHUT-DOWN: fault shutdown in progress
5.6	IMMOBILIZATION: temperature limit value exceeded
5.7	SHUT-DOWN: no operator command 'load demand'
5.8	IMMOBILIZATION: external temperature $T_{\text{ext}} > T_{\text{ext max}}$
5.9	IMMOBILIZATION: external temperature $T_{\text{ext}} < T_{\text{ext min}}$
5.10	IMMOBILIZATION: monitoring period exceeded; azimuth angle outside wind direction tolerance range
5.11	FAULT SHUT-DOWN: $\Delta\gamma_{\text{Win/Azim}}$ exceeded 10 times /d
5.12	FAULT SHUT-DOWN: maximum azimuth angle γ_{max} exceeded
5.13	IMMOBILIZATION: monitoring period for maximum azimuth angle range exceeded $(\gamma > \gamma_{\text{max}} - \Delta\gamma)$
5.14	IMMOBILIZATION: azimuth angle γ_2 exceeded
5.15	IMMOBILIZATION: monitoring period for azimuth angle range γ_2 exceeded $(\gamma > \gamma_2 - \Delta\gamma)$
5.16	IMMOBILIZATION: speed gradient $dn_{\text{ru}}/dt < 0$
5.17	IMMOBILIZATION: speed gradient $dn_{\text{ru}}/dt < dn_{\text{ru crit}}/dt$ (for $n_{\text{crit}} - \Delta x < n_{\text{act}} < n_{\text{crit}} + \Delta x$)
5.18	IMMOBILIZATION: speed gradient $dn_{\text{ru}}/dt > dn_{\text{ru max}}/dt$
5.19	FAULT SHUT-DOWN: $dn_{\text{ru}}/dt > dn_{\text{ru max}}/dt$ exceeded 10 times /d
5.20	IMMOBILIZATION: running-up time elapsed and synchronization not initiated
5.21	SHUT-DOWN: synchronization initiated, rotor speed $n_{\text{act}} < n_{\text{syn min}}$
5.22	SHUT-DOWN: rotor speed $n_{\text{act}} > n_{\text{syn max}}$
5.23	IMMOBILIZATION: synchronization time elapsed $t_{\text{syn}} > t_{\text{syn max}}$
5.24	FAULT SHUT-DOWN: five synchronization attempts reached /d

within the regulation reserve range by blade pitch adjustment. A small overload range is permissible in the case of gusts, so that the blades do not have to be adjusted as fast or as often. The overload range must, however, be of limited duration, depending upon the thermal behavior of the entire system. The termination conditions of this operating state are checked continuously and the necessary messages given out (Table 5.4). The tower nacelle is yawed in the direction of the wind.

5.6.1.8   Shut-down (transient)

From part-load operation, full-load operation and running-up, it must be possible at all times to shut the turbine down, bring it into the waiting mode operating state and report the appropriate states. To achieve this, after desired values have been set by the management system, the power output is reduced by the frequency converter and the turbine is decelerated by adjusting the

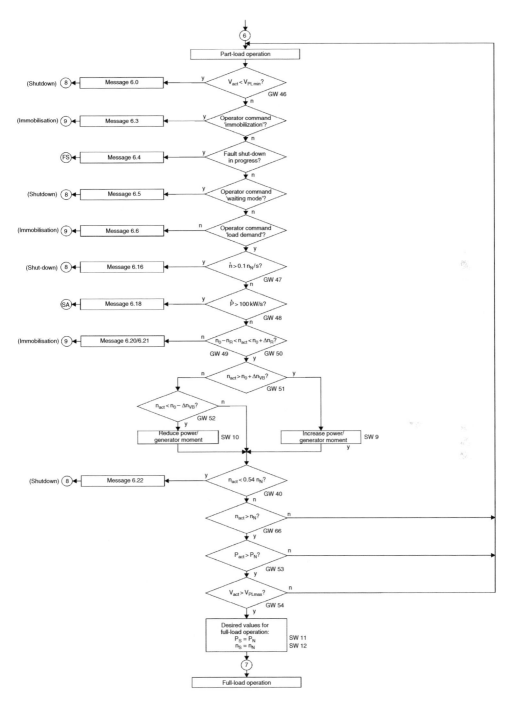

Figure 5.45 Management system structure and desired values in part-load operation

Table 5.3   Messages and desired values in part-load operation

Output	Meaning
6.0	PART-LOAD OPERATION
6.1	SHUT-DOWN: wind speed too low $v_{act} < v_{PL\,min}$
6.2	IMMOBILIZATION: supply grid disconnected
6.3	IMMOBILIZATION: temperature limit value exceeded
6.4	IMMOBILIZATION: operator command 'immobilization'
6.5	FAULT SHUT-DOWN: fault shutdown in progress
6.6	SHUT-DOWN: operator command 'waiting mode'
6.7	IMMOBILIZATION: no operator command 'load demand'
6.8	IMMOBILIZATION: external temperature $T_{ext} > T_{ext\,max}$
6.9	IMMOBILIZATION: external temperature $T_{ext} < T_{ext\,min}$
6.10	IMMOBILIZATION: monitoring period exceeded; azimuth angle outside wind direction tolerance range
6.11	FAULT SHUT-DOWN: $\Delta\gamma_{Win/Azim}$ exceeded 10 times /d
6.12	FAULT SHUT-DOWN: Maximum azimuth angle γ_{max} exceeded
6.13	IMMOBILIZATION: Monitoring period for maximum azimuth angle range exceeded $(\gamma > \gamma_{max} - \Delta\gamma)$
6.14	IMMOBILIZATION: azimuth angle γ_2 exceeded
6.15	IMMOBILIZATION: Monitoring period for azimuth angle range γ_2 exceeded $(\gamma > \gamma_2 - \Delta\gamma)$
6.16	IMMOBILIZATION: speed gradient $dn_{PL}/dt > dn_{PL\,max}/dt$
6.17	FAULT SHUT-DOWN: $dn_{PL}/dt > dn_{PL\,max}/dt$ exceeded 10 times /d
6.18	FAULT SHUT-DOWN: power gradient $dP/dt > 100\,kW/s$
6.19	IMMOBILIZATION: rotor speed below acceptable range $n_{act} < n_0 - 10\%$
6.20	IMMOBILIZATION: rotor speed above acceptable range $n_{act} > n_0 + 10\%$
6.21	IMMOBILIZATION: rotor speed too low $n_{act} < n_{PL\,min}$
6.22	FAULT SHUT-DOWN: rotor speed $n_{act} > n_{max}$

blade pitch towards the feathered position, so that values are reached that permit the generator system to be disconnected from the supply grid. The fault shut-down and braking conditions are checked in a recurring sequence. After a successful separation process, the turbine returns to the waiting mode operating state.

5.6.1.9   Immobilization (transient)

It must be possible to stop the turbine from any operating state. Immobilization is similar to shut-down. If the speed has fallen below the minimum value predetermined by the management system then the rotor and nacelle are braked and the turbine goes into the immobilization state. Again, the fault shut-down and braking conditions must be checked repeatedly and state messages displayed during the immobilization operating state.

5.6.1.10   Fault shut-down (transient)

Fault shut-down takes place in a similar manner to immobilization. The turbine management system can, however, impose steeper desired value ranges than are normally used when

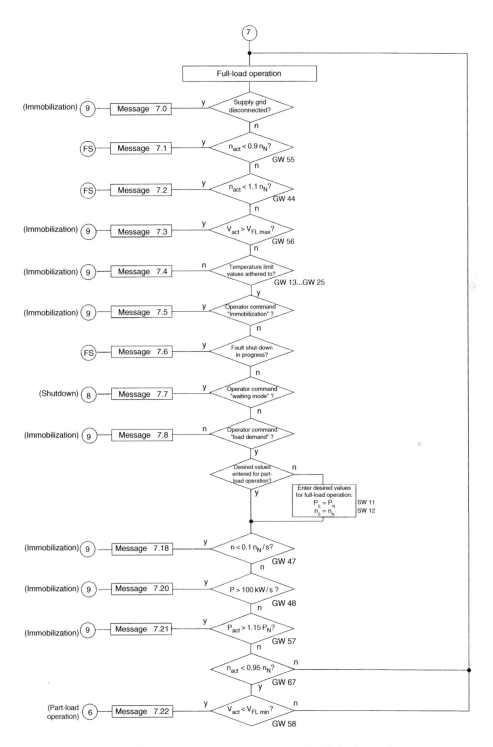

Figure 5.46 Management system structure in full-load operation

Table 5.4 Messages in full-load operation

Output	Meaning
7.0	FULL-LOAD OPERATION
7.1	IMMOBILIZATION: supply grid disconnected
7.2	IMMOBILIZATION: value below rotor-speed range $n_{act} < n_0 - 10\%$
7.3	IMMOBILIZATION: value above rotor-speed range $n_{act} > n_0 + 10\%$
7.4	IMMOBILIZATION: wind speed too high $v_{act} > v_{FL\,max}$
7.5	IMMOBILIZATION: temperature limit value exceeded
7.6	IMMOBILIZATION: operator command 'immobilization'
7.7	FAULT SHUT-DOWN: fault shutdown in progress
7.8	SHUT-DOWN: operator command 'waiting mode'
7.9	IMMOBILIZATION: no operator command 'load demand'
7.10	IMMOBILIZATION: external temperature $T_{ext} > T_{ext\,max}$
7.11	IMMOBILIZATION: External temperature $T_{ext} < T_{ext\,min}$
7.12	IMMOBILIZATION: Monitoring period exceeded; azimuth angle outside wind direction tolerance range
7.13	FAULT SHUT-DOWN: $\Delta\gamma_{Win/Azim}$ exceeded 10 times /d
7.14	FAULT SHUT-DOWN: maximum azimuth angle γ_{max} exceeded
7.15	IMMOBILIZATION: monitoring period for maximum azimuth exceeded $(\gamma > \gamma_{max} - \Delta\gamma)$
7.16	IMMOBILIZATION: azimuth angle γ_2 exceeded
7.17	IMMOBILIZATION: monitoring period for azimuth angle range γ_2 exceeded $(\gamma > \gamma_2 - \Delta\gamma)$
7.18	IMMOBILIZATION: speed gradient $dn_{FL}/dt > dn_{FL\,max}/dt$
7.19	FAULT SHUT-DOWN: $dn_{FL}/dt > dn_{FL\,max}/dt$ exceeded 10 times /d
7.20	IMMOBILIZATION: power gradient $dP/dt > 100\,kW/s$
7.21	IMMOBILIZATION: maximum permissible power exceeded $P_{act} > 1.15 P_N$
7.22	Wind speed too low for full-load operation $v_{act} < v_{FL\,min}$

stopping the turbine. This operating state can also be initiated from higher speeds, in which case the rotor brake may be applied in a controlled manner if the rotor is not decelerating quickly enough. If the rotor speed falls below a minimum value predetermined by the management system then the rotor and nacelle are braked and the turbine goes over into the immobilization state. During fault shut-down, the emergency shut-down and braking conditions must be checked continuously.

5.6.1.11 Emergency shut-down (transient)

An emergency shut-down is triggered if a normal immobilization procedure is not possible. This state lasts until the turbine is stationary. It can be initiated either by the turbine management system or by a higher safety system. As an emergency shut-down can take place from full speed, all braking systems should be used to bring the turbine to a standstill as safely as possible. The safest, but mechanically least favourable, method is the abrupt operation of brakes and blade-adjusting hydraulics. The rotor is locked as soon as it is stationary. Further operation of the turbine is prevented by the management system. Recommissioning is only possible after manual release.

5.6.2 Faults

To ensure the reliable operation of the turbine and its components, disturbances to normal operation must be recognized by the management system. The management system should cut in before the safety system, so that the latter is used as seldom as possible. The safety system can be made up of a redundant monitoring computer or a speed sensor connected either directly to the hydraulic system or to an electric adjustment mechanism.

5.6.2.1 Rapid auto-reclosure of grid

In the event of grid failures – even those of only short duration – it is necessary to prevent an excessive increase in speed. In the designs selected here, only the frequency converter can recognize grid failures. It must therefore shut down immediately and send a message to the management system. As the turbine's generator is no longer opposed by a load moment, speed increases. Using blade pitch adjustment and, if necessary, the brake (in the upper speed range), the turbine is decelerated into the waiting mode. As soon as all conditions (usually grid OK) are again fulfilled, running-up can be automatically initiated once again. If the grid is not available after a certain time, a fault shut-down (i.e. immobilization) must take place.

5.6.2.2 Short-circuits

Short-circuits bring about high currents that can damage or even destroy turbine components, circuitry and protective devices. To prevent damage, short-circuits must be recognized quickly and protective measures initiated. This process is concluded by the tripping of the main switch. At the same time, the frequency converter reports the short-circuit to the management system, which initiates a fault shut-down.

A generator short-circuit can result in a sudden reduction in voltage in one or more phases of the generator-side input to the frequency converter, despite adequate speed. As soon as the management system recognizes this error, a fault shut-down is initiated.

A short-circuit within the frequency converter must be recognized independently. The internal frequency converter electronics then switch off the power section and report a fault. The management system then initiates a fault shut-down.

5.6.2.3 Overspeed

When the turbine is at full load, i.e. at wind speeds above the nominal range, the speed is held within the control range by adjustments to the blade pitch angle. A regulation reserve allows delayed reaction to increases in speed. If the speed nevertheless climbs above the maximum permissible operating speed (e.g. 10% above the nominal range), a fault shut-down is initiated. If the rotor continues to run too fast despite the intervention of the management system and reaches the tripping speed, the safety system must work to limit the speed. In this case, the safety system immediately initiates the emergency shut-down procedure.

5.6.2.4 Overtemperature

All turbine components are designed such that in normal operation no impermissably high temperatures occur. If these temperature limits are exceeded, it can be assumed that there is a fault or overload in the system. Therefore the fault shut-down must be initiated.

5.6.3 Determining the state of system components

This description relates to a gearless wind turbine with a blade pitch adjustment system, fitted with a permanent-magnet synchronous machine and connected to the grid via a pulse-controlled a.c. converter. The turbine components will be described below and the options for ensuring the sensing of relevant states investigated.

For the blade adjustment device, the adjustment angle and adjustment speed must be monitored and limited. The pitch adjustment angle and direction, as well as the adjustment speed and locking of the blades, are predetermined. Moreover, the hydraulic or power supply to the device must be ensured.

Braking intervention and drive-train locking are predetermined for the generator. Output values are speed, rotational direction, temperature and electrical variables such as current, voltage, power, power factor and frequency.

The frequency converter takes on the functions of grid and generator monitoring, temperature limitation and grid synchronization. The power output or electrical moment of the generator are specified. As well as checking that the system is ready to be powered up and connected to the grid, the current and voltage in the d.c. link, and the output power are sensed.

For the azimuth drive, the adjustment of the nacelle and the adjustment speed and locking of the machine housing must be predetermined, and the nacelle angle and any cable twist reported. The drive and braking function and the brake lining thickness must be monitored. As well as the speed and direction of the wind, vibrations in the nacelle, tower and foundations should be sensed. On the grid side, the energy must be sensed, fuses monitored and an uninterruptible power supply provided for the management system, safety system, emergency lighting, etc.

For the control and monitoring of all operating states, the management system must receive not only operator commands but also all measured and monitored variables, so that it can stipulate the desired values for the frequency converter, blade pitch adjustment system and azimuth adjustment system, and display the status of all turbine components. Moreover, fault messages, remote interrogation and monitoring should be possible, and error diagnosis and fault prediction systems integrated to ensure reliable operation of wind turbines.

5.7 Monitoring and Safety Systems

Besides the normal turbine and operational management components, further monitoring and safety systems should be taken into account when considering the management and safety of the turbine. These may depend upon requirements relating to the turbine, grid or location.

Such systems cover measuring and monitoring systems for temperature, pressure, moisture, acceleration, oscillations, voltage, etc. Furthermore, illumination systems for the tower, nacelle and grid station, a system for the automatic rectification of cable twisting and navigation lights should also be considered. Measures to protect against lightning and other extreme effects such as earthquakes, tornadoes, etc., should also be taken into account. Aerodynamic, mechanical and electrical braking systems (see Sections 2.3.2.4 and 3.6.2.2 to 3.6.2.4) protect against overspeed and serve to bring the rotor to a standstill. Requirements and design notes for safety systems are listed in reference [5.102].

In addition to air density and humidity, the wind conditions at turbine sites are particularly important for determining the drive power of a wind turbine. These will be briefly described in what follows.

5.7.1 Wind measuring devices

Relevant flow conditions for wind turbines are determined by the air speed and its direction in relation to the horizontal. To determine this, individual or combined wind gauges can be fitted on the nacelle, comprising an anemometer (usually a cup anemometer) and a vane. A lightning rod can also be fitted on the wind gauge to protect against a direct lightning strike.

The measuring range of the device must cover the cut-in (v_{cut-in}), nominal (v_N) and shut-down (v_{shut}) wind speeds of the turbine. To determine the average values over 1, 3, 5, 10 or 15 minute periods, the minimum and maximum occurring values must also be determined, which means that the measuring range must cover at least

$$v_{meas} = (0-1.5)v_{shut}$$

Therefore, for example, at a shut-down wind speed of 25 m/s, a measuring range up to approximately 40 m/s is required.

5.7.2 Oscillation monitoring

In order to protect the turbine from severe jarring and high-amplitude movements in the nacelle, imbalance in the rotor system and similar effects, vibrations are monitored. If limit values are exceeded, the turbine is brought to a standstill.

Vibrations in the longitudinal and transverse (and, if required, vertical) directions can be determined as a vector variable with frequency and amplitude dependences by an acceleration sensor in the bottom of the nacelle. A reliable and robust design option for the acceleration sensor is offered by piezo elements. The measured values can be processed with the aid of charge amplifiers. Critical operating conditions, e.g. caused by natural resonances of the tower, rotor blade deflection, etc., must be terminated as quickly as possible, e.g. by ensuring that the turbine only passes through the speed range in question for a brief period. All vibrations are monitored, and if the limit value is reached a message must be sent to the turbine management system. At amplitudes of 50 to 60% of the applicable limit value a (delayed) fault shut-down should be initiated and at a maximum of 90% an immediate emergency shut-down should be initiated.

If the option of selecting different amplitudes and acceleration values for the initiation of shut-down procedures is rejected then much cheaper designs can be used. Mechanical systems offer very simple but effective options for the monitoring of vibrations. These are often used in small (and medium)-sized wind turbines. In most cases a ring-and-ball system is used, fitted in the nacelle or the top of the tower. In this case, the relative diameters of the ball and ring should be selected such that the free-lying ball falls from the ring if the acceleration limit value is reached. This trips an emergency stop switch that immediately brings the turbine to a standstill. Another design for the detection of vibrations in the nacelle is the fitting of a pendulum rod. This pendulum is made of an electrically conductive material and passes through a metal ring so that in the event of oscillation the pendulum makes contact with the ring and an electrical signal initiates the shut-down of the turbine. The frequency and amplitude of the vibrations can be set to suit the turbine parameters by the selection of the length and mass of the pendulum length and the internal diameter of the ring.

5.7.3   Grid surveillance and lightning protection

In the case of voltage or frequency deviations exceeding, for example, 10 or 5% of the nominal values, the turbine must be disconnected from the grid to prevent unwanted separate operation in grid branches. The turbine is protected from overvoltage damage caused by overvoltage at the generator or by direct or indirect lightning strikes by means of powerful coarse and fine protective devices in the measurement and control circuits, at the generator and at the supply mechanisms, etc.

Direct lightning strikes usually result in serious damage. Diverters in the rotor blades specifically designed to conduct current through connections to the shaft and tower and through an effective (low-resistance) foundation ground connection allow damage to be limited. For this purpose, metal caps are fitted on the blade tips and coarse copper mesh is fitted on to the blade surfaces to conduct lightning currents away without causing much damage.

5.7.4   Surveillance computer

Wind turbines are usually built at some distance from towns and the operator. Visual monitoring is therefore not usually possible. To keep the down-time of turbines low, remote diagnosis systems are necessary. These require suitable measuring, transmission and monitoring units for individual turbines and wind farms.

The analogue and digital data collected can include turbine states plus grid and meteorological conditions such as output, rotation speed, turbine position, temperature, etc. These data are generally processed to ensure fault-free transmission. To this end, the information to be transmitted is divided into blocks of information, provided with error protection, error checking and error correction, and synchronized in blocks. Physical signal preparation takes place by means of encoding and modulation [5.103].

The data collected can be used for control and management, for error checking and for statistical evaluation by the operator, maintenance company and manufacturer. Data transmission is therefore necessary for the transfer of statistical data or the immediate reporting of faults.

As shown in Figure 5.47, individual turbines can be connected to the monitoring computer by copper or fibre-optic cables, modem and telephone or radio connections at the turbine and computer. For data transmission, analogue equipment such as the telephone system or the C1 radio system, or digital equipment such as ISDN or the D1 or D2 network, can be used.

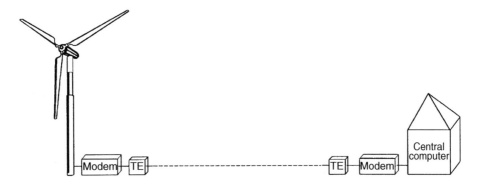

Figure 5.47   Individual wind turbine monitoring

Analogue transmissions are sometimes subject to serious interference. Digital data, on the other hand, can be checked and corrected using codes. Transmission errors can be significantly reduced in this manner and monitoring systems made relatively reliable.

Wind farm monitoring can take place in different ways. Depending upon factors such as the distance between turbines and the system configuration, the data from individual turbines may be transmitted via integral modems as shown in Figure 5.47, brought together as a group (Figure 5.48(b)) or transmitted to the central computer (Figure 5.48(a)). Moreover, with the use of a wind farm computer, it is possible to process, evaluate and compress data on site and send it to the central computer. In this case, the connection between the turbines can be made via cables or the local radio network. As well as cost and safety aspects, upgradability, e.g. with regard to fault prediction, should play a major role in system selection. This aspect will be described briefly in the following section.

5.7.5   Fault prediction

Fault prediction is taking on increasing importance in the field of quality assurance and in the monitoring of technical plant and equipment. By monitoring and evaluating relevant measured signals of a wind turbine, fault indications can be determined before visual, vibration or acoustic changes become apparent and serious damage is done to subcomponents or the system as a whole. In this manner, secondary damage can be avoided, subsequent costs reduced, maintenance intervals adjusted to the state of the turbine and necessary repair work planned in advance and carried out in periods of low wind for safety reasons. Such a system also permits remote monitoring and remote diagnoses to be carried out. Therefore the down-time of the turbine can be reduced, reliability and economic viability improved and the service life of the turbine increased. The most common causes of faults, which are listed in detail in reference [5.104], are defective components and the turbine control system. External effects due to storms, lightning strikes and grid faults, and turbine-specific effects caused by the loosening of components are also of significance. Significant causes of faults in the mechanical components of a wind turbine are the fatiguing of materials and wear and loosening of components. Changes observed in the event of such defects, e.g. in relation to vibration behavior, can generally be recognized before they become critical. It is thus possible to rectify expected faults in advance.

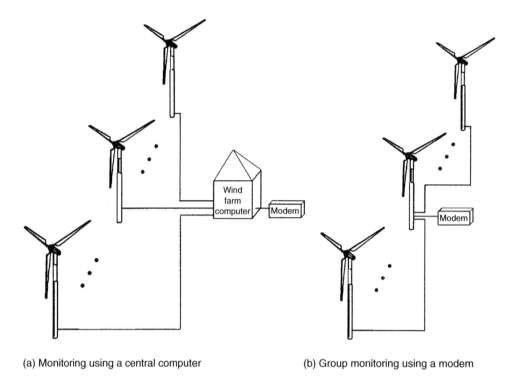

(a) Monitoring using a central computer   (b) Group monitoring using a modem

Figure 5.48   Different methods of wind farm monitoring

In fault-prediction systems, relevant measuring signals are continually captured and evaluated in relation to fault-related characteristics [Chapter 4, Reference 41]. The most expressive measuring variables are mainly used, which are always available in running operation. State-related information that is relevant to faults can be determined from electrical power (see Figure 4.34), generator currents, turbine rotation speed and the acceleration of vibration-monitoring systems. Furthermore, body noise and possibly also air noise measurements can be used for fault prediction.

Spectral analysis processes are particularly suitable for the evaluation due to the permanent random and periodic excitations to the turbine caused primarily by the wind and the rotation of the rotor. At this point, measuring signals are divided into deterministic and random components, broken down into sections of equal length and weighted using a window function. This allows direct components with occurring trends to be filtered out. Using a fast Fourier transform (FTT), the spectra of the filtered components are calculated and the mean taken. By comparison of the measurements with known spectra of fault-free turbines and turbines with faults, changes and the development of faults can be recognized.

Precise knowledge of the turbine behavior in normal operation and in fault states permits a detailed diagnosis to be made of the current turbine state, and allows necessary measures for fault diagnosis to be introduced [5.105–5.113]. In modern high-output turbines, fault predictions are expected to form a fixed component of the turbine monitoring system in the near future.

5.7.6   *Voltage limitation*

In order to guarantee safe operation of a wind turbine, its operating voltage must be confined at all times within a required nominal range. If a plant is operated in a voltage-controlled manner, then the maintenance of the operating limits in undisturbed operation poses no problems. If, however, the plant is operated with reactive power control, as, e.g. in a wind farm with central reactive power control (see also Section 4.6.3), then the maintenance of the voltage limits cannot always be guaranteed, as the voltage is the result of the given reactive power. Therefore, an additional voltage limitation must be implemented that will prevent the plant voltage moving beyond the nominal range. This must monitor the plant voltage and in the case of a threatening voltage injury must adapt the reactive power of the plant such that the voltage remains within the defined limits.

A corresponding algorithm can be implemented either in the central reactive power controller or in the control room of the plant. However, as this is a protective function, it should be carried out as near to the plant as possible in order to minimize the dependency on the communication arrangement. Therefore, the voltage limitation is normally not implemented in the central reactive power controller but in the control room of the plant (see also Figure 5.49)

Various methods can be used for the actual implementation of such an algorithm for voltage limitation. The most common one is the limitation of the voltage on the basis of a $Q(U)$-static that limits the reactive power with approximation to the voltage limits in the

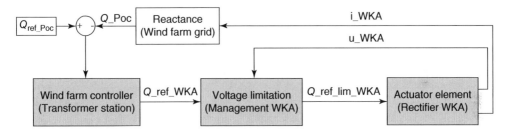

Figure 5.49   Reactive power control circuit of a wind farm with central reactive power control and local voltage limitation [5.114]

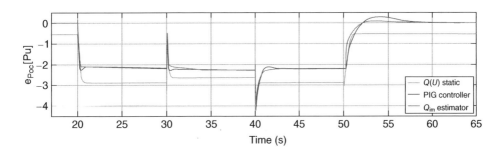

Figure 5.50   Comparison of control error in the voltage limitation by means of $Q(U)$ static, PID controller and Q_{lim}-estimator [5.114]. Investigations were carried out on grid voltage jumps ($t = 20$ s, $t = 55$ s) as well as active power jumps of the plants ($t = 30$ s, $t = 40$ s)

form of ramps. As such, a static cannot make use of the complete reactive power capacity of the plant; research is being carried out on improved algorithms for voltage limitation and voltage control [5.114–5.116]. In Figure 5.50, the reactive power control error is shown for three different control algorithms. It can be shown that although the $Q(U)$ static shows a better dynamic behavior at the PID controller or $Q_{\lim}$-estimator attempt in comparison to $Q(U)$ static, there is a smaller control error and thus a better utilization of the reactive power presentation. With rising requirements of the grid behavior of wind farms, the importance of such optimizing continues to increase, yet it must be seen whether such complex algorithms are feasible in practice.

References

[5.1] Albrecht, P., Cramer, G., Drews, P., Grawunder, M., Heier, S., Kleinkauf, W., Leonhard, W., Speckheuer, W., Thür, J., Vollstedt, W. and Wettlaufer, R., *Betriebsverhalten von Windenergieanlagen, Abschlußbericht zum BMFT-Forschungsvorhaben O3E-4362-A, BMFT-FB-T 84-154*, Part II, Department of Electrical Energy Supply Systems, Kassel University, and Institute for Control Technology Braunschweig Technical University, Karlsruhe, 1984.

[5.2] Heier, S. and Kleinkauf, W., Regelungskonzept für GROWIAN (Große Windenergieanlage) in Seminar-und Statusbericht Windenergie, Kernforschungsanlage Jülich GmbH, Projektleitung Energieforschung (Hrsg), October 1978, pp. 407–18.

[5.3] Heier, S., Regelungskonzepte für Windenergieanlagen, in Wind Energy Conference of the German Association for Wind Energy (DGW) and KFA Jülich, Oldenburg, 27–28 March 1987, pp. 123–40.

[5.4] Heier, S., Regelungskonzepte für Windenergieanlagen, *Elektrotechnik/Schweiz*, 1988, 9, 51–6.

[5.5] Heier, S., Generatoren für kleine Windkraftanlagen im Netzbetrieb unter Berücksichtigung der regelungstechnischen Konzeption, in VI Symposium on *Micromachines and Servosystems*, Warszawa, Poland, May 1988.

[5.6] Heier, S., Kleinkauf, W. and Sachau, J., Wind Energy Converters at Weak Grids, in European Community Wind Energy Conference, Herning, Denmark, May 1988, pp. 429–33.

[5.7] Durstewitz, M., Heier, S., Hoppe-Kilpper, M., Kleinkauf, W. and Sachau, J., *Elektrische Energieversorgung mit Windenergieanlagen. Auslegung und Regelung von verbraucher-orientierten Versorgungseinheiten und Inselnetzen*, BMFT Final Report, March 1992.

[5.8] Caselitz, P. and Krüger, T., Drehzahlvariable Windkraftanlagen mit Überlagerungs-getriebe, *Windkraft-Journal*, 1993, 13.

[5.9] Nigim, K.A., *Static Exciter for Wound Rotor Induction Machine*, IEEE, 1990, pp. 933–7.

[5.10] Holmes, P.G. and Nigim, K.A., A Stand-Alone Induction Generator with Secondary Control to Give Constant Frequency and Voltage, in 22nd UPEC, Sunderland Polytechnic, UK, 14–16 April 1987.

[5.11] Arsudis, D., *Doppeltgespeister Drehstromgenerator mit Spannungszwischenkreis-Umrichter im Rotorkreis für Windkraftanlagen*, Dissertation, Braunschweig Technical University, 1989.

[5.12] Kiel, E. and Schumacher, W., Der Servocontroller in einem Chip, *Elektronik*, April 1994.

[5.13] Körber, F., Besel, G. and Reinhold, H., *Meßprogramm an der 3 MW-Windkraftanlage GROWIAN. BMFT-Forschungsbericht Förderkennzeichen 03E-4512A*, Hamburg, 1988.

[5.14] Arafa, O. and Heier, S., *Accurate Modelling of Parallel Chopper-Controlled Induction Motor Drives*, Report DAAD, University Gh Kassel, 1996.

[5.15] Arafa, O., *A Study of the Performance Characteristics of the Asynchronous Cascade in the Driving Mode*, Final Report DAAD, University Gh Kassel, 1996.

[5.16] Ritter, P. and Rotzsche, L., *Schlupfsteuerung von Asynchrongeneratoren für Windkraftanlagen zur Minderung ihrer Leistungsschwankungen*, Thesis, University Gh Kassel, 1995.

[5.17] Ritter, P., *Schlupfregelung einer Windkraftanlage mit Asynchrongenerator*, Thesis, University Gh Kassel, 1996.

[5.18] Hawranke, I., *Betriebsführung eines Teststandes zur Nachbildung von Windkraftanlagen*, Thesis, University Gh Kassel, 1996.

[5.19] Schmid, Ch., Glasbrenner, M. and Hörmannskirchner, S., Regelung von Windenergieanlagen mit hydrodynamischem Überlagerungsgetriebe. Wind Turbine Control Using Hydrodynamic Variable Speed Superimposing Gear, *at – Automatisierungstechnik*, 2013, G1(5), 339–50.

[5.20] Leonhard, W., *Einführung in die Regelungstechnik. Nichtlineare Regelvorgänge*, 2nd Revised Edn, Friedrich Vieweg & Sohn, Braunschweig, 1977.

[5.21] Buxbaum, A. and Schierau, K., *Berechnung von Regelkreisen der Antriebstechnik*, Elitera-Verlag, Berlin, 1974.

[5.22] Dörrscheidt, F. and Latzel, W., *Grundlagen der Regelungstechnik*, 2nd Revised Edn, B.G. Teubner, Stuttgart/Leipzig, 1993.

[5.23] Pfaff, G., *Regelung elektrischer Auftriebe*, 2nd Edn, R. Oldenburg Verlag, Munich, 1984.

[5.24] Bothe, H.-H., *Fuzzy Logic. Einführung in Theorie und Anwendung*, 2nd Enlarged Edn, Springer-Verlag, Berlin, Heidelberg, New York, London, Paris, Tokyo, Hong Kong, Barcelona, Budapest, 1995.

[5.25] Kähny, H., *Eigenschaften eines Fuzzy-Reglers zur Leistungsregelung einer Windkraftanlage mit Blattverstellung*, Dissertation, Universität Gesamthochschule Kassel, 1994.

[5.26] Danesi, A. *et al.*, A Self Adaptive Pitch Blade Control of a Large Wind Turbine with Predictive Wind Velocity Measurements, in European Wind Energy Conference, Rome, Italy, 1986, p. 641.

[5.27] Arsudis, D. and Bönisch, H., Self-Tuning Linear Controller for the Blade Pitch Control of a 100 kW WEC, in European Community Wind Energy Conference, Madrid, Spain, 10–14 September 1990, pp. 564–8.

[5.28] Barton, R.S., Bowler, C.E.J. and Piwko, R.J., *Control and Stabilization of the NASA / DOE MOD-1 Two Megawatt Wind Turbine Generator*, American Chemical Society, 1979, pp. 325–30.

[5.29] Rothmann, E.A., The Effects of Control Modes on Rotor Loads, in Second International Symposium on *Wind Energy Systems*, 1978, pp. 107–17.

[5.30] Kos, J.M., Online Control of a Large Horizontal Axis Wind Energy Conversion System and Its Performance in a Turbulent Wind Environment, in Proceedings of 13th Conversion Engineering Conference, San Diego, California, 1978.

[5.31] Hinrichsen, E.N. and Nolan, P.J., Dynamics and Stability of Wind Turbine Generators, *IEEE Transactions on Power Apparatus and Systems*, 1982, PAS 101, 2640–8.

[5.32] Svensson, J.E. and Ulen, E. The Control System of WTS-3 Instrumentation and Testing, in 4th Symposium on *Wind Energy Systems*, BHRA, Stockholm, 1982.

[5.33] Hinrichsen, E.N., Controls for Variable Pitch Wind Generators, *IEEE Transactions on Power Apparatus and Systems*, 1984, PAS 103, 866–92.

[5.34] Liebst, B.S., *Pitch Control Systems for Large Scale Wind Turbines*, American Institute of Aeronautical Engineering and Mechanics, Vol. 7, 1982, pp. 182–92.

[5.35] Murdoch A., Winkelman, J.R., Javid, S.H. and Barton, R.S., Control Design and Performance Analysis of a 6 MW Wind Turbine Generator, *IEEE Transactions on Power Apparatus and Systems*, 1983, PAS102, 1340–7.

[5.36] Mattson, S.E., *Modelling and Control of Large Horizontal Axis Wind Power Plants*, Lund, Sweden, 1984.

[5.37] Grimble, M.J., Two and a Half Degrees of Freedom LQG Controller Solution and Wind Turbine Control Applications, in Proceedings of American Control Conference, 1992, pp. 676–80.

[5.38] Steinbuch, M., *Dynamic Modelling and Robust Control of a Wind Energy Conversion System*, Dissertation, Delft University of Technology, 1989.

[5.39] Steinbuch, M. and Bosgra O.H., Optimal Output Feedback of a Wind Energy Conversion System, in Proceedings of 9th IFAC on *Power Systems: Modeling and Control Applications*, Brussels, Belgium, 1989, pp. 313–9.

[5.40] Makila, P.M. and Toivonen, H.T., Computational Methods for Parametric LQ Problems – a Survey, *IEEE Transactions on Automatic Control*, 1987, AC-32, 658–71.

[5.41] Bongers, P.M.M and Schrama, R.J.P., Application of LQ-Based Controllers to Flexible Wind Turbines, in Proceedings of 1st European Control Conference, Grenoble, France, 2–5 July 1991, pp. 2185–9.

[5.42] Bongers, P.M.M. and Dijkstra, S., Control of Wind Turbine Systems Aimed at Load Reduction, in Proceedings of American Control Conference, Chicago, Illinois, 24–26 June 1992, pp. 1710–4.

[5.43] Bongers, P.M.M., *Modeling and Identification of Flexible Wind Turbines and a Factorizational Approach to Robust Control*, Dissertation, Delft University of Technology, The Netherlands, 1994.

[5.44] Kreißelmeier, G. and Steinhauser, R., *Systematische Auslegung von Reglern durch Optimierung eines vektoriellen Gütekriteriums, Regelungstechnik*, 1979, Book 3, pp. 76–9.

[5.45] Reck, T., *Untersuchung zur aktiven Dämpfung von Turmschwingungen an einer Windkraftanlage mit Blattverstellung*, Dissertation, University Gh Kassel, 1996.

[5.46] Adam, H., *Entwurf eines Mehrgrößenreglers für eine drehzahlvariable Windkraftanlage durch Gütevektoroptimierung*, Dissertation, Universität Gesamthochschule Kassel, 1996.

[5.47] Caselitz, P., Krüger, T. and Petschenka, J., Load Reduction by Multivariable Control of Wind Energy Converters–Simulations and Experiments, in European Union Wind Energy Conference, Göteborg, Sweden, 1996.

[5.48] Caselitz, P., Giebhardt, J. Krüger, T., Mevenkamp, M., Petschenka, J. and Reichardt, M., Neue Verfahren zur Regelung von Windkraftanlagen, in Forschungsverbund Sonnenenergie, Annual Conference, 1996.

[5.49] Krüger, T., *Regelungsverfahren für Windkraftanlagen zur Reduktion der mechanischen Belastung*, Dissertation, Universität Gesamthochschule Kassel, 1997.

[5.50] Kalman, R.E. and Bucy, R.S., New Results in Linear Filtering and Prediction Theory, *Transactions of the ASME, Series D, Journal of Basic Engineering*, 1961, 83, 95–108.

[5.51] Luenberger, D.G., An Introduction to Observers, *IEEE Transactions on Automatic Control*, 1971, 16(6), 596–602.

[5.52] Weinmann, A., *Regelungen – Analyse und technischer Entwurf*, Vol. 2: *Nichtlineare, abtastende und komplexe Systeme; modale, optimale und stochastische Verfahren*, Springer-Verlag, Vienna, 1984.

[5.53] Isermann, R., *Digitale Regelsysteme*, 2nd Revised and Enlarged Edn, Vol. 1, *Grundlagen, Deterministische Regelungen*, corrected reprint, Springer-Verlag, Berlin, Heidelberg, New York, London, Paris, Tokyo, 1988.

[5.54] Störzel, K., *Untersuchungen zur mechanischen Beanspruchung drehzahlvariabler Windkraftanlagen unter Mittelgebirgsbedingungen im WKA-Testfeld Vogelsberg*, LBF Report No. 7545, Frauenhofer Institute für Betriebsfestigkeit, Darmstadt, 1996.

[5.55] Changeux, J.-P., *Der neuronale Mensch*, Rowohlt-Verlag, Reinbeck, 1984.

[5.56] Eccles, J.C and Popper, K.R., *Das Ich und sein Gehirn*, R. Piper Verlag, Munich, 1989.

[5.57] Shatz, C.J., Das sich entwickelnde Gehirn, *Spektrum der Wissenschaft*, 1992, 11, 44–52.

[5.58] Kohonen, T., Self-Organized Formation of Topologically Correct Feature Maps, *Biological Cybernetics*, 1982, 43, 59–69.

[5.59] Kohonen, T., Analysis of a Simple Self-Organizing Process, *Biological Cybernetics*, 1982, 44, 135–40.

[5.60] Kohonen, T., *Self-Organization and Associative Memory*, Springer-Verlag, Berlin, 1984.

[5.61] Kohonen, T., Adaptive, Associative and Self-Organizing Functions in Neural Computing, *Applied Optics*, 1987,26(23), 4910–8.

[5.62] Ritter, H. and Schulten, K., Topology Conserving Mappings for Learning Motor Tasks, in Tagungsband zur AIP Conference, Snowbird, Utah, 1986, pp. 376–80.

[5.63] Ritter, H. and Schulten, K., Extending Kohonen's Self-Organizing Mapping Algorithm to Learn Ballistic Movements, in *Neural Computers*, Springer-Verlag, Heidelberg, 1988, pp. 393–406.

[5.64] Ritter, H. and Schulten, K., Convergence Properties of Kohonen's Topology Conserving Maps: Fluctuations, Stability, and Dimension Selection, *Biological Cybernetics*, 1988, 60, 59–71.

[5.65] Ritter, H., *Selbstorganisierende neuronale Karten*, Dissertation, Munich University, 1988.

[5.66] Ritter, H., *et al.*, *Ein Gehirn für Roboter*, mc, No. 2, 1989, pp. 48–61.

[5.67] Ritter, H., *et al.*, Topology-Preserving Maps for Learning Visuomotor-Coordination, *Neural Networks*, 1989, 2, 159–68.

[5.68] Ritter, H., *et al.*, 3D-Neural-Network for Learning Visuomotor-Coordination of a Robot Arm, in Tagungsband zur IJCNN-89 Conference, Washington, Vol. II, 1989, pp. 351–6.

[5.69] Ritter, H., *et al.*, *Neuronale Netze: Eine Einführung in die Neuroinformatik selbstorganisierender Netzwerke*, Addison-Wesley, Bonn, 1990.

[5.70] Stoll, M., *Ein Schätzverfahren über den inneren Zustand geschlossener Bleiakkumulatoren*, VDI-Verlag GmbH, Düsseldorf, 1994.

[5.71] Watschke, H., *Untersuchung zum Einsatz neuronaler Netze in der Regelung und Betriebsführung von Windkraftanlagen*, Dissertation, Universität Gesamthochschule Kassel, 1993.

[5.72] Shan, M., Fischer, B. and Brosche, P., Regelungsentwurf für Windenergieanlagen, *at – Automatisierungstechnik*, 2013, 61(5), Oldenbourg Wissenschaftsverlag, 305–17.

[5.73] Geyler, M. and Caselitz, P., Regelung von drehzahlvariablen Windenergieanlagen Automatisierungstechnik, *at – Automatisierungstechnik*, 2008, 56(12), 614–26.

[5.74] Johnson, K.E., Pao, L.Y., Balas, M.J. and Fingersh, L.J., Control of Variable Speed Wind Turbines, Standard and Adaptive Techniques for Maximizing Energy Capture, *IEEE Control Systems Magazine*, June 2006.

[5.75] Bianchi, D., De Battista, H. and Mantz, R.J., Wind Turbine Control Systems, Springer-Verlag, London, 2007, ISBN 1-84628-492-9.

[5.76] Krüger, T., *Regelung für Windkraftanlagen zur Reduktion mechanischer Belastung*, Dissertation, Universität Gesamthochschule Kassel, 1998.

[5.77] Stol, K.A., Disturbance Tracking and Blade Load Control of Wind Turbines in Variable-Speed Operation, Presented at the 2003 AIAA/ASME Wind Symposium Reno, Nevada, January 6–9, 2003.

[5.78] van der Hooft, E.L. and van Engelen, T.G., Feed Forward Control of Estimated Wind Speed, Technical Report, ECN, ECN-C-03-137, December 2003.

[5.79] Suarez, J., Azagra, E., Urroz, Y. and HdezMascarell, O., Application of Wind Speed Estimation for Power Production Increase, Proc. EWEA 2012, Copenhagen, DK.

[5.80] Bottasso, C.L. and Croce, A.C., Advanced Control laws for Variable-Speed Wind Turbines and Supporting Enabling Technologies, Scientific report DIA-SR 09-01, Dipartimento di ingegneria Aerospaziale, Politecnico di Milano, 2009.

[5.81] Schlipf, D. et al.: LIDAR Assisted Collective Pitch Control, UPWIND Project, Deliverable 5.2, February 2011.

[5.82] Dunne, F., Pao, L., Wright, A., Jonkman, B. and Kelley, N., Combining Standard Feedback Controllers with Feedforward Blade Pitch Control for Load Mitigation in Wind Turbines, Proc of the AIAA Aerospace Sciences Meeting, Orlando/FL 2010.

[5.83] Bossanyi, E.A., Un-Freezing the Turbulence: Improved Wind Field Modeling for Investigating Lidar-Assisted Wind Turbine Control, Proc. EWEA 2012, Copenhagen, DK.

[5.84] Andersen, P.B., Gaunaa, M., Bak, C. and Buhl, T., Load Alleviation on Wind Turbine blades Using variable Airfoil Geometry, Proc. EWEC 2006, Athens, Greece.

[5.85] Shan, M., Duckwitz, D., Choi, J., Jacobsen, J., Bauer, F. and Rösmann, T. and Adelt, S., Schlussbericht zum Verbundprojekt: Lastreduzierende Regelungs-verfahren für Multimegawatt Windkraftanlagen im Offshore-Bereich, Rev. 02, Verbundprojekt, Fraunhofer IWES, Areva Wind

GmbH, Moog Unna GmbH, gefördert vom Bundesministerium für Umwelt, Naturschutz und Reaktorsicherheit veröffentlicht, Juni 2012.

[5.86] Larsen, T. and Hanson, T., A Method to Avoid Negative Damped Low Frequent Tower Vibrations for a Floating, Pitch Controlled Wind Turbine. The Science of Making Torque from Wind, *Journal of Physics Conference Series*, Vol. 75 DTU, Copenhagen, DK, 2007.

[5.87] Fischer, B., Reducing Rotor Speed Variations of Floating Wind Turbines by Compensation of Non-Minimum Phase Zeros, Proc. EWEA 2012, Copenhagen, DK, 2012.

[5.88] Caselitz, P., Giebhardt, J. and Mevenkamp, M., On-Line Fault Detection and Prediction in Wind Energy Converters, Proc. EWEC 1994, Thessaloniki, Greece, 1994.

[5.89] García Márquez, F.P., Tobias, A.M., Pinar Pérez, J.M. and Papaelias, M., Condition Monitoring of Wind Turbines: Techniques and Methods, *Renewable Energy*, 2012, 46, 169–78.

[5.90] van Wingerden, J.-W., Hulskamp, A.W., Barlas, T.K., Marrant, B., van Kuik, G.A.M., Molenaar, D.-P. and Verhaegen, M., On the Proof of Concept of a "Smart" Wind Turbine Rotor Blade for Load Alleviation, *Wind Energy*, 2008, 11, 265–80.

[5.91] Isermann, R., Modellbasierte Überwachung und Fehlerdiagnose von kontinuier-lichen technischen Prozesse, *at – Automatisierungstechnik* 2010, 58, 291–305.

[5.92] Pourmohammad, S. and Fekih, A., Fault-Tolerant Control of Wind Turbine Systems – A Review, Green Technologies Conference (IEEE-Green), Baton Rouge, LA, 14–15 April 2011, pp. 1–6, 2011.

[5.93] Sloth, C., Esbensen, T. and Stoustrup, J., Robust and Fault-Tolerant Linear Parameter-Varying Control of Wind Turbines, *Mechatronics*, 2011, 21, 645–59.

[5.94] Odgaard, P.F. and Johnson, K.E., Wind Turbine Fault Detection and Fault Tolerant Control – A Second Challenge, http://www.kk-electronic.com/Files/Billeder/kk-electronic.

[5.95] Barlas, T.K. and van Kuik, G.A.M., Aeroelastic Modelling and Comparison of Advanced Active Flap Control concepts for Load Reduction on the Upwind 5MW Wind Turbine, Proc. EWEC 2009, Marseille, France, 2009.

[5.96] Berg, D.E., Wilson, D.G., Barone, M.F., Resor, B.R., Berg, J.C. and Paquette, J.A., Zayas, J.R., Kota, S., Ervin, G. and Maric, D., The Impact of Active Aerodynamic Load Control on Fatigue and Energy Capture at Low Wind Speed Sites, Proc. EWEC 2009, Marseille, France, 2009.

[5.97] Barlas, T.K. and van Kuik, G.M.A., Review of State of the Art in Smart Rotor Control Research for Wind Turbines, *Progress in Aerospace Sciences*, 2010, 46(1), 1–27.

[5.98] Marrant, B.A.H. and van Holten, T.H., Comparison of Smart Rotor Blade Concepts for Large Offshore Wind Turbines. Offshore Wind Energy and Other Renewable Energies in Mediterranean and European Seas, Civitavecchia, 2006.

[5.99] Mayda, E.A., van Da, C.P. and Yen Nakafuji, D., Computational Investigation of Finite Width Microtabs for Aerodynamic Load Control. AIAA 2005-1185, 2005.

[5.100] Wilson, D.G., Berg, D.E., Barone, M.F., Berg, J.C., Resor, B.R. and Lobitz, D.W. Active Aerodynamic Blade Control Design for Load Reduction on Large Wind Turbines, Proc. EWEC 2009, Marseille, France, 2009.

[5.101] Buhl, T., Gaunaa, M. and Andersen, P.B., Stability Limits for a Full Wind Turbine Equipped with Trailing Edge Systems, Proc. EWEC 2009, Marseille, France, 2009.

[5.102] Germanischer Lloyd, Vorschriften und Richtlinien. IV – *Nichtmaritime Technik*. Part 1 – *Richtlinie für die Zertifizierung von Windkraftanlagen*, Selbstverlag des Germanischen Lloyd, Hamburg, 1999.

[5.103] Adzic, L., *Fernüberwachung von Windkraftanlagen*, Thesis, Universität Gh Kassel, 1997.

[5.104] Institut für Solare Energieversorgungstechnik (ISET), *Windenergie Report Deutschland, Wissentschaftliches Meß- und Evalierungsprogramm zum Breitentest '250 MW-Wind'. Jahresauswertungen 1990 bis 2001*, Eigendruck Kassel.

[5.105] Caselitz, P., Giebhardt, J. and Mevenkamp, M., Fehlerfrüherkennung in Windkraftanlagen, in Kongreßband Husum Wind '95, Husum, 1995, pp. 143–51.

[5.106] Caselitz, P., Giebhardt, J. and Mevenkamp, M., *Verwendung von WMEP-Onlinemessungen bei der Entwicklung eines Fehlerfrüherkennungssystems für Windkraftanlagen, Jahresauswertung 1994, Wissenschaftliches Meß- und Evaluierungsprogramm zum Breitentest '250 MW Wind' im Auftrag des Bundesministeriums für Forschung und Technologie*, ISET, Kassel, 1995, pp. 155–61.

[5.107] Caselitz, P., Giebhardt, J. and Mevenkamp, M., Development of a Fault Detection System for Wind Energy Converters, in European Union Wind Energy Conference and Exhibition, Göteborg, Sweden, 1996.

[5.108] Morbitzer, D., *Simulation und meßtechnische Untersuchung der Triebstrangdynamik von Windkraftanlagen*, Dissertation I, ISET, Universität Gesamthochschule Kassel, 1995.

[5.109] Osbahr, S., *Untersuchung von Parameterschätzverfahren für die Fehlerfrüherkennung in Windkraftanlagen*, Dissertation, ISET, Hannover University, 1995.

[5.110] Eibach, T., *Untersuchung von Verfahren der Lager- und Getriebeüberwachung für die Fehlerfrüherkennung in Windkraftanlagen*, Dissertation I, ISET, Universität Gesamthochschule Kassel, 1995.

[5.111] Adam, H., *Implementation und Untersuchung Künstlicher Neuronaler Netze zur Fehlerfrüherkennung in Windkraftanlagen*, Thesis, ISET, Universität Gesamthochschule Kassel, 1995.

[5.112] Hobein, A., *Entwicklung eines Hardware-Moduls zur analogen Leistungsberechnung für ein PC-gestütztes Meßdatenerfassungssystem*, Thesis, ISET, Universität Gesamthochschule Kassel, 1995.

[5.113] Werner, U., *Entwicklung eines Hardware-Moduls zur Drehzahlmessung für ein PC-gestütztes Meßdatenerfassungssystem*, Thesis, ISET, Universität Gesamthochschule Kassel, 1995.

[5.114] Thurner, L., *Entwicklung eines Regelungsalgorithmus zur Begrenzung der Spannung an Windenergieanlagen im Parknetz*, Master thesis, University of Kassel, 2012.

[5.115] Fortmann, J. and Erlich, I., Blindleistungsregelung von WEA im Netzparallelbetrieb *at-Automatisierungtechnik*, 2013, 61, 351–8.

[5.116] Hau, M., Robuste Spannungsregelung von Windparks mit Q(U)-Kennlinie, *at – Automatisierungtechnik* 2013, 61, 359–73.

[5.117] Leonhard, W., *Einführung in die Regelungstechnik. Lineare Regelvorgänge*, 2nd Improved Edn, Friedrich Vieweg & Sohn, Braunschweig, 1972.

[5.118] Gockel, M., *Betriebsführung für einen getriebelosen Windkraftgenerator*, Thesis, Universität Gh Kassel, 1994.

[5.119] Appel, D., *Entwurf einer Betriebsführung für eine getriebelose Windkraftanlage*, Dissertation, Universität Gh Kassel, 1995.

6

Using Wind Energy

The use of wind power for the supply of electricity broadens the energy base and reduces environmental pollution. It is particularly practical if it can be made to be economically competitive with conventional energy sources [6.1, 6.2]. In countries such as Denmark or regions such as Schleswig-Holstein in Germany wind power already makes a significant contribution to the electricity supply.

Knowledge of system costs and the expected energy yield are of fundamental importance in this context [6.3, 6.4]. Good wind conditions at the planned site of a turbine or wind farm must be viewed as being the most important prerequisite for the economical exploitation of wind energy [6.5]. Moreover, in densely populated, coastal and offshore areas planning permission issues take on a critical role.

6.1 Wind Conditions and Energy Yields

The Earth is surrounded by an atmosphere in which various physical processes influence the weather, which includes the winds. The atmosphere is kept in motion primarily by differences between heating processes. The contours of the Earth's surface have a decisive influence on local wind speeds. Good conditions for the exploitation of wind energy can be expected near water and in smooth areas of land. Trees, buildings and hills in the immediate vicinity, on the other hand, impair the flow of air.

6.1.1 Global wind conditions

The rate at which the speed of the wind increases with increasing height above ground depends upon the roughness of the landscape (e.g. water, meadow, grassland with bushes, trees, buildings) (see Figure 2.9). Figure 6.1 illustrates the concentration of favourable wind conditions – with regard to their exploitation – in coastal areas. High inland areas can also offer similar conditions.

Grid Integration of Wind Energy: Onshore and Offshore Conversion Systems, Third Edition. Siegfried Heier.
© 2014 John Wiley & Sons, Ltd. Published 2014 by John Wiley & Sons, Ltd.

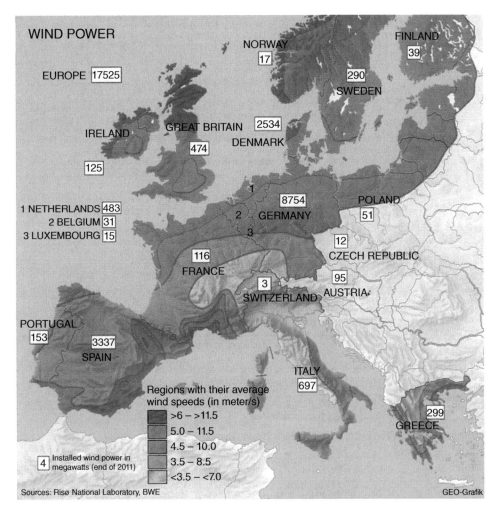

Figure 6.1 Wind speeds in Europe and installed wind turbine outputs in the countries in question in year 2011

The economic operation of wind turbines is completely dependent upon the local wind conditions. These can often deviate significantly from the values stated on wind maps. Statistically determined wind velocity and their wind velocity distributions are decisive for the expected energy yield. These indicate the percentage of the time or the number of hours per year for which every relevant wind speed occurs. Moreover, the daily and yearly wind-speed graphs, the height dependence of wind speeds, the contours and roughness of the landscape and the influence of obstacles must also be taken into account. The gustiness of the area, the degree of turbulence and the maximum wind speeds all impose requirements with regard to the stability of the structure and the control of the turbine. Large inland turbines in particular are often subject to dynamic loads that should not be underestimated.

Before wind turbines are erected, the expected energy yields should be predicted as precisely as possible to determine the economics of the project for the operator and to minimize the investment risk. Site surveys and energy yield predictions based on measurements and calculations are thus required. For cost reasons, measurements are generally only carried out for wind farm projects or in the case of sites for which insufficient reliable data are available.

6.1.2 Local wind conditions and annual available power from the wind

Relatively precise forecasts can currently be obtained by model calculations to determine the local wind potential and turbine-specific energy yields. The limitations of these must, however, be kept in mind.

Precise knowledge of local wind conditions is of fundamental importance for the assessment of a site since wind turbine output and energy yields are proportional to the cube of wind speed. As well as climatological factors such as the shape of the land (orography), surface roughness (topography) and obstacles near the location (mechanical turbulence) influence air density, temperature and sunshine (thermal turbulence) and the direction and strength of the wind [6.5].

Energy predictions based on local wind conditions measured at the hub height of a planned turbine give the most precise results. However, this involves an expensive and time-consuming process. For today's turbine sizes, measurement at the hub height (50 to 100 m) is barely feasible for cost reasons and due to the cumbersome nature of large measuring masts. Therefore, wind speed and direction are measured at lesser heights (10, 20, 30 and 40 m) and the measurements arithmetically extrapolated to the hub height (see Figure 6.2). A measured

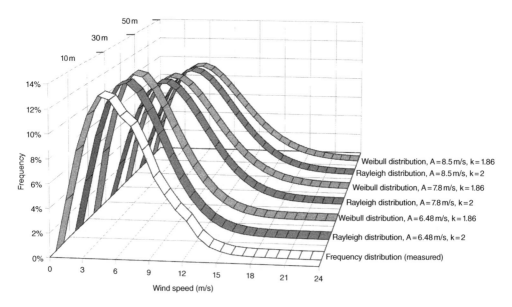

Figure 6.2 Measurement of the frequency distribution at 10 m and calculation of the Rayleigh and Weibull distributions at heights of 30 and 50 m (IWES)

or numerically determined frequency distribution of the wind speed is approximated by an analytical function. The Weibull frequency distribution of wind speeds is usually used for this. The frequency density

$$h_{\text{Weibull}}(v) = \frac{k}{A}\left(\frac{v}{A}\right)^{k-1} e^{-(v/A)^k} \qquad (6.1)$$

is completely determined, in addition to wind speed, by the dimensionless form parameter k with values from 1 to 3 and the scaling factor A with units of m/s. The mathematically simpler special case $k = 2$, known as the Rayleigh distribution function, is used to describe wind conditions in the event that more precise site data are not available, and is generally sufficiently precise

$$h_{\text{Rayleigh}}(v) = \frac{2v}{A^2} e^{-(v/A)^2}. \qquad (6.2)$$

In the Rayleigh distribution the form factor A is calculated directly from the average wind speed

$$A = v_m \frac{2}{\sqrt{\pi}}. \qquad (6.3)$$

A further special case, $k = 3.5$, represents the approximation of the Gaussian distribution.

To determine the relative frequency of a certain wind class the wind speed at the centre of the class is established and the calculated frequency density is multiplied by the breadth of the class (e.g. 1 m/s). The summed frequency of the wind speed

$$F(v) = 1 - e^{(-v/A)^k} \qquad (6.4)$$

is determined by the same parameters. For its calculation below a certain wind speed the upper limit of the last class to be included is used.

The main components of modern wind measurement systems are the anemometer, anemometer mast and measuring computer. Such systems permit fully automatic and maintenance-free operation if they are weatherproof and have internal lightning protection

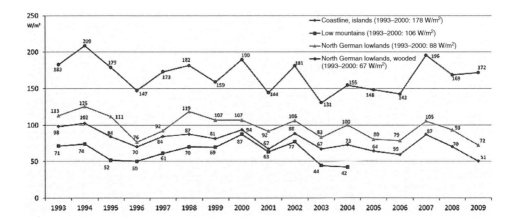

Figure 6.3   Gross available wind energy in the period 1993 to 2001. *Source*: WMEP-Messungen in 10m Höhe – Windenergie-Report, Germany 2001, IWES

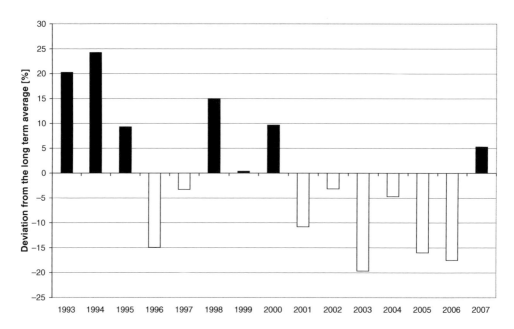

Figure 6.4 Wind index as a deviation from the long-term average of the wind energy supply Period 1993-2007, wind energy Report Germany 2008, ISET

and an effective power supply. Primarily cup anemometers are used for the measurement of wind speed. Ultrasound, hydrometric vane, hot-wire anemometers and venturi nozzles are also occasionally encountered. To record wind speeds and directions over longer periods of time, automatic recording devices (so-called 'data loggers') are required, which facilitate a computerized evaluation of the data. The system often also incorporates a radio modem for the remote interrogation of the measurements.

6.1.3 Calculation of site-specific and regional turbine yields

Using various calculation procedures, relatively precise energy yield forecasts are drawn up for certain wind turbines based upon measured or calculated average wind speeds and frequency distributions for wind speed and wind direction. Figure 6.5 illustrates this. As an example, a Weibull distribution at the hub height is extrapolated from the frequency distribution of wind speeds measured at a height of 10 m (bottom left in Figure 6.5). Based upon the relative frequencies, the duration of time of each individual wind speed, which will prevail each year, can be determined. With the multiplication of this figure by the output of the wind turbine at the wind speed in question we find the so-called class yields. Summing these gives the annual energy yield in the form of the cumulative curve (top right in Figure 6.5).

In order to determine the power values for each wind speed from the wind velocity distributions of the wind value in question, aerodynamic turbine behavior, system design (generator power, rotor transmission), the operational characteristics of the turbine and the influences of control and management must be taken into consideration. Since energy yields can be derived

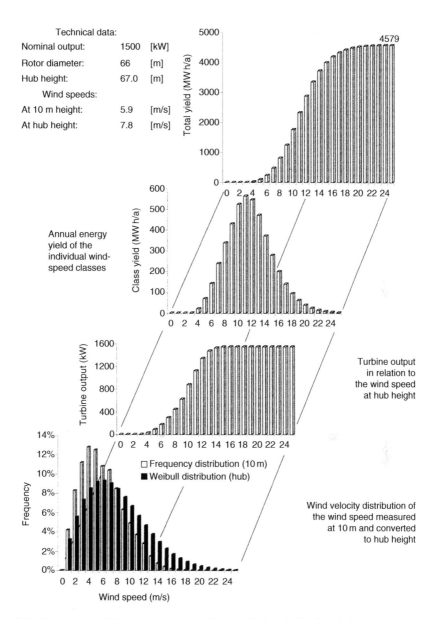

Figure 6.5 Calculation of the annual energy yield (cumulative yield) of a wind energy converter from the measured (or calculated) frequency distribution of the wind speed at a height of 10 m

from the power characteristic (third graph from the top in Figure 6.5) in connection with the distribution of the wind speed in question, these procedures permit the determination of the available power or energy during a period of one year. In newer large turbines in particular, turbulence, gustiness and the unevenness of the wind speed in relation to the entire swept area of the rotor also play an important role. The quality and details of turbine control can also

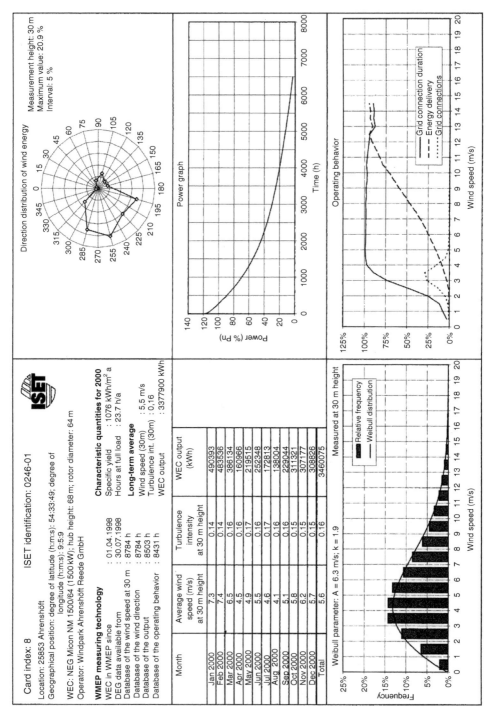

Figure 6.6 Individual results of data evaluation in the large-scale test programme 250 MW Wind (ISET)

influence the economics. Practical experience in the operation of wind turbines is therefore indispensable both when making preliminary calculations and when drawing up estimates of economic viability [6.6]. In the framework of large-scale investigations [6.7], measurements were carried out on wind turbines of different design and size throughout the whole of Germany. The results of site-specific evaluations for the duration of one year for a 1500 kW turbine in northern Germany are shown as an example in Figure 6.5. In addition to the monthly average wind speeds, the figure also shows the wind velocity distribution of wind speeds and arithmetic values such as the turbulence intensity and Weibull parameter. Monthly energy yields, the energy-weighted wind direction distribution and the graph showing the number of hours per year for which the various power levels are achieved are also shown. According to this, the turbine may, for example, deliver nominal output or above for 300 hours per year and around half this value for 1700 hours, with the turbine operating for around 7000 hours.

With the aid of the large-scale tests important findings could be obtained about the wind conditions and energy yields at various sites and in various regions. Figure 6.7 shows the wind velocity distribution of wind speed in the site categories of coastal and islands, North German lowlands and low mountain ranges. This clearly shows that in relation to lowlands and low mountain ranges, windy coastal sites exhibit lower frequencies for low wind speeds and greater distribution for higher wind speeds.

The specific annual energy yield in kilowatt hours per square metre turbine rotor swept area per year (kW h/m² a) is drawn upon in order to be able to compare energy yields from different wind turbines. Investigations in the large-scale testing program have shown that at

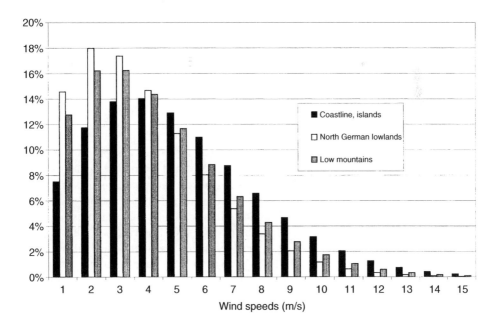

Figure 6.7   Wind-speed wind velocity distribution at various site categories (ISET)

coastal sites wind turbines of the 50 kW class achieve specific annual energy yields of around 500 kW h/m^2 a, turbines of the 100 kW class achieve 600 kW h/m^2 a, turbines of the 200 kW class achieve 900 kW h/m^2 a, turbines of the 300 kW class achieve 1000 kW h/m^2 a and turbines of the MW class achieve approximately 1200 kW h/m^2 a. Inland turbines achieve values of around half these, whereas turbines in low mountain ranges are significantly better, currently achieving values of a good 60% of those quoted above. Since roughness and orography have a particularly great influence upon the wind conditions nearer to the ground, greater hub heights should be chosen for inland sites in comparison to coastal areas.

In site-related economic considerations, the annual energy yield calculations can firstly be oriented towards the considerations of different turbine configurations and models (fixed or variable-speed systems, geared or gearless designs, etc.), regulation systems used (stall/pitch), tower heights, etc. Secondly, calculations for various turbine sizes can also be of great importance, e.g. in order to achieve the most economically favourable construction density for a site. Figure 6.7 aims to highlight the basic procedure on the basis of four relatively diverse turbine sizes, since this helps to illustrate the large differences between the individual power and yield values. As in Figure 6.5, based upon the wind-speed distribution at a height of 10 m (bottom left in Figure 6.8) the Weibull distribution of wind speed at hub height is determined for each turbine (not shown in the figure). Based upon the wind velocity distribution in question, we find the class and cumulative yields per year associated with the turbine outputs in question (top right in the figure).

Wind conditions at the site have a decisive influence upon energy yields. Figure 6.9 shows how wind conditions affect the annual yield of a 2 MW plant. Based upon the three frequency distributions of wind speed at hub height for coast, low mountain ranges and lowlands (bottom left in the figure) the annual class and cumulative yields (top right in the figure) are determined with the aid of the associated power values. These illustrate the great differences in the annual energy yields to be expected, with values varying from almost 1300 MW h in the lowlands to around 1700 MW h in low mountain ranges and approximately 2700 MW h in coastal areas.

6.1.4   Wind atlas methods

For locations for which no measurements are available, model calculation procedures have been developed that allow the potential of the wind to be estimated with a reasonable degree of precision. Such calculations can be performed using commercial programs on standard PCs. This process is based on the 'European Wind Atlas' compiled at the Danish Risø Research Centre on behalf of the European Union and the so-called 'Wind Atlas Analysis Application Programme (WASP)' [6.8], which uses the wind Atlas method (Figure 6.9). In this approach, specific measurements recorded over many years [6.9] are incrementally standardized – taking into account local conditions such as obstacles, surface roughness and orography – to standard environments (flat land, no obstacles, etc.) (left-hand side in Figure 6.10 with the arrow pointing upwards). Taken together, these data represent the European Wind Atlas, and reflects the regional wind conditions disregarding landscape influences. One hundred and seven sites have as yet been recorded throughout the whole of Germany.

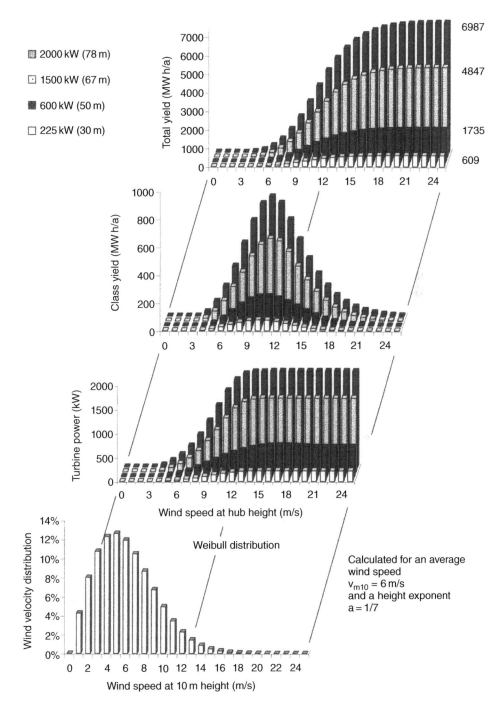

Figure 6.8 Calculation of the annual energy yield on the basis of wind measurements at a height of 10 m for various turbine sizes

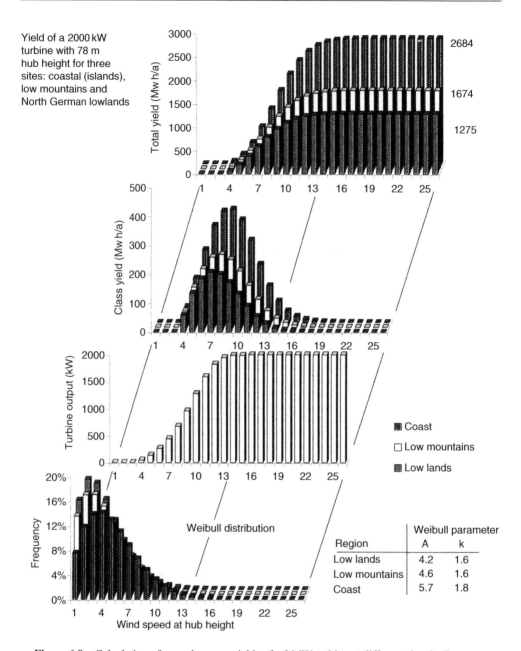

Yield of a 2000 kW turbine with 78 m hub height for three sites: coastal (islands), low mountains and North German lowlands

Region	Weibull parameter A	k
Low lands	4.2	1.6
Low mountains	4.6	1.6
Coast	5.7	1.8

Figure 6.9 Calculation of annual energy yields of a 2 MW turbine at different sites in Germany

In order to calculate the conditions at the site under consideration, we now move in the opposite direction (see the right-hand side of Figure 6.10 with the arrow pointing downwards). Regional statistics are included in the WASP programme using local parameters. Wind climatological factors such as the structure of the landscape (see Figure 6.11), surface

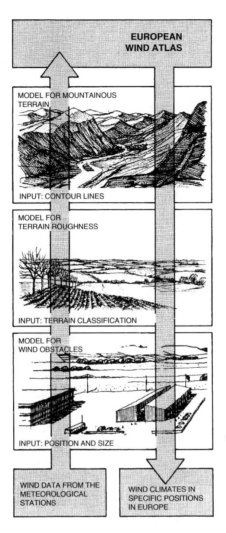

Figure 6.10   Procedure for resource analysis according to the wind atlas method. Reproduced by permission of National Laboratory Risø

texture or roughness and obstacles at the site are particularly important. Shadowing losses in wind farms can also be taken into account.

The WASP programme was developed for use in areas without complex orography. It therefore provides reliable information about local wind conditions for the analysis of areas in coastal regions. However, in highly structured landscapes inland and in low mountain ranges, such calculation procedures are of limited use. However, more complicated methods, e.g. the so-called Mesoskala model which takes into account factors like jet effects in the landscape, do supply relatively good predictions even for complex landscape structures, but they are much more expensive. The scientific measurement and evaluation programme 250-MW Wind was

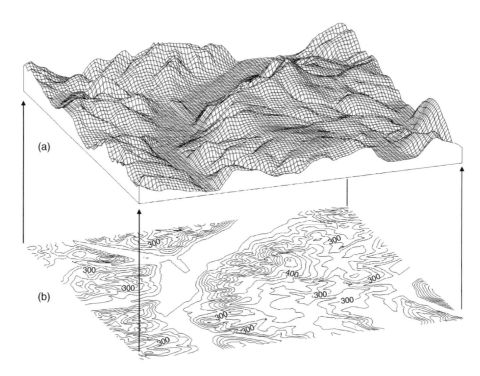

Figure 6.11   Surface structure (orography) of (a) a location based upon (b) map contours

a long-term evaluation test. In the whole of Germany, more than 1600 wind turbines that were installed in the 1990s were tested for ten years in almost the whole of the multitude of their systems, plant size and manufacture, especially their site-specific power output, the operating and failure times up to the grid coupling duration as well as the servicing and repair effort. After the completion of this, more than decade and a half test in the year 2006, this unique comprehensive measuring and evaluation could no longer be carried out completely so that often no new pictures of various depictions were available. The new edition of a similar evaluation test programme was not able to provide comparable results for manufacturer-independent and not-yet built multi-megawatt plants and substantially improve the perspectives of these systems.

The suitability of the landscape along with wind speed plays a decisive role for wind energy utilization. In coastal areas, locations directly next to the water are preferable. Surface roughness is lowest in such areas. Turbines installed at a distance of 5 km from the coastline achieve significantly lower energy yields than turbines located directly on the coast. Inland, exposed locations are of particular interest. Elevated plains and mountain ranges, with as little woodland as possible, and in which the wind can flow freely in the most common wind direction (usually south-west in Germany), are the preferred locations. There should be no other hills or obstacles in the immediate vicinity. It is difficult to obtain planning permission for sites close to nature reserves or national parks and buildings or towns. The distance from the supply grid should be kept as short as possible for cost reasons. Other factors that should be particularly taken into consideration in the selection of a location – again primarily for cost reasons – are land ownership, roads that are available or under construction and the solidity of the building land.

6.2   Potential and Expansion

In the context of wind energy exploitation, we must differentiate between the site and economic potential. The former covers the conditions for the erection of a turbine, which depend upon meteorology, topography, buildings and conditions relating to the permissible operating methods. Economic potential, on the other hand, relates to those locations that offer the possibility of profitable operation under the prevailing economic conditions for energy. In Germany, the coastal locations dominate this category.

6.2.1   Wind energy use on land

From the middle of the 1970s, some potential estimates [6.36–6.42] of power generation from wind energy especially for land use had already been undertaken. These produced widely differing results. Locality analyses carried out at the start of the 1990 were based on Lower Saxony [6.42] and Schleswig-Holstein [6.43] also arrived at differing expectations of potentials. Exclusion criteria played an important role in this. Today, conservative estimates have in part been widely exceeded [6.44–6.47].

The development of the installed wind turbine power of the German States, which in the past few years have become the most advanced in the world, are shown in Figure 6.12. It is remarkable that the frontrunners are not the States that have excellent wind conditions and large expansion that have the greatest technical and commercial potential. Here, the energy policy of the respective State is the deciding factor.

Figure 6.15 shows clearly that the USA after the first great wind energy boom in California in the middle of the 1980 had no perceivable growth for the following more than ten years. Only at the end of the 1990s were new installations erected. Meanwhile, the USA with approximately 9.9 GW in 2009 has achieved the largest growth figures. In the 1990s, Germany overtook the USA and at the end of 2002, 44% of the worldwide wind turbine power was installed there. This proportion decreases continuously due to the global expansion of wind energy and was 14% in 2010. The plants in 2010 in Germany produced around 40 TWh wind current. The breakthrough for this success was achieved due to a more than twenty-five-year intensive development which helped since 1990 by the necessary political and economical framework conditions.

The stormy development of the 1990s and at the beginning of the twenty-first century is flattening out at the present and a form of saturation is being observed to the end of this decade. The extension rates dropped in 2010 to 5.6% of the overall installed power of the previous year.

A similar development is seen in Spain with a time displacement. Also, in 2010, the USA, a country with gigantic wind energy potentials and enormously large available spaces was overtaken by China with the highest wind plant power in the world.

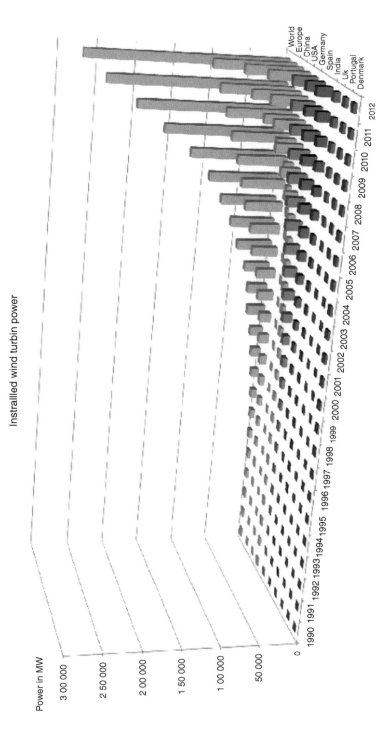

Figure 6.12 Development of the installed wind turbines in the world and Europe as well as in the most important developing countries (BWE-EWEA data, etc.)

India, Great Britain and China have made enormous efforts in the past years in the use of wind energy on a large scale. In this, they were able to establish themselves fully and leave the classic 'wind lands' of Denmark and The Netherlands far behind.

6.2.2 Offshore wind energy use

An important expansion of the energy potential on land can be achieved by the erection of wind energy plants in the ocean (offshore). Investigations for this were already being carried out in the 1980s [6.48, 6.49] and continued to the present with worldwide [6.50] as well as country-specific studies of the potentials (e.g. [6.51]). Further investigations have been carried out throughout Europe and Germany [6.52, 6.53]. They showed the technical feasibility and concentrated on the economical aspects. Large areas must be excluded from their use as they affect sea lanes, military restricted areas, pipelines and sea cables as well as nature reserves. *Offshore* is defined as anything that is at least 12 nautical miles from the coast. At these distances, the depth of water in the North Sea is already 15 m to 30 m deep. The resulting high costs for foundations and the much greater effort and servicing required for grid connection in comparison to coastal or land plants must be compensated by higher yields and larger farms. However, at present, there is no sufficient experience on loading due to water currents [6.54], waves, ice, etc. or of wind data for heights of approximately 60 to 100 m above the surface of the water. Also, countries such as Denmark [6.55], The Netherlands [6.56] or Sweden [6.57–6.59] only have experience in the coastal region. These will only be used to a limited extent even for future large wind farms due to their visibility from the coast. As offshore sizes in the region of 5 MW will be achieved, there is great potential for substantial cost reductions – similarly as in the past on land. The first projects were carried out in Germany in 2009 (Figures 6.13 and 6.14). Optimistic prognoses [6.60] in Germany are for an offshore wind energy utilization of 45 GW in the time period to 2020 and 85 GW by 2050.

Already in the 1990s, investigations were carried out by other European countries with – from the present point of view – relatively small wind turbines and mostly 'mini farms' to gather first offshore experience. Only in the new millennium was there a noticeable increase in offshore installations. These have meanwhile exceeded the gigawatt limit all over the world. Thus, wind turbines were first erected in water depths between 4 and 11 m and at distances of between 5 and 10 km from the coast.

In Germany, the offshore wind farms are erected in water depths of more than 15 m and distances of more than 10 km from the coast in order to ensure the smallest possible effect on the Wattenmeer National Park. The costs for foundations and grid connections for the plants are correspondingly higher. As the feed-in subsidy in Germany up to now was insufficient for the erection and economical operation of offshore wind farms or to cover the additional high risks involved, no offshore wind farm were erected up to 2008. With the new feed-in subsidy according to the renewed extension of the Renewable Energy Law (EEG-4) from 01.04.2012, an economical use of wind turbines at sea is expected.

In order to limit the investment risks of offshore wind farms which move in a framework of 100 million into the billions of euros, there were erected in the Germany sea area, several research platforms (FINO 1 to 3) with up to 100-m high instrumentation masts. Besides the tests of the foundations, the investigations include the measurement of the wind velocity at various heights as well as the movements of the water, such as currents, wave paths, etc. In addition, comprehensive physical, hydrological, chemical and biological measurements are being carried out as well as the behavior of various birds is observed with the aid of videos. In this way, the approval authorities and wind farm operators expect to gain important knowledge before offshore erection.

The first German wind farm, Alpha Ventus, was erected in the measuring area of the FI-NO platform (Figure 6.15) and measurement results have been available since 2003. The test field is being further extended 45 km North-West of the Island Borkum. It is situated outside the Wattenmeer National Park and the 12-nautical mile zone in the exclusively industrial zone (EIZ). However, the underwater cables had to be laid in part through the national park of the Lower Saxony Wattenmeer. The wind form consists of 12 plants of the 5-MW class. Respectively six turbines of Repower (see Figure 6.16) and Areva Multibrid were erected in 30 m of water in two different models of foundations.

6.2.2.1   North Sea: Offshore wind farms

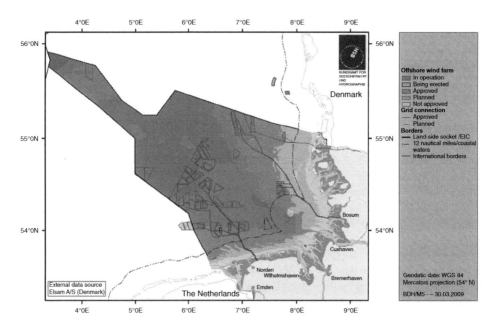

Figure 6.13   Wind farms in the German North Sea [6.61]

6.2.2.2   Baltic Sea: Offshore wind farms

All the estimates show that wind energy despite, for instance, already built-up areas still has substantial expansion potential. The present growth will continue especially by means of enlargement of individual plants, the replacement of smaller plants by larger ones (so-called Repowering) as well as offshore extension.

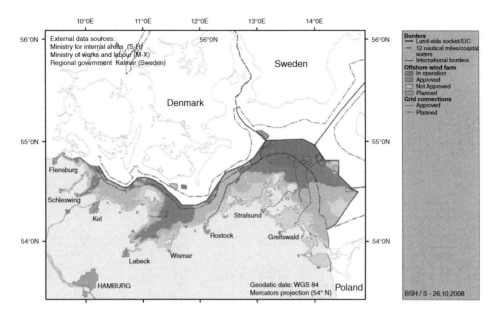

Figure 6.14   Wind farms in the German Baltic Sea [6.62]

6.2.3   *Repowering*

The replacement of older mostly smaller wind turbines by new, larger units has many advantages. The operation of more modern plants is more efficient. They also feed more electrical energy into the grid. In addition, the landscape is improved as wind farms are quasithinned out [6.63, 6.64]. A study, in which the coastal countries were investigated [6.65], shows that within fifteen years the wind turbines in Germany could be modernized. This would result – on the basis of 2005 – with 17 GW installed wind turbine capacity – with an increase of 2.5× to 42.5 GW and a 3× current yield with around 90 TW/h.

Individual results of the study are the following.

- With repowering, the energy yield increases by a factor of 2.2 to 4.3.
- The installed capacity rises by a factor of 1.5 to 3.5.
- The plant capacity (referenced to the average loading) is improved by 13 to 45%.
- The number of plants is reduced by half to a fifth.
- The height of the towers is approximately doubled.

An analysis of potential of repowering measures for the addition of wind energy in the most populated German State of North Rhine Westphalia [6.66] shows how the aim of the State government of covering 15% of the current by 2020 with wind energy can be carried out. With approximately the same demand, it will be necessary to feed in approximately 23 TWh per year compared to today's approximately 5 TWh/a or 18 TWh/a from wind energy. This will result in a need of around 6.8 GW of additional wind turbine power. With approximately 7 ha/MW area requirement, in additional areas, 5.6 GW can be added and 1.2 GW can be won from repowering. With comprehensive repowering of all possible plants in the first step with

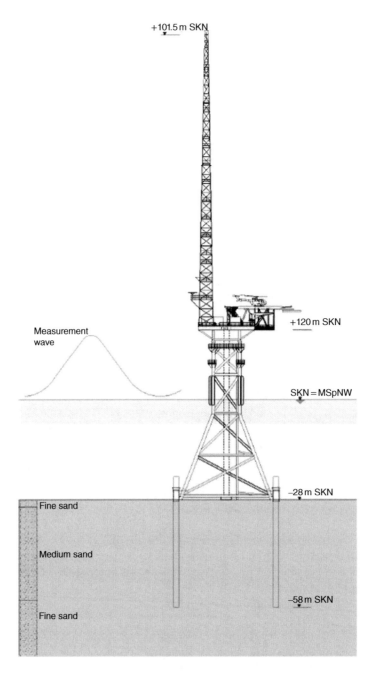

Figure 6.15   Research platform FINO 1 (Germanic Lloyd)

Figure 6.16   Repower-offshore plants

3 MW and in the second step with 5 MW plants, an additional 6.5 GW can be installed with the same number of plants. With repowering of plants to 600 kW by 3 MW ones and larger than 1.5 MW with 5 MW systems, it is possible with 10% less plants to add 4.6 GW capacity. By means of doubling the installed power and distributing to 3 and 5 MW plants, it is still possible to achieve around 2 GW with a halving of the number of plants.

The tight height limitations and spacing regulations of the States are a disadvantageous factor for repowering wind turbines. Adhering to them brings only minimum efficiency advantages. Also, this type of perspective makes large-scale new investments more difficult. With their regulations, the States obstruct not only industrial growth and increased clean power generation, but also the desired improvement of the appearance of the landscape. Legal framework conditions of repowering are given in reference [6.67].

At present, already four German States (Schleswig-Holstein, Saxony-Anhalt, Mecklenburg-Vorpommern and Brandenburg) cover their power requirements by 50% and more from wind energy. Throughout Germany, the proportion in 2013 was around 9%.

Thus, wind energy has already made a significant contribution to the declared aim of the White Book of the EU 1997 [6.68]; the regenerative portion of gross primary energy requirement by 2010 has doubled from 6% to 12%.

6.3   Economic Considerations

Renewable energy systems can only compete with conventional installations in the long term if their use brings about significant advantages in the

- technical,
- political,

- economic,
- labour-market and
- ecological

fields. Such wide-ranging improvements cannot, however, be achieved in all these aspects compared with new and established systems. Partial advantages, on the other hand, are certainly possible. Particular importance must therefore be assigned to the weighting of individual aspects. In the future, in addition to operational economic considerations, social prospects for the future and ecological effects will increase in importance in this weighting. Special consideration should be given to the fact that the wind energy industry is relatively labour-intensive in comparison to the conventional power sector, as shown in Figure 6.17. For every direct job approximately another two indirect jobs are created in supplying industries. Danish manufacturers also contribute to this by importing a significant quantity of German components (gearboxes, generators, etc.). Moreover, the disturbance of the landscape will have to be given more consideration with regard to its ecological effects, such as the effects associated with brown-coal power stations with their mines and open-cast workings compared with wind and solar farms. From this point of view, energy sources that are still considered to be uneconomic today could bring many social and ecological advantages in the short, medium and long term.

These aspects are evaluated very differently in different countries. Accordingly, there are also large differences in the evaluation and payment systems for electrical energy from conventional

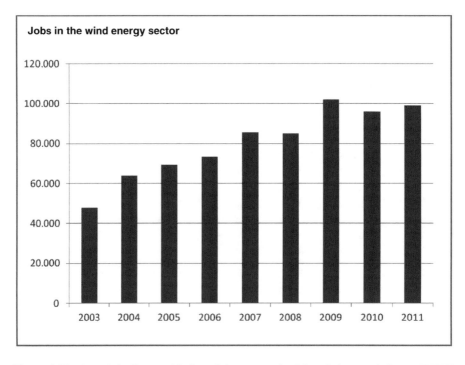

Figure 6.17   Growth in direct and indirect jobs as a result of the wind power industry (DEWI)

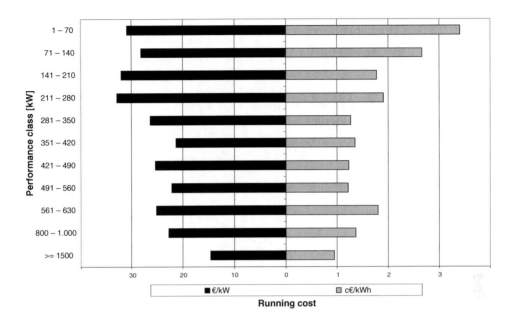

Figure 6.18   Growth in direct and indirect jobs as a result of the wind power industry (DEWI)

installations with fossil or nuclear conversion units and renewable systems. A unified tariff system within Europe or internationally cannot be expected in the foreseeable future.

6.3.1   Purchase and maintenance costs

Turbine costs (reference values are listed in Figures 2.78 to 2.80) can be determined after the compilation of a comprehensive profile of requirements and the definition of usage details, including technical turbine data and predicted costs for maintenance and repairs by the manufacturer or supplier. It is relevant to the cost calculation whether the converter will be used in isolated operation or combined with other power generation units (in grids, with diesel generators, etc.) and what demands will be made of the safety-related component groups (e.g. for installation in water protection areas). Costs for local transport, foundation, cabling and connection should also be taken into account. The total secondary costs lie somewhere between 15–30% of the cost of the turbine itself. If investment cost subsidies are granted in the framework of promotional measures, these can be drawn into the equation as a reduction of the acquisition costs.

6.3.2   Power supply and financial yields

The 'Energiewirtschaftsgesetz' [Act for the Promotion of the Fuel and Electricity Industries] obliges the operator of a wind turbine to inform the responsible power supply utility (PSU) of the fact. In addition to technical requirements to preclude the possibility of damage to the pubic supply grid, the provisions of the VDE [Verband der Deutscher Elektrotechniker [Association of German Electrical engineers]] must be adhered to. The Technical Connection Conditions

(TCC) can differ from one supply area to the next. The precise matching of the protective devices and instrumentation must be carried out in consultation with the PSU.

Financial yield – found by taking the product of the energy supplied and the tariff at which it is sold – is of fundamental importance for economic operation and thus also for the long-term use of power supply plants. This aspect will be considered further in the following.

The 'Erneuerbare Energien Gesetz (EEG)' (Renewable Energy Act) was introduced in April 2000 and January 2009 and extended in August 2011. This replaced the 'Stromeinspeisungs-gesetz' (Electricity Supply Act) that had been in force since 1991. The EEG sets the minimum prices that the PSU has to pay to the supplier of the power from renewable energy (Figure 6.19). The tariff for electricity from wind turbines on land that were put into service in 2013 was at least 8.80 eurocent per kilowatt hour for the period of five years calculated from the time of commissioning onwards. Thereafter, the tariff for turbines that in this time have achieved 150% of the yield calculated for the reference turbine according to the Annex to the law is at least 4.80 eurocents per kilowatt hour. At sites that do not achieve the reference yields, the period of higher remuneration is extended. For additional grid services such as voltage support, etc. that modern wind turbines can provide, an additional bonus amounting to 0.47 eurocent (for IBN 2013) is awarded.

For offshore feeds, there is a payment from 2013 of twelve years of starting subsidy of 15 eurocent per kilowatt hour, which is substantially higher than the onshore subsidy. The degression begins in 2018 and is 7% per years. In this way, the starting subsidy is lifted to a comparable subsidy with other EU countries. In a counteraction, the base subsidy was substantially reduced to 3.5 eurocent per kilowatt hour.

As an alternative to a fixed subsidy, the Renewable Energy Law (EEG) since 2012 has offered the possibility of direct marketing of the onshore or offshore generated electrical energy. The differential amount between the stock exchange yield and the regular EEG subsidy is thus paid by means of a market premium. In addition to this a management premium of 1 eurocent per kilowatt hour (for IBN 2013) is paid and is meant to cover the additional costs and risks of this form of marketing (Figure 6.20).

Spain is also currently using a minimum price arrangement. Other countries are taking different routes, but despite the fact that these countries often have a greater exploitable potential

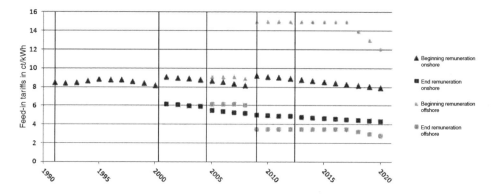

Figure 6.19 Average operating costs per kW rated power and specific operating cost per kWh of annual energy yield, Wind Energy Report Germany 2008, ISET

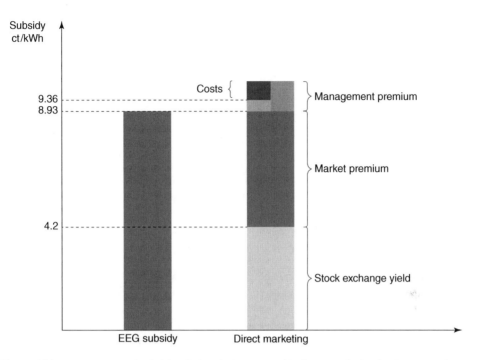

Figure 6.20   (a) Example of subsidy of electrical energy with direct marketing in the sense of EEG 2012 (yy1)

than Germany these approaches have not led to such large increases in wind power genera-tion as the German minimum price system. Great Britain has carried out tendering rounds in connection with its Non-Fossil Fuel Obligation (NFFO) Denmark has recently gone over from a minimum price system to a certificate trade in green power, which has led to a sharp decrease in growth (see Figure 6.12)

These descriptions show that the systems of remuneration for wind-generated electricity are very differently managed even in Western Europe, Investigations regarding the achievable yields were carried out for differently sized turbines at selected sites or the corresponding wind conditions for a few Western European countries under the applicable conditions [6.69]. These studies used yield data for different wind turbines at different times and locations determined by the 250 MW Wind Scientific Measurement and Evaluation Programme' (WMEP), so that the labour and service components of the payment in the countries under consideration could be accounted for according to the particular conditions The considerations thus relate to wind conditions at German sites. Under the conditions mentioned, the investigations showed that Switzerland, Luxembourg and Germany pay the highest tariffs, while Sweden, France and Ireland pay the lowest. Results from the so-called large-scale testing programme (WMEP) show the dependency of the monetary annual yields upon the turbine size and in particular upon the site category in Germany. In order to be able to undertake a comparison between different sizes, etc., the monetary yields for the year 2000 were related to the installed turbine output. Figure 6.18 shows, e.g. that for turbines of the 50 kW class the yields amount to around 180 euros per kW installed power in the medium mountain ranges and in the North German lowlands or 320 euros per kW a. 2 to 3 MW plants generally exceed these values.

The annual financial yields from wind turbines and wind farms can be estimated from the specific values or calculated from the energy yields and tariffs. They are balanced by the electricity generation costs, which will be considered in what follows.

6.3.3 Blue section

The actual costs of power and wind energy ion euro per PWh are shown in Figure 6.22 in dependence of the specific investment costs in euro per kW of installed capacity and for the utility duration of the plants in full-load hours per years. In addition, the present feed-in subsidies that can be expected after erection of the plants until reaching the reference feed for offshore and onshore can be seen as two parallel surfaces (0.09 and 0.15 euro/kWh). In addition, the current actual costs for annual operating costs are supported by between 1 and 6 ct/kWh as six parallel surfaces. In this, a plant life of twenty years and a present rate of interest of 7% or annuity factor of 0.0944 was estimated.

Figure 6.21 shows that offshore plants with, for instance, high specific investment costs of approximately 3000 euros per kW and annual operating costs of 6 ct/kWh requires about 3200 full-load hours. For low investment costs, lesser full-load hours are sufficient for an economical offshore operation.

In the onshore utilization, in contrast, with low specific investment costs of approximately 1200–1500 euro/kW operating costs can be economically achieved already with 1500 full-load hours.

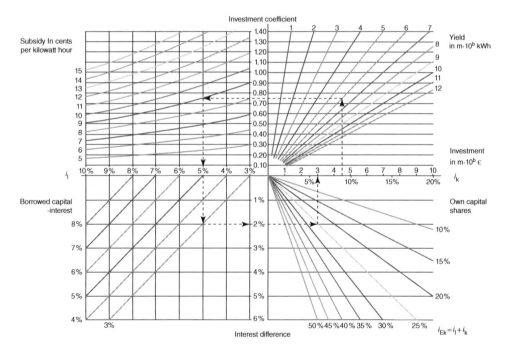

Figure 6.21 Nomogram for determining the economic viability of wind turbines or wind farm as an investment object with m as a scaling factor and b as exponent [6.70]

6.3.4 *Commercial calculation methods*

Static as well as dynamic calculations can assess the economic viability of wind turbines. In the annuity method, returns and costs are assumed to remain constant (static) over the entire depreciation period. In contrast to this, the capital value method of calculation takes into account the loss of value of the loan resulting from inflation and increasing return due to increasing remuneration for supply. Moreover, promotion schemes in the European Community, Germany and the individual States play an important role in the question of economic viability.

6.3.4.1 Annuity method

The annuity reflects the annual percentage portion of interest and repayments for externally financed loans. Capital costs can be easily determined from the relationship

where $k = p + \dfrac{p}{\left(1 + \frac{p}{100}\right)^{z} - 1}$

p = Interest in percentage
Z = Repayment in years
K = Annuity in percentage

Annual operating costs, capital costs and taxes must be included in the calculation. The operating and additional costs incorporated should be approximately 1.25%, insurance approximately 0.9% and approximately 4 to 0.6% for excess and other costs. For a fifteen-year duration and an interest rate between 4 and 6%, an annuity of 9 to 10% can be expected.

Dynamic calculations using the capital value method can provide a long-term point of view in the financial evaluation of wind turbines.

6.3.4.2 Capital value method

For commercial evaluation of wind turbines, the dynamic method of calculation with the aid of the capital value method is suitable for a long-term point of view
The starting point is the equation

$$C_0 = -I + \sum_{t=1}^{T} \frac{Z_t}{(1 + i)^t}$$

where
C_0 = capitalvalueint = 0
t = year
T = life
I = investment sum
Z_t = $E_t \cdot (p - k)$, payment excess in year t
E_t = Generated energy in year t
p = Subsidy
k = Periodic costs per kilowatt hour

Iterative solution methods permit the determination of the redemption period *AZ*, i.e. the year *t* in which the capital value is zero or the internal interest level, i.e. the rate of interest at which the capital value is zero. The following figures show a graphic depiction of the linkages described by the equation.

where

i_I	= interest on the investment amount
i_K	= interest on the capital structure
$i_{EK} = i_I + i_K,$	= interest on own capital

where

$$I_{ges} = \frac{i_{EK}}{I_{FK}} + i_{FK} \cdot \frac{I}{I_{FK}} \text{capital costs}$$
i_{EK} = own capital sum
I_{FK} = i_{EK}, foreign capital sum
i_{FK} = borrowed capital interest

In order to use the nomograms also in the border regions, a scaling factor *m* can be introduced that is to be selected as an equal value for the investment sum as well as the annual energy production. The nomograms are based on an assumption of a constant periodic cost of 3 eurocents per kilowatt hour (operating and safety surcharge) as well as a plant life of twenty years. Various results can be read off the selected parameter regions of the nomogram. The suitable nomogram must be selected [6.70] depending on the point of view.

The first nomogram (Figure 6.22) shows the viewpoint of a wind turbine as a pure investment object. The interest on the capital invested as well as the interest of the own capital can be read off this nomogram.

The second nomogram (Figure 6.23) is suitable for showing the economic viability of a wind turbine for own use of the generated electricity. The final economic coefficient found here is the dynamic amortization time. The view taken here is the political economic relevance of the effects of rising energy costs that lend the nomogram its strength also from global political economic points of view. Both nomograms have in common the so-called investment coefficient that can be determined also on the basis of the mathematical relationship:

$$\text{Investment characteristic} = \frac{I}{E}$$

With an example of a 3-MW plant that is operated in parallel with the grid and is to serve as an investment object (choose the first nomogram for this). The estimated investment costs are 4.5 Mio, $\Delta = 4.5 \times 10^6 \Delta$. The annual energy yield is assumed to be 6MiokWh = 6×10^6 kWh. This gives an investment coefficient of 0.75. For a subsidy of 9 eurocents per kilowatt hour, there is an interest on the overall capital of 5%. If one considers loan capital of 3% as well as an own capital portion of the investment sum of 25%, then one obtains additionally from the capital structure of the investment an interest of 6%. The interest on the own capital is the sum of the calculated interest, in this case of 11% ($i_{EK} = i_I + i_K = 5\% + 6\%$).

If, however, a plant with the same costs and yields is operated as a plant for own use of the generated electricity (choose the second nomogram), whereby the energy costs with 9 eurocents per kilowatt hour is determined, then the result with an assumed annual energy cost rise of 4% as well as capital costs of 5% is a dynamic amortization time of approximately

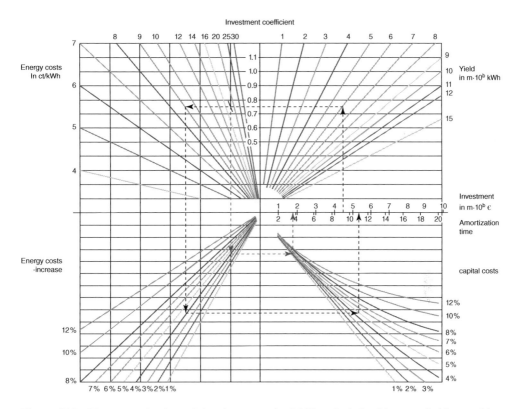

Figure 6.22 Nomogram for determining the economic viability of wind turbines or wind farms taking increasing energy costs for own use or global views into account with *m* as a scaling factor and *b* as exponent [6.70]

twelve years. If, on the other hand, one views an island supply with double fuel-conditioned energy costs of 18 eurocents per kilowatt hour under the otherwise same limiting conditions, then the result is an amortization time of 3.8 years.

Also other feedbacks such as, for instance, minimum subsidies or maximum bearable loan capital interest can be calculated with the aid of this nomogram that provides the influence of various parameters in a clear manner.

6.4   Legal Aspects and the Installation of Turbines

The densely populated nature of Germany and the enormous growth in the use of wind power over recent years mean that it is crucial to harmonize the interests and objectives of turbine planners and those living in the vicinity of the planned site. The main considerations here are the casting of shadows, the propagation of noise and the long-distance visibility of the now usually very large turbines. Beneficial effects of development have been that rotor speeds have fallen continuously as turbine diameters have risen and that large turbines give a much more peaceful impression than the older, smaller types in the 100 kW class.

The realization of the EU directive on environmental testing [6.71] should lead to the above-mentioned problem areas being countered and the interests of the environment being taken into account. Formal environmental compatibility tests are required for groups of more than 20 turbines. For groups of up to 19 turbines there must be a general preliminary test to determine the necessity for an environmental compatibility test. For three to six turbines only a site-specific preliminary test is required. The precise implementation of the environmental compatibility test for wind turbines has not yet been set down in the statutes in Germany. Therefore the regulations vary between different regions and often between individual communities. Schleswig-Holstein, however, is taking the lead in harmonizing the regulations [6.72]. Even without unified regulations, many planners have calculations of the noise emissions, shadow casting and the visibility of their turbines performed in order to rule out from the outset conflicts with those living nearby.

6.4.1 Immission protection

In Germany, the Bundesemissionsgesetz (Federal Emission Act) forms the legal basis for emission testing [6.73]. Where wind turbines are planned close to residential areas, the main considerations are noise emissions and the casting of shadows by the turbines. The previously much-discussed disco effect, caused in particular by reflective coatings on the rotor blades, has now been countered by the use of further-developed matt coatings.

6.4.1.1 Noise propagation

The propagation and effects of noise are broken down into the following categories:

- emission [6.74],
- transmission [6.75] and
- immission [6.76].

The DIN and ISO standards mentioned and the VDI directives deal with the category in question. Further standards and directives are concerned with the minimization of noise emissions [6.77].

The effects of noise are ultimately judged on the basis of immissions. Values for immissions are given in the Technischen Anleitung Lärm (TA Lärm) (Technical Directive on Noise Control) [6.78] and in the Baunutzungsverordnung (Ordinance on the Use of Buildings) [6.79]. Furthermore, an immission protection ranking is specified in this, adherence to which is evaluated by the immission protection authorities as part of the Gewerbeaufsichtsamtes (Trade Supervisory Agency) or the Umweltamtes (Environment Agency).

For night-time hours, when stricter measures are specified, the following maximum noise levels apply:

- 35 dB(A) for purely residential, leisure or health resorts;
- 40 dB(A) for general residential areas and small estates (primarily homes);
- 45 dB(A) for central, mixed and village areas where no usage type predominates;
- 50 dB(A) for industrial areas (primarily industrial plant).

The logarithmic dB(A) scale reflects to some degree the specific noise-sensitivity of the human ear. If individual values are exceeded, but not the limit values for the day, an order may be imposed for the shutting down of the turbine overnight. However, this would result in a severe sacrifice in terms of returns.

In order to ascertain the level of noise emissions with a reasonable level of certainty in the planning phase, commercially available programmes are now used that permit a relatively precise prediction of the risk areas. To this end, the orography of the surrounding area, as described in Section 6.1.2, is first recorded and relevant immission areas, such as estates, mapped out together with their limit values. The wind turbine is entered at the planned site and the noise immission calculated according to the standards [6.74-6.78]. The emission of the wind turbine is determined by the manufacturer according to a standardized procedure [6.80] and calculated at the wind speeds available at the site. Tones and pulses subjectively perceived as annoying, such as those caused by gearboxes, are evaluated by adding a supplement to the noise level. Existing noise levels due to roads, etc., are also taken into account, as are reflections and absorption, for example, due to the ground, the air or obstacles.

The result of such a calculation is a map as shown in Figure 6.23, in which along with the wind turbine and the relevant immission areas the ISO lines of the above-mentioned noise

Figure 6.23   Noise immission map for wind turbines (drawn up using the forecasting program [6.84])

limit values are drawn in. Vibrations in the ground give rise to so-called 'infrasound', which can now be classified as harmless [6.81, 6.82].

6.4.1.2   Shadow casting

Wind turbines are usually installed at exposed sites, e.g. on a hill. As a result they cast long shadows. This effect can be precisely determined by means of standard programs [6.83].

To determine the shadows cast, the orography data that are available from energy and noise predictions are drawn into the calculation along with the path of the sun (which can be determined by the geographic coordinates) and the turbine dimensions. This information is then represented in the form of a map (Figure 6.24). The characteristic path of the areas in shadow usually has a butterfly-like shape, which is caused in particular by the low positions of the sun in the morning and evening.

The calculations can be based upon the worst case for the time of day using the assumption of permanent sunshine. This yields the maximum possible shadow casting periods. For a more realistic estimation, meteorological data on the statistical sunshine hours at the site in question is drawn upon. The prediction can be refined still further by including turbine stationary times. Intelligent turbine control systems can position the rotor with one blade downwards during the stationary period so that the total height of the turbine including rotor, and thus the shadow casting, is minimized. If predetermined limit or guide values are exceeded, the turbine can be switched off at critical times of the day.

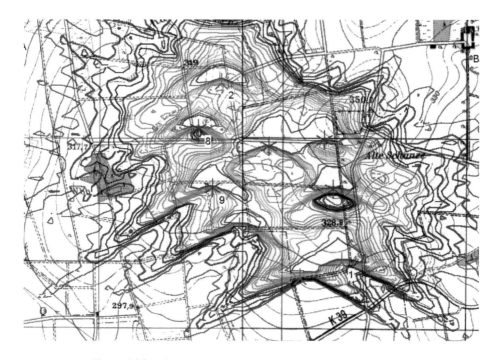

Figure 6.24   Shadowed area of a wind farm of eight turbines [6.83]

Since rotating turbines of this order of magnitude, which can exceed 150 m blade tip height, are new to a field of building regulations that usually relates to buildings, statutory guidelines and limit values have yet to be revised. Up until now, limit values of 30 hours per year or 30 minutes per day have normally been used in worst-case calculations. Furthermore, the criterion developed at the Technical University of Kiel, according to which a brightness change with less than 20% coverage of the visible disc of the sun by the rotor blade is not subjectively perceived, is also used. Thus the distance of the immission point from the turbine takes on increasing importance.

6.4.2   Nature and landscape conservation

In Germany, in contrast to many other countries, nature and landscape conservation is assigned particular importance. In this connection, disruption of the landscape due to visibility of turbines and of the migration and breeding of birds must be considered as well as effects upon the ecology of land and water animals, plants and the water balance (pollution caused by hydraulic and gear oil, etc.). Often compensatory landscaping measures are required. The following brief descriptions should, however, be limited to the two first and last areas mentioned.

Wind turbines, due to their size, represent a disruption to the landscape, although in terms of time this is limited due to the service life of approximately 20 years.

6.4.2.1   Visibility

First of all, the question of where the planned wind turbines are actually visible from must be answered during the planning phase. In addition, computers are used to check whether the view to the topmost point of a turbine, the rotor tip, is interrupted by landscape elements (mountains) and obstacles (buildings). A visibility map drawn up in this manner shows both visibility and shadowing areas around the turbine. In the case of wind farms, the number of turbines visible at the location in question is determined. The evaluation of this map is, however, subject to subjective points of view. Figure 6.25 shows an example of such a visibility map in which the shadowing of individual obstacles is indicated by means of dark specks.

The visual impression of a planned wind turbine can be illustrated with the aid of commercial photo-processing programs [6.85], as shown in Figure 6.26. At this point three-dimensional models of the planned turbines are incorporated into a photo and visually displayed. Furthermore, the simulated rotor movement can also be considered on the computer. Visualizations are particularly common in public presentations of projects.

6.4.2.2   Compensatory measures

In Germany, if the landscape is disrupted compensatory or replacement measures must be put in place. To this end, a survey is drawn up in which the environment surrounding the wind turbines are first of all divided into so-called landscape-aesthetic units, which are then classified according to their value. This is then used to evaluate the intensity and importance of the disruption for each unit. Additional factors, such as perception coefficients and compensation area values ultimately determine the size of the compensation area that must be upgraded

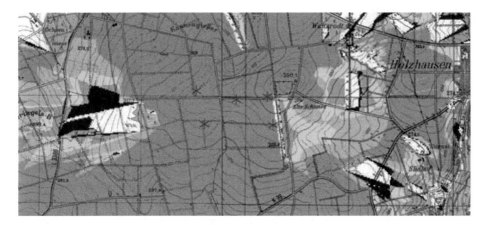

Figure 6.25 Results map of a visibility analysis [6.86] (visibility shadows of obstacles are shown as dark specks)

Figure 6.26 Visualization of a wind farm showing existing turbines with planned turbines superimposed with the aid of a computer [6.85]

in aesthetic terms by special measures such as tree planting. Several procedures have been developed for this calculation, e.g. the landscape compensation method by Dr W. Nohl.

6.4.3 Building laws

With regard to building laws and the planning of wind turbines, the Baugesetzbuch (BauGB) (Building Code) [6.87] and Baunuzungsverordnung (BauNVO) (Ordinance on the Use of Buildings) [6.88] must be observed within the jurisdiction of Germany, and the relevant Landesbauverordnungen (LBO) (Regional Building Ordinances) must be observed within the jurisdiction of the German regions.

In Germany, basic requirements concerning planning permission for turbine sites are regulated on a national level. According to the German Baugesetzbuch:

1. A general development plan must be submitted, with a construction and area utilization plan covering constructional and other uses.
2. The constructional use must be specified within the general development plan together with the BauNVO, taking into account areas that can be built upon and those that cannot be built upon and the permissible constructions on the construction areas.

According to the Baugesetzbuch, planning competence is transferred to the local planning authority (usually the municipal authority). In built-up areas (local turbines) the turbine must blend in with existing buildings and may not detract from the view of the area.

According to section 35 of the German Baugesetzbuch, projects in outlying areas are only permissible if the project is not detrimental to the public interest, if the public is adequately informed and if the project supplies the public with electricity, heat, water, etc. Due to the particular environmental requirements of wind turbines, e.g. favourable wind conditions, the possibility of connection to the grid and the like, the prerequisites for planning permission may be granted. In the updating of the Baugesetzbuch (BauGB) 1997 [6.89] wind turbines were categorized as so-called privileged plant, which makes it much easier to obtain planning permission. Permission can only be refused if it is not in the public interest. Encroachment alone is not sufficient. Due to this privileged position the allocation of prime sites for wind turbines should be promoted by local authorities as part of a space utilization plan.

6.4.4   Planning and planning permission

The rules of procedure for plant construction planning permission stipulate that it is compulsory to obtain planning permission for the erection of wind turbines. There are no guidelines that specify a maximum tower height and rotor diameter at which wind turbines can be erected without permission. For smaller turbines, however, the rules of procedure and inspection procedures are normally relaxed and turbine construction is simplified.

Therefore, in principle, permission must be requested for all projects. It is advisable to submit an application for outline planning permission to the responsible inspector of works through the municipal authorities. This application should include a description of the entire site, site plan, ground plan and views, possibly with a drawing or photograph of the wind turbine. The authorities will reply, explaining the basic erection options and giving notes on the further procedure with regard to changes to the construction plan or to the turbine. They will also indicate any additional licensing procedures that are required, together with the responsible authorities (e.g. nature conservation or countryside preservation authorities) or the requirement for a regional planning order (if not already initiated).

In the application for planning permission, differentiation must be made between

- private or commercial users, who must direct their application to the responsible building supervisory board (local authority, District Office, president of the administrative district) and

- authorities, who must initiate a consent procedure via the president of the administrative district to the responsible state building supervisory board (e.g. the University Building Office).

The quantity and scope of documentation to be submitted are not regulated in a unified manner. They are, however, precisely specified by the local authorities. The following must normally be submitted in triplicate:

- building specifications;
- layout plan or copy of the cadastral map (1:1000 or 1:5000) showing the location of the turbine;
- construction drawings with view, ground plan and at least one sectional drawing (1:100), which will normally be provided by the manufacturer;
- static calculations for the tower and foundations to prove its stability, and a certificate of operating safety;
- type approval is normally necessary for mass-produced plants – if this is not available then the following individual licences are required:
  - appraisal report, certificate, etc., for construction and components, according to the applicable guidelines;
  - certificate covering the turbine's safety equipment;
  - technical certification of the nacelle and rotor;
  - test results concerning noise measurements and possibly turbine vibration behavior;
  - instructions for the operator;
- maps for
  - noise immission,
  - shadow impact,
  - visibility and the
  - visualization of the turbine or wind farm.

This completes the documentation. If the application is rejected, an appeal can be lodged, which leads to an investigation of the process.

Procedures and guidelines for checking the compatibility of wind turbines with nature and the countryside are not currently regulated in a unified manner. This can significantly extend the licensing procedure. Therefore, a time schedule incorporating all stages up to the operation of the turbine must incorporate certain planning uncertainties.

6.4.5   Procedure for erecting a wind turbine

Figure 6.27 shows the sequence for planning, approval and erection of wind turbines. This flowchart will clarify the important steps and decisions on the path to plant operation in a coarsely structured and greatly simplified manner.

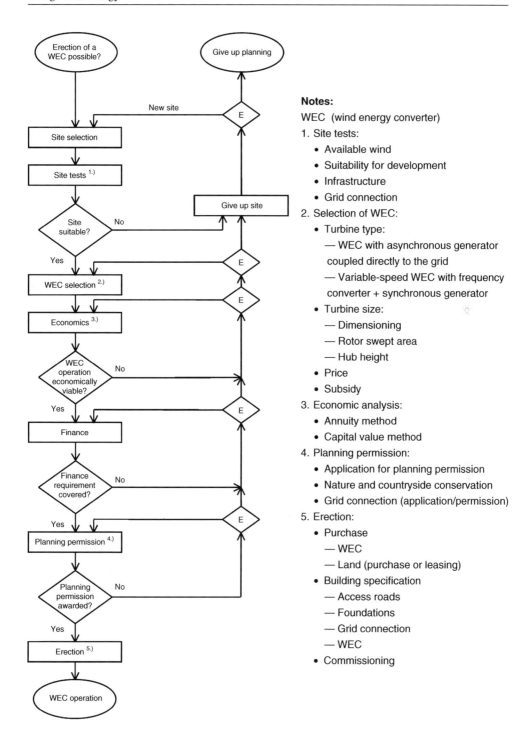

Notes:

WEC (wind energy converter)

1. Site tests:
   - Available wind
   - Suitability for development
   - Infrastructure
   - Grid connection

2. Selection of WEC:
   - Turbine type:
     - WEC with asynchronous generator coupled directly to the grid
     - Variable-speed WEC with frequency converter + synchronous generator
   - Turbine size:
     - Dimensioning
     - Rotor swept area
     - Hub height
   - Price
   - Subsidy

3. Economic analysis:
   - Annuity method
   - Capital value method

4. Planning permission:
   - Application for planning permission
   - Nature and countryside conservation
   - Grid connection (application/permission)

5. Erection:
   - Purchase
     - WEC
     - Land (purchase or leasing)
   - Building specification
     - Access roads
     - Foundations
     - Grid connection
     - WEC
   - Commissioning

Figure 6.27   Procedure for the planning and erection of wind turbines

6.4.6   Offshore utilization of wind energy

The large-scale utilization of wind energy on land will reach its limits in the next decades, especially in heavily populated countries. On the one hand, the availability of windy onshore localities will become increasingly scarce. On the other hand, the plant dimensions will become ever larger so that the transport of components will become increasingly difficult. This development on land can be seen as the trailblazer for offshore wind energy utilization. Only the steps taken on land for turbines in the 5 MW range and extensive operating experience will permit an economically feasible use at sea. A condition for this, however, is the safe operation of the plants with a high degree of availability (of, for instance, 98%). Its Figure 6.27 realization on land simplifies advances at sea greatly. With production facilities of the component and plants in the vicinity of harbours, there are practically no transport limitations for gigantic rotor blades of 60 to 80 m or large generators and machine house dimensions, for instance, up to 12 m diameter and 25 m length as well as for heavy component weights up to 1000 tonnes.

In Germany, the technical and economically usable offshore potentials are around 11–18%, in The Netherlands 20–30%, in Great Britain around 10–15% and in Denmark, half the present current usage. At a political level, these impressive wind energy potentials at sea have increased enormously in importance with respect to an independent energy supply. On the one hand, planning safety was created for investors in Germany with the fourth version of the Renewable Energy Law (EEG 2012). On the other hand, the legal framework conditions were improved. With the Infrastructure Acceleration Law, it is defined, for instance, at what point the grid connection starts for the grid operator.

Germany will only be able to maintain its position as one of the leading countries on the world market for construction and export of wind turbines in the medium and long term when, here too, offshore wind farms are erected and own experience is gained. The research platforms at sea (FINO 1 to 3) and the wind farm ALPHA VENTUS were erected for the primary purpose of gaining comprehensive experience with research and with the feed-in subsidy of 2012; important framework conditions were created that offer the best conditions for successful installations at sea. In addition, higher and more constant wind conditions than on land are available that become ever more advantageous with increasing distance from the coast but they also result in mostly deeper water depths and higher foundation and grid connection costs.

The erection possibilities of wind farms at sea are mainly limited by the depth of the water. Shipping lanes as well as the influence of the turbines on the environment, fish and birds are further considerations. Visual effects are generally not to be expected due to the great distances from the coast,

Besides the influences of the wind turbines on the environment, also the effects on the turbines must be taken into account. Here, among others, weather conditions, the salt content in the air and sea water, intensive UV radiation, waves and currents as well as ice formation and ice forces [6.90] on plant and foundation must be considered as well as influences on the ocean floor and the anchoring of the foundation. First, experiences were gathered with long-term studies for research platforms [6.91]. From this, it became clear that the effects on the turbines can be fully mastered by technically acceptable means. Further, it has been shown that no serious damage for the animal world need be expected.

A special challenge for erection and operation of offshore wind turbines are the foundations. The sea water resistance to the materials used and the selected structure is a precondition for safe long-term operation. Foundation variants must be selected in accordance with the water depth and the properties of the ocean floor and these must be able to be installed without great preparations [6.92]. Thus, effects on the environment are kept to a minimum.

Basically, the foundations can be divided into different categories. Monopile, tripile, tripod, jacket and bucket models as well as gravity and floating foundations are the most important variants and will be briefly discussed in the following.

Monopile consists of a steel pipe that is rammed into the ocean floor by means of a pile driver. This variant can be installed without ground preparation and was used in the Danish wind farm 'Horns Rev' for eighty 2 MW wind turbines. However, it is limited mostly to small wind turbines in shallow water. With increasing plant size and water depth, the material required increases and the costs rise disproportionally.

Tripile and tripods consist of three steel tube stands that combine in different forms into a pipe in the centre. Tripiles have a simple star connection above the water. Tripods, in contrast, are reinforced into the central tube by means of three additional braces. This protrudes out of the water, does not touch the bottom and is provided with ice protection. Anchoring is by means of three steel tube stands that, for a tripod, can be rammed up to 30 m into the ground. Thus, applications in 20 to 50 m water depths are possible. No great preparation is necessary as these foundations are only fixed to the ground at three places. The deep anchoring also protects against scouring, washing away or undercutting on the ocean floor. However, the ramming work is intensive and loads the environment. Tripods for 5 MW turbines have a weight of approximately 300 tonnes.

Jackets have their origins in oil platform technology. They are very similar in construction to lattice masts. Their anchoring to the ocean floor is carried out as for tripods. However, they use less material and thus the costs are substantially lower. A decisive disadvantage is that they offer little safety in the case of a collision by a ship.

Buckets have a bucket shape that is placed on the ocean floor in an inverted form. Then, the water inside the foundation is drawn out. The vacuum pulls it into the ocean floor and anchors it. Thus, no ramming work is required, which permits an environmentally friendly installation. However, the ocean floor must possess a consistency such that the anchoring to the floor is ensured.

Gravity foundations originate in bridge building. Hollow prefabricated foundations are manufactured on dry ground, floated to the site and sunk there. For this, almost no ground preparation and no ramming is required. In addition, little steel is required for manufacture. However, the costs rise strongly below 10 m depth. This variant can hardly be used due to the great distances of the offshore wind farm from the German coast.

Floating foundations are anchored to the ocean floor by means of a flexible connection at great water depths with comparable little effort. A particular challenge is posed by the static and dynamic forces acting on the wind turbine due to wave movements. In addition, the floating base must be provided with a corresponding counterweight that keeps the wind turbine in all wind and operating conditions in the atmosphere and protects it from the sea water.

Offshore wind turbines will probably be dimensioned in the 5 to 10 MW region and will have a turbine diameter of approximately 120 to 140 m. Plants and component with these dimensions can no longer just be brought from inland factories to the coast or the harbour. Their manufacturers must therefore move their production facilities to the coast in order to load all possible products directly to ships. Harbours, storage halls, loading systems and ships [6.93] must then also be dimensioned to suit the wind turbine sizes and masses, and sea lanes must be correspondingly deepened in order that even at low tide expensive delays in transport, installation and servicing is avoided.

In the transport of wind turbines and in sea transport, in general, it must under no circumstances occur that a ship is so damaged by a foundation or a wind turbine due to collision that persons and the environment are endangered. The safety against collision that can occur due to foundation-specific hazard potentials is thus a further important decision criterion for foundation design. As German offshore wind farms are only erected at approximately 30 km from the coast only foundations for greater water depths are considered.

The financial and economical framework conditions for erection and operation of offshore wind farms far exceed those of onshore installations. Small investor groups and medium-sized companies are generally not in a position to raise sums amounting to billions for offshore installations and offer financial reserves for plant operation. Here, large lines of credit from banks are required. These normally demand correspondingly high security. Lack of experience with offshore wind turbine investments and this branch of the technology make the completion of projects more difficult. With the aid of rating processes, attempts are made to evaluate the risks of these systems [6.94]. In order to obtain even more security, the banks demand much longer guarantee times from the manufacturers.

Safe framework conditions, however, can only be obtained by as many as possible different plants, foundations and linkage concepts at different sites by means of long years of operating experience.

6.5   Ecological Balance

Changes caused by people in the landscape, nature, climate and the animal world on land, at sea and in the air have effects on the overall ecosystem. An estimate of comparison of this for wind turbine technology with other power stations will be undertaken in the following sections. Good comparison possibilities are provided by figures for energy amortization and the yield factor.

6.5.1   Contribution to climate protection

The presently dominating fossil energy technologies cause the largest CO_2 emissions in Germany. These are expected to fall from 313 million tonnes in 1990 to around 290 million tonnes in 2020 in order to limit the increasing warming of the earth and climate and natural catastrophies.

Germany has set itself the target of reducing the CO_2 emissions from 1990 to 2020 by 25% and by 2050 by 80%. This extension of renewable energies, especially wind energy, is expected to play a key role at the present and in the next years. In the conversion of energy, the following CO_2 emission can be expected: For brown coal 280 000, for black coal 260 000 and for natural gas 130 000 tonnes CO_2 per petajoule (PJ) [6.95]. In comparison to this, the emissions caused by renewable energies can be seen as very low. They are caused mainly in the manufacture of the plant. There are practically no emissions during the operation. The CO_2 equivalent of wind energy can be assumed to be 20 and that of black coal with 950 g/kWh of electrically generated energy. Thus, wind energy contributes substantially to the protection of the climate.

6.5.2   Landscape utilization

The surroundings of wind turbines and the areas of wind farms can be used almost completely as agricultural and grazing land. Only the area of the foundations (e.g. 15 × 15 m) is lost to agriculture. Access, on the other hand, can have a double use. With own operation of wind turbines, or leasing of wind farm localities, farmers earn a substantial second income or a so-called 'second harvest'.

Although the energy densities of renewable compared to conventional conversion processes are relatively small, the result of the continuous regeneration of energy offer an advantageous effective surface relationship. In comparison to the brown coal dismantling, with wind energy utilization, less than 7.5% of the area of around 480 km^2 is occupied and is used up by surface construction in order to generate 150 billion kWh [6.96].

6.5.3   Bird strike

Numerous ornithological investigations of the occurrence of resting, breeding and migrating birds in the vicinity of wind farms have come to the result that only a few types are permanently affected in their behavior [6.97]. With older studies that take mostly smaller plants into account approximately 0.5 to 2 bird strikes per plant per year were discovered. Newer investigations give a figure of an average of five dead birds per plant and year. This corresponds at present statistically to 120 000 dead birds a year. More than 20 birds per wind turbine and year occurred according to [6.98] only at wet areas and the crests of mountains.

Compared to approximately 5 to 10 million birds that die per year in road traffic and high tension lines, this type of bird strike occurs relatively seldom in wind turbines. Thus, according to [6.99], bird strikes have no influence on population sizes or the local number of birds. Indeed, at some localities with many wind turbines, the population has risen. An example of this is the brood quantity of the cranes that was 1700 pairs in 1990 and have grown today to 7000 pairs [6.100] recently.

6.5.4   Bats

The effects of wind turbines on bats have only been investigated. Depending on the type of bat, the danger of collisions has a rising tendency. As the nocturnals fly higher, they collide more often than other types. Information on numbers is only conditionally available and for this reason an evaluation is not possible. The bat population is normally ignored in the planning of wind farms. According to [6.100], this can lead to problems with approvals or to limitation of operation.

6.5.5   Recycling of wind turbines

Wind farms represent a relatively large structure with a high mass. The cost of manufacture and the technical reliability are not the onyx factors under consideration in the selection of the material used, but also the recyclability at the end of the utility phase. An environmentally friendly energy technology must also work on this solution at an early stage. Conditioned by the life of the plants, it is foreseeable that sometime around the year 2020, the number of plants going out of service will grow. Wind turbines with drives and a steel tube tower (including foundation) consist 82% of steel. Fibreglass and carbon fibre-reinforced plastics comprise 8%; aluminium, electrical components and operating fluids are distributed over the remaining 7%. After disassembling the wind turbine, the concrete of the foundation (more than 1.5 m below the ground can be left behind) and, possibly the tower can be used as additive in road building. Metals such as steel, cast iron, aluminium and copper are re-melted in foundries and electrical waste is separated into the various materials in separating works and used further. While around 20% remains as unusable scrap. Except for the rotor blades, modern wind turbines can be reused to almost 100%. The composite material used in rotor blades cannot be recycled and is thermally utilized [6.101].

6.5.6 Energetic amortization time and harvest factor

The energetic amortization time is the time a system requires to again generate the primary energy that was required for its own manufacture.

The harvesting factor, on the other hand, gives the relationship of the electrical energy generated during the overall utilization period to the primary energy consumed. Both values form an important base for ecological considerations. This means that the smaller the energetic amortization time and the larger the harvest factor the more energetic effective is the power generation from wind turbines.

Investigations of examples of two gearless wind turbines (Enercon E-40 with 500 kW and E^{e}66 with 1500 kW rated power) showed, according to [6.102], energetic amortization times between three and six months and harvest factors of approximately 70 for the large and approximately 40 for the small plant with a life of twenty years.

For its plants, the manufacturer Enercon mentions a harvesting factor of 35.4 for inland sites and 51 for coastal sites for the twenty years operating life of its plants. The primary energy use for production, erection, operation, dismantling and disposal of an E-82 plant is given as 2880 MWh [6.103].

Conventional power stations have a much lesser result as, during their operation, energy in the form of raw materials must be provided continuously.

References

[6.1] Haas, O., Heier, S., Kleinkauf, W. and Strauß, P., Zukunftsaspekte regenerativer Energien und die Rolle der Photovoltaik. Fortschrittliche Energiewandlung und -anwendung. Schwerpunkt: Dezentrale Energiesysteme, in VDI Conference, Bochum, 13–14 March 2001, VDI reports 1594, pp. 3–16, ISBN 3-18-091594-3.

[6.2] Heier, S., Situation und Perspektiven der Windenergie. Fortschrittliche Energiewandlung und -anwendung. Schwerpunkt: Dezentrale Energiesysteme, in VDI Conference, Bochum, 13–14 March 2001, VDI reports 1594, pp. 523–37, ISBN 3-18-091594-3.

[6.3] Heier, S., *Nutzung der Windenergie*, 4th Edn, TV Rheinland, Cologne, 2000.

[6.4] Berlipp, A., *Standort- und anlagenspezifische Energieerträge von Windkraftanlagen*, Thesis, Universität Gh Kasel, 1996.

[6.5] Döpfer, R. and Otto, K., Untersuchung eines Mittelgebirgsstandortes im Hinblick auf die Eignung zur Windenergienutzung, in *Abschlußarbeit Energie und Umwelt*, Kassel University, 1994.

[6.6] Durstewitz, M., Enlin, C., Hahn, B., Hoppe-Kilpper, M. and Rohrig, K., Ausgewählte Betriebserfahrungen mit Windkraftanlagen im Binnenland, in *Wind Energie Aktuell*, Book 10, January 1994, pp. 22–6.

[6.7] Institut für Solare Energieversorgungstechnik (ISET), *Wissenschaftliches Meß- und Evaluierungsprogramm zum Breitentest '250 MW-Wind', Jahresauswertungen 1990–2001*, Eigendruck, Kassel.

[6.8] Mortensen, N.G., Landberg, L., Troen, I. and Petersen, E.L., *Wind Atlas Analysis Programme (WASP)*, Risø National Laboratory, Roskilde, Denmark.

[6.9] Troen, I. and Petersen, E.L., *European Wind Atlas*, Risø National Laboratory, Roskilde, Denmark.

[6.10] Kleinkauf, W., Meliß, M., Molly, J.-P., *et al.*, *Energiequellen für Morgen? Part III:* Nutzung der Windenergie, BMFT-Study, Umschau, Frankfurt, 1976.

[6.11] Windheim, R., *Nutzung der Windenergie*, KFA Jülich, 1980.

[6.12] Selzer, H., *Solar Energy R & D in the European Community*, D. Reidel, 1986.

[6.13] Bierbrauer, H. von, *et al.*, *Darstellung realistischer Regionen für die Errichtung insbesondere großer Windenergieanlagen in der BRD*, BMFT-FB-T 85-053, Lahmeyer International, Frankfurt.

[6.14] Fichtner Development Engineering, *Abschätzung des wirtschaftlichen Potentials der Windenergienutzung in Deutschland und des bis 2000/2005 zu erwartenden Realisierungsgrades sowie der Auswirkung von Fördermaßnahmen*, BMFT Bonn/Forschungszentrum Jülich GmbH, Stuttgart, 1991.

[6.15] Consulectra, *Wind Power Penetration Study of the European Commission*, Federal Republic of Germany, 1991.

[6.16] European Wind Energy Association, *Time for Action/Wind Energy in Europe*, CEC DG XVII, October 1991.

[6.17] DEWI, *Feststellung geeigneter Flächen als Grundlage für die Standortsicherung von Windparks im nördlichen Niedersachsen*, German Wind Energy Institute Commissioned by the Lower Saxony Ministry for the Environment, Wilhelmshaven, January 1993.

[6.18] Glocker, S., Richter, B. and Schwabe, J., Methoden und Ergebnisse bei der Ermittlung von Windenergiepotentialen und Flächen in Mecklenburg-Vorpommern, Hamburg und Schleswig-Holstein, in German Wind Energy Conference 92, Wilhelmshaven, 1992, pp. 93–9.

[6.19] Durstewitz, M., Heier, S., Hoppe-Kilpper, M. and Kleinkauf, W., Entwicklung der Windenergietechik in Deutschland, in *Plenarvortrag Windenergie*, Internationales Sonnenforum, Cologne, 26–30 July 1998.

[6.20] Durstewitz, M., Heier, S. and Hoppe-Kilpper, M., Ausbaustrategien für die Windenergienutzung in Deutschland, in *Forschungsverbund Sonnenenergie-Jahrestagung*, Bonn, Forschungsverbund Sonnenenergie Themen 98/99, Cologne, 8–9 September 1998, pp. 40–5.

[6.21] Heier, S., Kleinkauf, W., Durstewitz, M. and Hoppe-Kilpper, M., Anwendung der Windenergie in Deutschland, in VDI-Tagung Regenerative Energien (Hungary), Budapest, Hungary, 14–16 September 2000.

[6.22] Östergrad, C., *Randbedingungen zur Seeaufstellung großer Windkraftanlagen*, Report STB No. 924, Germanischer Lloyd, Hamburg, 1982.

[6.23] Pernpeitner, R., Offshore Siting of Large Wind Energy Converter Systems in the North Sea and Baltic Sea Regions, *Wind Engineering*, 1985, 203–313.

[6.24] Leutz, R., Akisawa, A., Kashiwagi, T., Ackermann, T. and Suzuki, A., World-Wide Offshore Wind Energy Potential, in Second International Workshop on *Transmission Networks for Offshore Wind Farms*, Stockholm, Sweden, 29–30 March 2001.

[6.25] Smith, K. and Hagermann, G., The Potential for Offshore Wind Development in the United States, in Second International Workshop on *Transmission Networks for Offshore Wind Farms*, Stockholm, Sweden, 29–30 March 2001.

[6.26] Matthies, H. G. and Garrad, A. D., *Study of the Offshore Wind Energy in the EC, JOULE I (JOUR 0072)*, Verlag Natürliche Energie, Brekensdorf, 1985.

[6.27] *Windenergienutzung auf See*, Position paper by the Bundesministeriums für Umwelt, Naturschutz und Reaktorsicherheit on the use of wind power in the offshore area, 2001.

[6.28] Henderson, A.R. and Camp, T.R., Hydrodynamic Loading of Offshore Wind Turbines, in European Wind Energy Conference, Copenhagen, Denmark, 2–6 July 2001, pp. 561–6.

[6.29] Sorensen, H.C., Hansen, J. and Volund, P., Experience from the Establishment of Middelgrunden 40MW Offshore Wind Farm, in European Wind Energy Conference, Copenhagen, Denmark, 2–6 July 2001, pp. 541–7.

[6.30] van Bussel, G.J.W. and Zaaijier, M.B., DOWEC Concepts Study, Reliability, Availability and Maintenance Aspects, in European Wind Energy Conference, Copenhagen, Denmark, 2–6 July 2001, pp. 557–60.

[6.31] Stalin, T., Utgrunden Offshore Wind Energy Project, Sweden, in Second International Workshop on *Transmission Networks for Offshore Wind Farms*, Stockholm, Sweden, 29–30 March 2001.

[6.32] Kühn, M. and Sievers, T., Utgrunden Offshore Windfarm – First Results of Design Verification by Measurements, in Second International Workshop on *Transmission Networks for Offshore Wind Farms*, Stockholm, Sweden, 29–30 March 2001.

[6.33] Loman, G., The Offshore Wind Farm at Lillgrund, Application and Environmental Impact Assessment, in Second International Workshop on *Transmission Networks for Offshore Wind Farms*, Stockholm, Sweden, 29–30 March 2001.

[6.34] Rehfeldt, K., Gerdes, G.J. and Schreiber, M., *Weiterer Ausbau der Windenergienutzung im Hinblick auf den Klimaschutz – Teil 1*, Bundesumweltministerium/DEWI, 2001, p. 104.

[6.35] EU Commission, *Energy for the Future: Renewable Sources of Energy*, White Paper for a Community Strategy and Action Plan, 26 November 1997.

[6.36] Kleinkauf, W., M. Meliß, J.-P. Molly *et al.*: Nutzung der Windenergie, Band III der Reihe Energiequellen für morgen? Umschau, Frankfurt, 1976.

[6.37] Windheim, R.: Nutzung der Windenergie. KFA Jülich, 1980.

[6.38] Selzer, H.: Potential of wind energy in the European Community. An Assessment Study, Band 2 der Reihe Solar Energy R&D in the European Community, Series G. Wind Energy. D. Reidel Publishing Company for the European Community, Dordrecht, Boston, Lancaster, Tokyo, 1986, ISBN 90-277-2205-6.

[6.39] Ritter, H. und K. Schulten: Topology conserving mappings for learning motor tasks. In: Proceedings of the AIP Conference, Snowbird/Utah (USA), 1986. Seiten 376–380.

[6.40] Brakelmann, H. und M. Jensen: Neues sechsphasiges Übertragungssystem für VPE-isolierte HVAC-See- und Landkabel hoher Übertragungsleistung. *ew*, 105(4):34–43, 2006.

[6.41] Consulectra: Wind power penetration study, the case of the Federal Republic of Germany. Report EUR 14249 EN, Commission of the European Communities, Brussels/Luxembourg, 1991.

[6.42] European Wind Energy Association (EWEA): Time for Action/Wind Energy in Europe. CEC DG XVII, October 1991.

[6.43] Consulectra: Wind power penetration study, the case of the Federal Republic of Germany. Report EUR 14249 EN, Commission of the European Communities, Brussels/Luxembourg, 1991.

[6.44] Heier, S.: Situation und Perspektiven der Windenergie. In: VDI-Gesellschaft Energietechnik (Herausgeber): Fortschrittliche Energiewandlung und -anwendung – Schwerpunkt: Dezentrale Energiesysteme, Düsseldorf, März 2001. Band 1594 der Reihe VDI-Berichte, Seiten 523–537, VDI-Verlag, ISBN 3-18-091594-3.

[6.45] Durstewitz, M., S. Heier, M. Hoppe-Kilpper und W. Kleinkauf: Entwicklung der Windenergietechnik in Deutschland, Plenarvortrag Windenergie. In: Deutsche Gesellschaft für Sonnenenergie e.V. International Solar Energy Society – German Section (Herausgeber): Tagungsband 11. Internationales Sonnenforum, Köln, 26.–30. Juli 1998. Solar Promotion GmbH.

[6.46] Durstewitz, M., S. Heier und M. Hoppe-Kilpper: Ausbaustrategien für die Windenergienutzung in Deutschland. In: *Forschungsverbund Sonnenenergie (Herausgeber): Themen*

98/99 – Nachhaltigkeit und Energie, Köln, 1999. Seiten 40–45, Forschungsverbund Sonnenenergie. Forschungsverbund Sonnenenergie-Jahrestagung, Bonn 8–9. September 1998.

[6.47] Heier, S., W. Kleinkauf, M. Durstewitz und M. Hoppe-Kilpper: Anwendung der Windenergie in Deutschland. In: VDI/GET-Tagung Regenerative Energien in Ungarn und Deutschland, Budapest, Ungarn, 14.–16. September 2000. ISBN 3-931384-32-2.

[6.48] Östergrad, C.: Randbedingungen zur Seeaufstellung großer Windkraftanlagen. Bericht STB 924, Germanischer Lloyd, Hamburg, 1982.

[6.49] Pernpeitner, R.: Offshore siting of large wind energy converter systems in the north sea and baltic sea regions. *Wind Engineering*, 9:203–313, 1985.

[6.50] Leutz, R., A. Akisawa, T. Kashiwagi, T. Ackermann und A. Suzuki: World-wide offshore wind energy potential. In: Second International Workshop on Transmission Networks for Offshore Wind Farms, Stockholm, Sweden, March 29–30, 2001.

[6.51] Smith, K. und G. Hagermann: The potential for offshore wind development in the United States. In: Second International Workshop on Transmission Networks for Offshore Wind Farms Stockholm, Sweden, March 29–30, 2001.

[6.52] Matthies, H. G. und A. D. Garrad: Study of the Offshore Wind Energy in the EC, JOULE I (JOUR 0072). Verlag Natürliche Energie, Brekensdorf, 1985.

[6.53] Bundesministerium für Umwelt, Naturschutz und Reaktorsicherheit (BMU): Positionspapier zur Windenergiennutzung im Offshore-Bereich: Windenergienutzung auf See, 2001.

[6.54] Henderson, A. R. und T. R. Camp: Hydrodynamic loading of offshore wind turbines. In: European Wind Energy Conference, Copenhagen, Denmark, July 2–6, 2001. Seiten 561–566.

[6.55] Sorensen, H. C., J. Hansen und P. Volund: Experience from the establishment of Middelgrunden 40MW offshore wind farm. In: European Wind Energy Conference, Copenhagen, Denmark, July 2–6, 2001. Seiten 541–547.

[6.56] Bussel, G. J. W. van und M. B. Zaaijier: DOWEC concepts study, reliability, availability and maintenance aspects. In: European Wind Energy Conference, Copenhagen, Denmark, July 2–6, 2001. Seiten 557–560.

[6.57] Stalin, T.: Utgrunden offshore wind energy project, Sweden. In: Second International Workshop on Transmission Networks for Offshore Wind Farms, Stockholm, Sweden, March 29–30, 2001.

[6.58] Kühn, M. und T. Sievers: Utgrunden offshore windfarm – first results of design verification by measurements. In: Second International Workshop on Transmission Networks for Offshore Wind Farms, Stockholm, Sweden, March 29–30, 2001.

[6.59] Loman, G.: The offshore wind farm at Lillgrund, application and environmental impact assessment In: Second International Workshop on Transmission Networks for Offshore Wind Farms Stockholm, Sweden, March 29–30, 2001.

[6.60] Kühn, P., V.D. Brune, D. Callies, S. Faulstich, G. Füller, A. Lopez, P. Lyding und R Rothkegel: Windenergie Report Deutschland 2010. Fraunhofer Institut für Windenergie und Energiesystemtechnik IWES, Kassel, 2011.

[6.61] Bundesamt für Seeschifffahrt und Hydrographie (BSH): Contis-Seekarte Offshore-Windparks (Pilotgebiete) in der deutschen AWZ (Nordsee). PDF-Dokument im Internet, 2008. http:// www.bsh.de/de/Meeresnutzung/Wirtschaft/CONTIS-Informationssystem/ContisKarten /NordseeOffshoreWindparksPilotgebiete.pdf

[6.62] Bundesamt für Seeschifffahrt und Hydrographie (BSH): Contis-Seekarte Offshore-Windparks (Pilotgebiete) in der deutschen AWZ (Ostsee). PDF-Dokument im Internet, 2008. http://www.bsh.de/de/Meeresnutzung/Wirtschaft/CONTIS-Informationssystem/ContisKarten /OstseeOffshoreWindparksPilotgebiete.pdf

[6.63] Bundesverband Windenergie e.V.: Themenseite Repowering. Internet-Seite, 2008. http://www.wind-energie.de/de/themen/repowering

[6.64] Bundesverband Windenergie e.V.: Repowering – weniger ist mehr. Elektronisches Dokument, Juni 2007. http://www.wind-energie.de/fileadmin/dokumente/Hintergrundpapiere/Wirtschaft _und_Strompreise/HG_Repowering.pdf Hintergrundinformation.

[6.65] Bundesverband Windenergie e.V.: Die Studie zum Repowering. Internet-Seite, 2005. http://www .wind-energie.de/de/themen/repowering/bwe-studie/, Studie im Auftrag des BWE, erstellt vom Hermann-Föttinger-Institut für Strömungsmechanik der TU Berlin in Zusammenarbeit mit der Firma Ecofys.

[6.66] Bundesverband Windenergie e.V.: Recycling. Internetseite, 2009. http://www.wind-energie.de /index.php?id=253

[6.67] Maslaton, M. und D. Kupke: Rechtliche Rahmenbedingungen des Repowerings von Windenergieanlagen. Verlag für alternatives Energierecht, Leipzig, 2005, ISBN 3-9809815-3-3.

[6.68] European Commission: Energy for the future – renewable sources of energy. White Paper for a Community Strategy and Action Plan, November 26, 1997

[6.69] Eischen, T.: Wirtschaftliche und technische Behandlung von Eigenerzeugungsanlagen in Europa Diplomarbeit II, Universität Gh Kassel, 1996.

[6.70] Küthe, Ch., *Wirtschaftlichkeit von Windenergieanlagen*, Bachelor thesis, University of Kassel, 2013.

[6.71] European Union, Directives 97/11/EC and 85/337/EEC on environmental compatibility testing.

[6.72] Amtsblatt für Schleswig-Holstein, 2001, No. 16/17, p. 216ff.

[6.73] Bundesimmissionsschutzgesetz BImSchG 1974, 1990.

[6.74] DIN 45635, VDI 2571.

[6.75] DIN ISO 9613-2, VDI 2720.

[6.76] DIN 45641, DIN 45645, TA Lärm, VDI 2058.

[6.77] VDI 2570, VDI 3720, DIN 4109, VDI 271.

[6.78] Technische Anleitung zum Schutz gegen Lärm, 1998.

[6.79] Baunutzungsverordnung BauNVO, 1990.

[6.80] Fördergesellschaft Windenergie e.V., *Technische Richtlinien zur Bestimmung der Leistungskurve, der Schallemissionswerte und der elektrischen Eigenschaften von Windenergieanlagen*, 1 April 1998.

[6.81] Ising, H., Markert, B., Shenoda, F. and Schwarze, C., Infraschallwirkungen auf den Menschen, in *Bundesministerium für Forschung und Technologie*, VDI Verlag, 1982.

[6.82] Buhmann, A., *Keine Gefahr durch Infraschall*, Neue Energie 1/98.

[6.83] EMD Deutschland GmbH, *WindPRO, Modul SHADOW*, Kassel, 2002.

[6.84] EMD Deutschland GmbH, *WindPRO, Modul Decibel*, Kassel, 2002.

[6.85] EMD Deutschland GmbH, *WindPRO, Modul VISUAL*, Kassel, 2002.

[6.86] EMD Deutschland GmbH, *WindPRO, Modul ZKI*, Kassel, 2002.

[6.87] Baugesetzbuch (BauGB), Bekanntmachung vom 8.12.86 (BGBl. I p. 2253), Änderung durch Artikel 24 Jahressteuergesetz 1997 vom 20.12.96 (BGBl. I p. 2049).

[6.88] Baunutzungsverordnung (BauNVO), Bekanntmachung vom 23.1.90 (BGBl. I p. 132), Änderung durch Artikel 3 Investitionserleichterungs- und Wohnbauland-Gesetz vom 22.4.94 (BGBl. I p. 466).

[6.89] Änderungsgesetz vom 30.07.1996 (BGBl. I p. 1189).

[6.90] Windenergie-Agentur Bremerhaven/Bremen e.V.: Der Wind, das Meer und die Zukunft der Energieversorgung. Magazin Offshore Windenergie, 2009. http://www.windenergie-agentur.de /deutsch/PDFs/OffshoreMagazin_reduced.pdf

[6.91] Germanischer Lloyd Industrial Services GmbH, Geschäftsbereich Windenergie: FINO 1 – Forschungsplattformen in Nord- und Ostsee. Internetseiten, 2009. http://fino1.de/

[6.92] Deutsche Energie-Agentur GmbH (dena): Willkommen bei www.Offshore-Wind.de. Internetseiten, 2009. www.Offshore-Wind.de

[6.93] BARD Engineering GmbH: BARD Engineering GmbH. Internetseiten, 2009. http://www.bard -offshore.de/

[6.94] Energiewende Verlag & Vertrieb: solarenergie.com. Internetseiten, 2009. http://www. solarenergie.com/

[6.95] Wagner, H.-J., M.K. Koch, J. Burkhardt, T. Große Böckmann, N. Feck und P. Kruse: CO2-Emissionen der Stromerzeugung – ein ganzheitlicher Vergleich verschiedener Techniken *Wasser und Abfall*, 59(10):44–52, 2007.

[6.96] Schmidt, J.: Erneuerbare Energien in der Fläche. Potenziale 2020 – Wie viel Flächen brauchen die Erneuerbaren Energien? Renews Spezial, 2010(23), 2010.

[6.97] Bundesverband Windenergie e.V.: Vogeltod durch WEA vergleichsweise selten. Internetseite, 2009. http://www.wind-energie.de/de/themen/mensch-umwelt/vogelschutz/vogeltod-durch -wea-vergleichsweise-selten/

[6.98] Hötker, H.: Auswirkungen des ≫Repowering≪ von Windkraftanlagen auf Vögel und Fledermäuse Oktober 2006.

[6.99] Dürr, T.: Auswertung der Zentralen Fundkartei der Staatlichen Vogelschutzwarte im Landesumweltamt Brandenburg, 2011.

[6.100] Arbeitskreis Naturschutz, BWE e.V.: POSITIONSPAPIER ZU NATURSCHUTZ UND WINDENERGIE DES BUNDESVERBANDS WINDENERGIE E.V., Mai 2011.

[6.101] Albers, H., Greiner S., Seifert H. und Kühne U.: Recycling of wind turbine rotor blades – fact or fiction? *DEWI Magazin*, 2009(34), 2009.

[6.102] Wagner, H.-J.: Ganzheitliche Energiebilanzen von Windkraftanlagen: Wie sauber sind die weißen Riesen? maschinenbauRUBIN, 2004. http://www.ruhr-uni-bochum.de/rubin/ maschinenbau/pdf/beitrag1.pdf, Sonderheft, 11–S. zgl. veröffentlicht im Internet.

[6.103] Enercon GmbH: Windblatt – Enercon Magazin für Windenergie, April 2011.

Index

Grid Integration of Wind Energy: Onshore and Offshore Conversion Systems, Third Edition. Siegfried Heier.
© 2014 John Wiley & Sons, Ltd. Published 2014 by John Wiley & Sons, Ltd.

Printed and bound by CPI Group (UK) Ltd, Croydon, CR0 4YY

17/04/2025

14658874-0001